Study Guide

Chemistry:
The Molecular Science

FOURTH EDITION

Moore

Stanitsky

Jurs

Prepared by

Michael Sanger
Middle Tennessee State University

BROOKS/COLE
CENGAGE Learning™

Australia • Brazil • Japan • Korea • Mexico • Singapore • Spain • United Kingdom • United States

For product information and technology assistance, contact us at **Cengage Learning Customer & Sales Support, 1-800-354-9706**

For permission to use material from this text or product, submit all requests online at **www.cengage.com/permissions** Further permissions questions can be emailed to **permissionrequest@cengage.com**

ISBN-13: 978-1-4390-4964-8
ISBN-10: 1-4390-4964-5

Brooks/Cole
20 Davis Drive
Belmont, CA 94002-3098
USA

Cengage Learning is a leading provider of customized learning solutions with office locations around the globe, including Singapore, the United Kingdom, Australia, Mexico, Brazil, and Japan. Locate your local office at: **www.cengage.com/global**

Cengage Learning products are represented in Canada by Nelson Education, Ltd.

To learn more about Brooks/Cole, visit **www.cengage.com/brookscole**

Purchase any of our products at your local college store or at our preferred online store **www.CengageBrain.com**

Printed in the United States of America
1 2 3 4 5 6 7 14 13 12 11 10

Table of Contents

Preface

This study guide is intended to accompany the 4th edition of the textbook *Chemistry: The Molecular Science*, written by John W. Moore, Conrad L. Stanitski, and Peter C. Jurs. (Brooks/Cole, 2011). The material in this study guide was revised based on the study guide for the 3rd edition, written by Michael J. Sanger. One of the major changes in the third edition of the study guide was that the chapters were organized around the instructional objectives appearing in the 'In Closing' section at the end of each chapter in the Moore, Stanitski, and Jurs textbook instead of the sections that appear within these chapters. This edition of the study guide continues this format. This study guide also includes many examples of worked problems in each chapter, including several examples involving molecular-level pictures that ask students to think about chemical processes at the nanoscale level.

The goal of this study guide is to assist students in increasing their understanding of the material presented in the Moore, Stanitski, and Jurs textbook and to make studying this material more efficient. The written text and the worked examples emphasize, in a concise form, the most important information in each chapter. Important terms and phrases that appear in the 'Key Terms' section at the end of each chapter in the textbook are highlighted in boldface, and other important terms or phrases are highlighted in italics. Because solving mathematic problems and answering conceptual questions is an important and necessary part of learning chemistry, this study guide offers several examples of conceptual and algorithmic (mathematical) questions in the 'Practice Test' section at the end of each chapter. The study guide also contains answers to the 'Practice Test' questions that not only shows the answers to these questions, but also provides comments and explanations regarding how these problems are solved.

I would like to thank Kevin Vaughn, a 7-12 chemistry teaching major at MTSU, for proofreading the chapters in the 3rd edition for typographical errors, and for making several suggestions for improving the clarity and readability of the text in these chapters.

Michael J. Sanger
Middle Tennessee State University
Murfreesboro, Tennessee

Chapter 1
The Nature of Chemistry

The goal of this chapter is to introduce you to the world of **chemistry** from the perspective of a chemist. Although chemists think about chemistry using macroscale, nanoscale, and symbolic techniques, chemistry is often referred to as the molecular science. It is this molecular (nanoscale) interest that makes chemistry distinct from other scientific fields. This chapter introduces several macroscale (physical properties, types of mixtures, states of matter), nanoscale (structural formulas, ball-and-stick models, space-filling models, atomic theory, kinetic-molecular theory), and symbolic (balanced equations, density calculations, dimensional analysis, periodic table) representations commonly used by chemists and non-chemists alike. It also explains how chemists perform scientific experiments based on hypotheses, theories, and laws and introduces one of the most important tools used by chemists—mathematics and mathematical calculations.

The Power of Chemistry to Answer Intriguing Questions (1.1)

The properties of many chemicals are governed by their molecular structures. This is why scientists from other fields including biology, engineering, geology, physics, and medicine need to study chemistry and why they work with chemists in interdisciplinary teams. Chemical principles are also used to answer many important (and difficult) questions that arise in non-scientific fields like music, art, and history.

Using a Scientific Approach to Solve Problems (1.2 & 1.3)

Chemistry, like most other sciences, is *empirical* in nature. This means that the laws, theories, and hypotheses proposed in chemistry are largely based on data collected in the laboratory ('empirical' simply means 'based on observation and experiment', and is the opposite of 'theoretical' which means 'based on theory or speculation'). Ultimately, if there is a conflict between the empirical data and a scientific theory, the theory must be adapted to fit the data. The field of chemistry represents the work done by scientists to make sense of the world at the molecular (atomic) level.

Hypotheses, Theories, and Laws (1.3)

Hypotheses, theories, and laws are the result of the work done by scientists in their attempts to understand the world. In contrast to most students' view of the scientific method, hypotheses are not a scientists' first step. **Hypotheses** (which are usually described as 'educated guesses') are based on data collected by scientists and usually represent the scientists' first attempt to explain their observations. The results of a hypothesis that appear to work after several experiments can be referred to as a **theory**, a unifying principle used to explain many facts. Although theories can still be proven wrong based on new data, theories are usually pretty solid because they are based on large amounts of data. Theories that are of extremely broad significance, and have withstood the test of time and many experiments, can be 'upgraded' to laws. **Laws** are usually very well established and rarely change, although this happens from time to time.

Quantitative and Qualitative Observations (1.3)

Quantitative observations are based on numerical values that are often determined through experimental measurements. **Qualitative** observations, on the other hand, involve non-numerical information collected using the five senses: Color or luster (sight), odor (smell), pops or bangs (hearing), texture or temperature (touch), etc. You should never use the sense of taste in a chemistry laboratory.

Physical Properties of Matter and Physical Changes (1.4)

Physical properties are characteristics of a **substance** that describe the object. These properties are usually measured non-destructively (i.e., the object is the same before and after the physical property has been measured). **Melting point** (which is also called the **freezing point**), **boiling point**, density, color, and shape are all examples of physical properties. **Physical changes** are changes that occur in which the chemical composition of a substance has not changed. The most common physical changes seen in a

1

chemistry laboratory are those involving changes in the state of matter (gas, liquid, solid, aqueous). A liquid boiling into a gas or a solid dissolving in water are common examples of physical changes.

Estimating Celsius Temperatures (1.4)

Although there is a formula for converting *Fahrenheit temperatures* to the **Celsius temperature scale** (which can be found in the back of the textbook), it is often useful to make quick qualitative comparisons of two or more **temperatures** (hotter/colder) without performing the mathematical conversion to get them in the same scale. In order to make these qualitative comparisons, you need to know the numerical values of certain temperatures in both scales. Some of the values include: $-40°C = -40°F$ (the temperature where these two values are the same), $0°C = 32°F$ (melting/freezing point of water), $37°C = 98.6°F$ (normal human body temperature), and $100°C = 212°F$ (boiling point of water).

Density Calculations (1.4)

The definition of **density** as 'the mass per unit volume' of an object is a symbolic definition (based on the mathematical symbols for the mass and volume: $d = m/V$). The nanoscale definition of density is the "crowdedness" of **matter** (the more crowded the matter is in a given volume the higher the density, and vice versa). The density of most solids and liquids fall in the range of 0.5-20 g/mL, while most gases have densities that are much lower than that of solids and liquids.

When performing density calculations, density can be viewed as a mathematical ratio (**proportionality factor**) of two properties related by a mathematical formula ($d = m/V$) or as a **conversion factor** that converts a mass measurement into its equivalent volume or a volume measurement into its equivalent mass.

Example: What is the volume (in mL) of a sample of 514 g of liquid bromine (Br_2)? The density of bromine is 3.12 g/mL.

Solution: If you are more comfortable with using the density formula, you can solve for V (volume) and plug the numbers that were given into the equation:

$$d = \frac{m}{V}, \ V = \frac{m}{d} = \frac{514 \text{ g}}{3.12 \text{ g/mL}} = 165 \text{ mL}$$

If you would prefer to think of density as a conversion factor that relates the mass and volume of bromine (i.e., 3.12 g = 1 mL), then you can start with the mass of Br_2 and convert it to volume of Br_2:

$$514 \text{ g} \times \frac{1 \text{ mL}}{3.12 \text{ g}} = 165 \text{ mL}$$

Both methods give you the same answer, and it is perfectly acceptable to use either method.

Chemical Properties of Matter and Chemical Changes (1.5)

Chemical properties are characteristics of a substance describing how it will react with other chemicals. A **chemical reaction** is considered a destructive process because once the **reactants** have been converted into the **products**, they are gone and cannot be recovered under normal conditions. The ability to bubble when placed in acid, to burst into flames when placed in water, and to turn cloudy when another substance is added are examples of chemical properties. **Chemical changes** are changes that occur in which the chemical composition of a substance is changed, and are often accompanied by changes in color, heat, etc.

Homogeneous and Heterogeneous Mixtures (1.6)

Most substances in the real world are *mixtures* (combinations of two or more chemicals). Mixtures can be classified as heterogeneous or homogeneous. **Heterogeneous mixtures** do not appear to be completely

2

mixed and differences in the mixture can be seen by the unaided eye. **Homogeneous mixtures** (also called **solutions**), on the other hand, appear to be completely mixed so that every sample of the mixture contains the same substances and in the same ratios. An example of a heterogeneous mixture would be Raisin Bran cereal (where differences between the raisins and bran flakes can be seen); an example of a homogenous mixture would be Kool-Aid (where every sample of the liquid looks the same).

At the nanoscale, the different molecules in a homogeneous mixture can be seen but these molecules are completely and randomly mixed. In a heterogeneous mixture, the different molecules in the mixture are not randomly mixed and different types of molecules are clustered together in different regions.

Separation, Purification, and Analysis (1.6)

Separation and purification are methods that can be used to convert mixtures into pure substances. This is important because a substance must be pure if chemists are to measure and identify its physical and chemical properties. Analysis is used to determine the amount of any chemical present, and the quality of these techniques tends to improve over time. It is important to know the total amount of a substance present, because human toxicity greatly depends on amount.

Chemical Elements and Chemical Compounds (1.7 & 1.9)

In their attempts to understand matter, early chemists focused on separating mixtures into their individual components, and on separating these components into more and more components. When they identified a substance that could not be separated further, they classified this substance as a **chemical element**. Pure substances that could be separated further (but were still pure) were called **chemical compounds**. The difference between pure compounds and mixtures was that the properties of the pure compound were always the same while the properties of mixtures could be different if different relative amounts of the constituents were present (i.e., salt water can be very salty or only slightly salty, so it represents a mixture).

At the nanoscale, pure elements have only one kind of atom present. Pure compounds have more than one atom present, but they are always combined in the same way (i.e., pure water has H and O in it, but always as H_2O and never in any other form). At the nanoscale, mixtures have two or more different substances present—for example, sugar water would have H_2O (water) molecules and $C_{12}H_{22}O_{11}$ (sucrose) molecules.

Classifying Matter (1.7)

The chart in Figure 1.13 of the textbook (p. 17) describes how matter may be classified into heterogeneous mixtures, homogeneous mixtures, pure substances, pure compounds, and pure elements. Separating mixtures can often be done using physical processes, although chemical processes can be used as well. Separating pure compounds into pure elements, on the other hand, requires the use of chemical methods.

Properties of Gases, Liquids, and Solids (1.8)

One of the most important physical properties of an object is its *state of matter* (gas, liquid, or solid). At the macroscale, **gases** expand to fill the entire volume of any container they are placed in, and they can also be compressed into smaller volumes. Unless the gas sample is colored, gases are also invisible. The macroscale properties of **liquids** are that they have fixed volumes but fluid shapes and motions. Liquids conform to the shape of the bottom of any container in which they are placed. **Solids** can be recognized by their rigid shape and fixed volume. Solid objects move as one object and there is no fluidity in the motion of a single solid—powdered solids like salt and sugar consist of many solid particles and while the powder may appear to have fluid motion, each particle in the solid moves as a single object.

Estimating Macroscale, Microscale, and Nanoscale Sizes (1.8)

When discussing macroscale, microscale, and nanoscale properties, it is useful to look at the origin of these words. The prefixes '*macro-*' means 'large', '*micro-*' means 'small or requiring magnification to be seen' (as in a microscope), and '*nano-*' means 'extremely small' and is about the size of atoms and molecules.

Scientists also use the prefixes '*micro-*' to mean 'one millionth' ($1/1,000,000 = 10^{-6}$) and '*nano-*' to mean 'one billionth' ($1/1,000,000,000 = 10^{-9}$). **Macroscale** observations are those that can be made using the human eye, and most everyday objects that you can see have sizes in the macroscale region. **Microscale** observations cannot be made by the unaided eye, and require the use of magnification devices like simple optical microscopes. Microscale objects are about 1 micrometer (μm) in size—the size of a single living cell. **Nanoscale** observations have only recently been made possible with the aid of powerful microscopes like the electron microscope and the scanning tunneling microscope. Nanoscale objects are about 1 nanometer (nm) in size—the size of single atoms or **molecules**.

Kinetic-Molecular Theory (1.8)

One of the most important nanoscale theories used in chemistry is the **kinetic-molecular theory**, which explains how atoms or molecules behave within gas, liquid, and solid samples. The kinetic-molecular theory will be described in more detail in the chapters on the properties of gases and the atmosphere (Chapter 10) and the properties of solids, liquids, and materials (Chapter 11).

At the nanoscale, gas molecules are very far apart from each other and there is a large amount of empty space between gas molecules. Gas molecules also have large amounts of **energy** and move very quickly in constant, random motions. Molecules in liquids and solids are close enough to touch each other. In a liquid, the molecules appear disorganized (no pattern) and have independent freedom of motion to move past each other. The molecules within a solid have an organized pattern and do not have the freedom to move from their positions. Solid molecules still have some motion, but it is usually in the form of simple vibrations around their fixed position in the solid.

Example: Consider the nanoscale pictures to the right.
(a) Do these pictures depict gases, liquids, or solids?
(b) Do these pictures depict pure substances, heterogeneous mixtures, or homogeneous mixtures?
(c) Do these pictures depict atoms, molecules, or both?
(d) Do these pictures depict elements, compounds, or both?

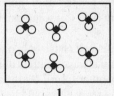 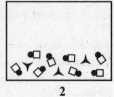

1   2

Solution:
(a) Picture 1 depicts a gas because the particles are occupying the entire container volume; Picture 2 depicts a liquid. These particles occupy the bottom of the container but do not show the ordered pattern that would be expected from a solid.
(b) Picture 1 depicts a pure substance; even though we can see two different atoms (circle and diamond), there is only one kind of chemical present (diamond with three circles). Picture 2 is a mixture because there are two different chemicals present (triangle and circle-square); this is a homogeneous mixture because the particles are randomly mixed and are not separated in different regions.
(c) Picture 1 depicts molecules made up of four atoms. Picture 2 contains atoms of triangle and molecules made up of circle and square atoms.
(d) Picture 1 depicts a compound made of two different kinds atoms. The atoms of triangle in Picture 2 represent an element. The circle-square molecules are compounds made of two different atoms.

Modern Atomic Theory (1.9)

Another important nanoscale theory used in chemistry is the **atomic theory**. Originally proposed by John Dalton in 1803, the atomic theory states that all matter is made of indivisible, indestructible particles called **atoms** (*atomos* means 'indivisible' in Greek). Chemical reactions are simply the rearrangements of atoms within the reactants to form new products. Each atom of the same element has very similar properties and reactivities, and atoms of different elements can have very different chemical and physical properties.

Dalton's atomic theory became popular because it could explain several puzzling yet reproducible observations made at the time. The **law of conservation of mass** could finally be explained because the atomic theory says that in chemical reactions, atoms are neither created nor destroyed, simply rearranged; if every atom that was present before the reaction is still present, then the total mass of the sample should not change. The atomic theory also explains the **law of constant composition**. The atomic theory says that each chemical is always made of the same number and kinds of atoms. Since each atom within the chemical has specific properties (masses) that don't change, then the ratio of masses of the atoms within this compound won't change either. The example below describes how the atomic theory explains the **law of multiple proportions**.

Example: When cuprous bromide is decomposed into its elements, it is found to be 55.7% bromine and 44.3% copper. When cupric bromide is decomposed into its elements, it is found to be 71.5% bromine and 28.5% copper. How do these compounds demonstrate the law of multiple proportions?

Solution: In order to test the law of multiple proportions, you need to determine the proportion of the mass of one element divided by the other in these two compounds. If you assume there is 100.0 g of each compound, you have 55.7 g of bromine and 44.3 g of copper in cuprous bromide, and 71.5 g of bromine and 28.5 g of copper in cupric bromide. The mass ratios of bromine to copper are calculated below.

$$\text{cuprous bromide:} \quad \frac{mass\ Br}{mass\ Cu} = \frac{55.7\ \text{g Br}}{44.3\ \text{g Cu}} = 1.26$$

$$\text{cupric bromide:} \quad \frac{mass\ Br}{mass\ Cu} = \frac{71.5\ \text{g Br}}{28.5\ \text{g Cu}} = 2.51$$

The ratios of the masses of bromine to copper are 1.26:1 in cuprous bromide and 2.51:1 in cupric bromide. Although these ratios do not tell you the **chemical formulas** of either compound, it does show that for the same amount of copper present, the total mass of bromine in cupric bromide must be exactly twice that present in cuprous bromide.

So, if the formula for cuprous bromide were Cu_xBr_y (where x and y are whole numbers), then the formula for cupric bromide would have to be Cu_xBr_{2y} (the actual formulas for cuprous bromide and cupric bromide are CuBr and $CuBr_2$, respectively).

Metals, Nonmetals, and Metalloids (1.10)

This chapter also introduced the single most important tool used by chemists—the periodic table. Atoms are arranged in the periodic table based on similarities in their physical and chemical properties. **Metals**, which are known to be good electrical conductors and have the ability to be deformed or reshaped without breaking (*ductile* and *malleable*), appear in the bottom left portion of the periodic table. Most atoms in the periodic table are metals.

Nonmetals have properties that are very different from the metals: They are generally poor electrical conductors, and while most metals are solids at room temperature, many of the nonmetals are gases or liquids at room temperature. Nonmetals appear in the top right portion of the periodic table.

Metalloids (also called *semimetals*) are elements that fall in between the metals and nonmetals on the periodic table, and their chemical and physical properties also fall somewhere between the properties of metals and nonmetals—many are lustrous like metals, and they conduct electricity but not as good as metals (this is why they are also referred to as *semiconductors*).

Molecular Elements and Allotropes (1.10)

Unlike metals and metalloids, most nonmetal elements are made of small molecules. Exceptions to this rule are the noble gases (column 8A in the periodic table), which exist as single atoms in the gas phase, and some solids like carbon (C) and selenium (Se). Most of the other nonmetals exist as simple **diatomic**

molecules—two atoms of the same element combined together. These diatomic molecules include hydrogen (H_2), nitrogen (N_2), oxygen (O_2), and the elements in column 7A (F_2, Cl_2, Br_2, and I_2).

Some elements can exist in two or more stable forms; these different forms are referred to as **allotropes** of each other. Classic examples of this are oxygen (O_2) and ozone (O_3), and graphite (solid C that is black and soft), diamond (solid C that is colorless and hard), and buckminsterfullerene (C_{60} molecules that look like soccer balls).

Macroscale, Nanoscale, and Symbolic Representations (1.11)

When chemists think about a chemical reaction, they commonly use three distinct but clearly related representations to describe the same chemical process—the macroscale, nanoscale, and symbolic representations. The *macroscale representation* is based on physical observations made using the five senses (color changes, odors, heat changes, changes in states of matter, etc.). The macroscale representation is most commonly used by students when performing experiments in the chemistry laboratory, or when an instructor performs chemical demonstrations in class.

The *symbolic representation* uses symbols to stand in the place of more abstract objects. Examples of symbolic representations are chemical symbols (H for hydrogen, O for oxygen), chemical formulas (H_2O for water), balanced chemical equations, units of measurement (like g for gram or L for liter), mathematical formulas, graphs, and tabulated data. The symbolic representation is used extensively in the chemistry classroom and laboratory in the form of balanced equations and mathematical formulas.

The *nanoscale representation* describes the chemical process in terms of the interactions of atoms and molecules. This representation often poses a problem for students because atoms and molecules are too small to be directly seen or touched. Some chemistry classes tend not to focus on the nanoscale representation—not because the instructor thinks that it's unimportant, but because it can be difficult to teach students about the interactions of particles that they cannot directly see or touch.

Chapter Review — Key Terms

The key terms that were introduced in this chapter are listed below, along with the section in which they were introduced. You should understand these terms and be able to apply them in appropriate situations.

allotrope (1.10)	dimensional analysis (1.4)	molecule (1.10)
atom (1.9)	energy (1.5)	multiple proportions, law of (1.9)
atomic theory (1.9)	freezing point (1.4)	nanoscale (1.8)
boiling point (1.4)	gas (1.8)	nonmetal (1.10)
Celsius temperature scale (1.4)	heterogeneous mixture (1.6)	physical changes (1.4)
chemical change (1.5)	homogeneous mixture (1.6)	physical property (1.4)
chemical compound (1.7)	hypothesis (1.3)	product (1.5)
chemical element (1.7)	kinetic-molecular theory (1.8)	proportionality factor (1.4)
chemical formula (1.10)	law (1.3)	qualitative (1.3)
chemical property (1.5)	liquid (1.8)	quantitative (1.3)
chemical reaction (1.5)	macroscale (1.8)	reactant (1.5)
chemistry (1.1)	matter (1.1)	solid (1.8)
conservation of mass, law of (1.9)	melting point (1.4)	solution (1.6)
constant composition, law of (1.9)	metal (1.10)	substance (1.4)
conversion factor (1.4)	metalloid (1.10)	temperature (1.4)
density (1.4)	microscale (1.8)	theory (1.3)
diatomic molecule (1.10)	model (1.3)	

Practice Test

After you have finished studying the chapter and the homework problems, the following questions can serve as a test to determine how well you have learned the chapter objectives.

1. A scientist collects data to test an established theory, but the collected data contradicts the theory.
    (a) What is the first step the scientist should take? Why?
    (b) If the data and the theory are found to be incompatible, should the data discarded or should the theory be changed?

2. Classify each statement below as qualitative or quantitative.
    (a) A sample of ice has two hydrogen atoms for every oxygen atom.
    (b) A sample of ice is cold to the touch.
    (c) A sample of ice weighs 35 grams.

3. Classify the changes listed below as chemical or physical changes. Explain your answer.
    (a) Plants take in carbon dioxide and release oxygen.
    (b) Liquid gasoline evaporates after being spilled on the pavement.
    (c) When placed on an icy road, salt melts the ice.

4. Which temperature is higher? Explain your answer.
    (a) 80°C or 240°F   (c) 20°C or 20°F
    (b) 37°C or 73°F   (d) −40°C or −40°F

5. What weighs more—a pound of lead or a pound of feathers? Explain your answer and relate it to the concept of density.

6. Which weighs more 313 mL of ethanol or 286 mL of benzene?

7. Classify each of these samples as an element, a compound, a heterogeneous mixture, or a homogeneous mixture. Explain your answer.
    (a) Cheerios cereal   (c) milk
    (b) Froot Loops cereal   (d) sugar

8. Benzene was first isolated from coal tar (a by-product of the distillation of coal) in the late 1800s. While studying the properties of benzene, von Baeyer noticed that a sample of benzene turned blue when treated with isatin and sulfuric acid (called the *indophenine test*). Later, Meyer used this test on a sample of benzene isolated in a different way (from calcium benzoate instead of coal tar), but the sample did not turn blue.
    (a) How does this story demonstrate how science is done?
    (b) Can you explain why the second sample of benzene did not turn blue?

9. Use the kinetic-molecular theory to explain these nanoscale properties of gases, liquids, and solids.
    (a) Colorless gases are invisible, but we can see colorless liquids.
    (b) Liquids and solids cannot be easily compressed into smaller volumes, but gases can be.
    (c) When liquid food coloring is added to liquid water, they slowly mix but when solid salt is poured onto a sample of solid sand, the samples do not mix.

10. Consider the nanoscale picture on the right.
    (a) Does this picture depict a gas, liquid, or solid? Explain your answer.
    (b) Does this picture depict a pure substance, a heterogeneous mixture, or a homogeneous mixture? Explain your answer.
    (c) Does this picture depict atoms, molecules, or both? Explain your answer.
    (d) Does this picture depict elements, compounds, or both? Explain your answer.

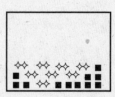

7

11. Use the atomic theory to explain why elements cannot be decomposed into two or more other substances.

12. A chemist has performed four experiments to analyze different samples of chemicals known to contain only nitrogen and oxygen. Can any of these compounds be the same chemicals? Explain your answer.
    (a) Compound A contained 18.4 g of nitrogen and 42.0 g of oxygen.
    (b) Compound B contained 23.8 g of nitrogen and 40.8 g of oxygen.
    (c) Compound C contained 42.6 g of nitrogen and 97.3 g of oxygen.
    (d) Compound D contained 52.5 g of nitrogen and 60.0 g of oxygen.

13. Do the following statements describe metals, nonmetals, metalloids, or a combination of the three?
    (a) The sample is a gas at room temperature.
    (b) The sample conducts electricity very well.
    (c) The sample has a shiny, lustrous appearance.

14. Classify the following atoms as metals, nonmetal, or metalloids based on their positions in the periodic table.
    (a) silicon, Si   (c) sodium, Na
    (b) silver, Ag   (d) sulfur, S

15. Classify the following statements as macroscale, nanoscale, or symbolic. Explain your answers.
    (a) A sample of water contains two hydrogen atoms for every one oxygen atom.
    (b) A sample of water is colorless and has no odor or taste.
    (c) The molecular formula for water is H_2O.

8

Chapter 2
Atoms and Elements

The goal of this chapter is to introduce the structure of the atom and the periodic table. The chapter describes the modern view of the atom and the historical processes by which this view developed. It also introduces two mathematical concepts important to chemists—conversion factors and significant figures. In addition, the concept of the mole and its usefulness in connecting the nanoscale view of the atom to the macroscale view of chemistry laboratory are explained in the chapter. Finally, this chapter describes the periodic table, the history of its development, and some of its interesting and useful features.

Atomic Structure and Subatomic Particles (2.1 & 2.2)

The historical description of how our understanding of the **atomic structure** has changed over time is a fine example of how science is done. With each new experiment and unexpected result, scientists were forced to reevaluate their view of the atom's structure and modify it (when needed) to fit the experimental data. The cathode-ray tubes told scientists that **electrons** (e⁻), the negatively charged objects in the atom, were on the outside of all atoms and were easily lost compared to the other atom components. While all atoms lost the same negative particles, the resulting positive atoms (**ions**) were different from atom to atom. The simplest (lightest) of these positive atoms was hydrogen, which was later shown to be a single **proton** (p^+), the positively charged objects in the atom. Further experiments showed that the nucleus of an atom is very small compared to the overall size of the atom and that, in addition to protons, the **nucleus** of an atom contains neutral objects called **neutrons** (n^0).

Using Conversion Factors (2.3)

Most scientific measurements are made using **metric system** units (also referred to as **SI units**, based on the international standards used to make scientific measurements). These units have the advantage that conversions from one metric value to another are based on units of ten (10 mm = 1 cm and 10 cm = 1 dm, compared to 12 in = 1ft and 3 ft = 1 yd in the English system). The metric units are grams (g) for **mass** and meters (m) for *length*. The metric system also uses prefixes to modify their units. Some of the prefixes commonly used include '*kilo-*' (k-, 10^3), '*milli-*' (m-, 10^{-3}), '*micro-*' (μ-, 10^{-6}), and '*nano-*' (n-, 10^{-9}). Conversion factors (which were introduced in Chapter 1) are very useful in converting a measured value from one unit to another, both within a system of units (English or metric) and between systems.

Example: A baseball weighs 145 g. What is its mass in kilograms (kg) and ounces (oz)?

Solution: For the conversion from grams to kilograms, you need to use the definition of '*kilo-*' (1 kg = 10^3 g). When writing conversion factors, many students are not sure whether the factor 10^3 belongs with the g or kg. One way to remember this is that 'k-' and '10^3' mean the same thing, so 'kg' is the same as '10^3 g'.

$$145 \text{ g} \times \frac{1 \text{ kg}}{10^3 \text{ g}} = 0.145 \text{ kg}$$

For the conversion from English to metric, you need an equality statement relating the two unit systems. In this case 16 oz = 454 g. This value was taken from Table 2.2 (p. 48) of the textbook.

$$145 \text{ g} \times \frac{16 \text{ oz}}{454 \text{ g}} = 5.1 \text{ oz}$$

Significant Figures of Numbers and in Calculations (2.4)

Significant figures (also called *significant digits*) are very important for communicating the degree of uncertainty associated with a measured value. To a mathematician, the numbers 5 and 5.00 are the same, but to a scientist they are different. The first number has an uncertainty of ±1 (and could be as low as 4 or as high as 6) while the second number has an uncertainty of ±0.01 (and could be as low as 4.99 or as high as 5.01). When reporting a measurement, scientists always keep the numbers they are confident about and one number that has some uncertainty. For a value of 0.12542935 in which the uncertainty is ±0.003 (it could be as low as 0.122 or as high as 0.128), the scientist is very certain about the numbers in the tenths and hundredths places but there is some uncertainty about the number in the thousandths place. So, the scientist should write this value as 0.125, keeping only one number with uncertainty.

When determining significant figures in a number, all non-zero numbers are significant. Zeroes between two non-zero numbers are significant (the zero in 501 is significant), and zeroes to the left of the first non-zero number are not significant (the zeroes in 0.00169 are not significant). Zeroes to the right of the last non-zero number may or may not be significant; zeroes to the right of the decimal place are significant (the zero in 3.60 is significant) but zeroes to the left of the decimal place aren't significant (the zeroes in 421,000 aren't significant) unless the decimal place is there (the zeroes in 421,000. are significant).

Example: Determine the number of significant figures in the following numbers:
(a) 4,050, (b) 0.00700, (c) 1500.00.

Solution:
(a) There are 3 significant figures—the zero between the 4 and 5 is significant but the zero at the end is not.
(b) There are 3 significant figures—the three beginning zeroes are not significant but the last two are.
(c) There are 6 significant figures—the rules described above might seem to suggest that the last two zeroes are significant because they are to the right of the decimal place but the first two are not because they are to the left of the decimal place. Based on the uncertainty in this number, it has an error of ±0.01 so they are all significant (the first two zeros are between two significant numbers and that makes them significant).

Although scientists are interested in sharing their measured data with other scientists, sometimes they need to perform mathematical calculations with their data and report these results. So, rules for determining significant figures after mathematical calculations have been performed are needed. For addition and subtraction, the final answer should be reported to the same decimal place as the number with the fewest decimal places reported. For multiplication and division, the final answer should be reported to the same number of significant figures as the number with the fewest number of significant figures reported.

Example: You want to find the density of aluminum, so you weigh a piece of aluminum and determine its volume by water displacement. The sample of aluminum weighs 180.134 g. Before the sample is added, you have a graduated cylinder filled with 35.3 mL of water; after the sample was added, the volume is now 102 mL. What is the volume (in mL) and density (in g/mL) of the aluminum sample?

Solution: The volume of the sample is: 102 mL – 35.3 mL = 66.7 mL. However, you should report this number as 68 mL (to the one place). The density is determined from mass and volume:

$$d = \frac{m}{V} = \frac{180.134\ \text{g}}{66.7\ \text{g/mL}} = 2.70066\ \text{g/mL}$$

In this calculation, you should use the unrounded value for the mass. In general, it is best to round numbers only at the end of the calculation to minimize errors in your answer. Since the mass has 6 significant figures (s.f.) and the volume has 2 s.f., the answer should have 2 s.f.: 2.7 g/mL.

Conversion factors within a unit system (like 1 kg = 10^3 g or 12 in = 1 ft) have infinite significant figures because these are definitions (scientists are infinitely confident that 12 in equals 1 ft). The normal rules for determining significant figures apply to conversion factors from one unit system to another because they are based on measurements and not definitions (so, the conversion factor 1 lb = 454 g would have 3 significant figures but the conversion factor 1 lb = 453.6 g would have 4 significant figures).

Isotopes and Mass Numbers (2.5)

The identity of an atom is based on the number of protons in the nucleus (the **atomic number**); if the number of protons in the nucleus changes, then the atom identity changes. This kind of transformation (where one atom changes into another) only occurs in nuclear reactions, like radioactive decay. The atomic number is also called the *nuclear charge* (Z). Since neutrons have no charge and protons have a +1 charge, the total charge of the nucleus is simply the number of protons in the nucleus ($Z = \#p^+$).

Since the mass of an atom in grams is very small (about 10^{-22} g), scientist have defined a smaller unit, the **atomic mass unit** (amu). The amu is defined so that an atom of carbon with six protons and six neutrons weighs exactly 12 amu. The mass of a proton and a neutron are both very close to 1 amu while the mass of an electron is much smaller (about 0.0005 amu). The **mass number** (A) of an atom is the mass of the atom to the nearest whole number (in amu), and is equal to the number of protons plus the number of neutrons in the atom ($A = \#p^+ + \#n^0$). The number of protons and neutrons can be determined by the atom's atomic number and mass number: $\#p^+ = Z$, and $\#n^0 = A - Z$.

We can describe the composition of an atom's nucleus using a *nuclide symbol*. Nuclide symbols have the general formula $^A_Z X$, where A is the mass number ($\#p^+ + \#n^0$), Z is the atomic number ($\#p^+$), and X is the chemical symbol for the atom (determined by Z). The value of Z is often left out of nuclide symbols because it is redundant information (the X symbol also tells you the value for Z).

Example: What is the nuclide symbol for an atom with 79 protons and 118 neutrons? How many protons and neutrons are present in an atom of tin-118 (^{118}Sn)?

Solution: The periodic table lists the atomic numbers for each atom (the whole number). Scanning the chart shows that gold has an atomic number of 79, so $Z = 79$ and X = Au. The mass number is $A = \#p^+ + \#n^0 = 79 + 118 = 197$. Therefore the nuclide symbol is $^{197}_{79}$Au or ^{197}Au. For tin-118, the Z value is not given. Looking at the periodic table, and remembering that $Z = \#p^+$, you can see that the number of protons for tin is 50. The number of neutrons is: $\#n^0 = A - Z = 118 - 50 = 68$.

Isotopes are two different atoms of the same element (same $\#p^+$) having different masses (different $\#p^+ + \#n^0$). Therefore, isotopes have the same number of protons but different numbers of neutrons. Because isotopes are atoms of the same element, they also have nearly identical physical and chemical properties.

Calculating Atomic Weights (2.6)

Very few elements exist as only one isotope (e.g., F and P); most have two or more isotopes that exist in nature. When weighing a macroscale sample of these elements, you will be weighing both isotopes and the atomic weight of the *element* will not match the atomic weight of any of the isotope *atoms*. The atomic weight of an element is determined by calculating a *weighted average* using the **atomic weights** and **percent abundances** of each of the naturally-occurring isotopes.

Example: The ^{28}Si isotope weighs 27.977 amu and is 92.23% abundant, the ^{29}Si isotope weighs 28.966 amu and is 4.67% abundant, and the ^{30}Si isotope weighs 29.974 amu and is 3.10% abundant. What is the atomic weight of elemental silicon?

Solution: A weighted average is determined by multiplying the mass of each isotope by its fractional abundance (percent abundance divided by 100) and adding these values together for each isotope.
$$(0.9223)(27.977 \text{ amu}) + (0.0467)(28.966 \text{ amu}) + (0.0310)(29.974 \text{ amu})$$
$$25.80 \text{ amu} + 1.35 \text{ amu} + 0.929 \text{ amu} = 28.08 \text{ amu (rounded to the hundredth place)}$$

Atomic Numbers and Atomic Weights (2.5 & 2.6)

The periodic table lists two numbers for each element. The whole number in the periodic table is the element's atomic number Z (the number of protons in the atom), and the number with decimals is the atomic weight of the element (the weighted average of the individual isotope masses).

The Mole and Molar Masses (2.7)

While scientists think about atoms at the nanoscale level, they must make laboratory measurements at the macroscale level. It is currently impossible to measure the mass of a single atom, and when scientists weigh a sample of an element, they are weighing a huge number of atoms. The periodic table lists atomic masses of elements in amu per atom, but it would be tremendously helpful if these numbers in the chart could be used when measuring masses in grams. Scientists defined a number of particles (the **mole**) so that a mole of carbon-12 atoms weighs exactly 12 g. This is similar to the definition used for the amu (1 atom of carbon-12 weighs exactly 12 amu). So, the atomic masses of the elements listed in the periodic table describe the mass of an element in units of amu/atom <u>and</u> in units of g/mol. The book uses the term atomic mass for the 'amu/atom' values and **molar mass** for the 'g/mol' values. The numerical value of a mole (Avogadro's number) is 6.022×10^{23} 'things'/mol (the 'thing' can be anything—atoms, molecules, ions).

Converting Between Mass and Mole Values (2.8)

Because atomic masses and molar masses are mathematical equalities (e.g., 1 atom Al = 26.98 amu Al, and 1 mol Ca = 40.08 g Ca), they can be used as conversion factors.

Example: In July of 2003, astronomers from the Australian National University reported that the current estimate of the number of stars in the universe (visible to modern telescopes) is 70 sextillion (7×10^{22}).
(a) How many moles of stars is this?
(b) What is the mass (in g) of the same amount of hydrogen atoms?

Solution:
(a) To convert number of stars into moles of stars, you need to use Avogadro's number.

$$7 \times 10^{22} \text{ stars} \times \frac{1 \text{ mol stars}}{6.022 \times 10^{23} \text{ stars}} = 0.1 \text{ mol stars (1 significant figure)}$$

(b) You can determine the mass of 70 sextillion hydrogen atoms in grams using Avogadro's number and the molar mass of hydrogen in g/mol.

$$7 \times 10^{22} \text{ atom He} \times \frac{1 \text{ mol He}}{6.022 \times 10^{23} \text{ atom He}} \times \frac{1.008 \text{ g H}}{\text{mol H}} = 0.1 \text{ g H}$$

The Periodic Table (2.9)

The **periodic table** is an important and useful tool for studying chemistry. It was originally organized by Mendeleev based on the atomic weights of the elements. When Mendeleev placed the atoms in order of increasing atomic weights, he noticed a periodic trend (repeating pattern) among the elements. He placed elements that had similar properties in columns, giving the familiar two-dimensional chart used today.

Example: When Mendeleev organized the periodic table, the reported atomic weights were 128 amu for tellurium (Te) and 127 amu for iodine (I). However, Mendeleev recognized that the physical and chemical properties of Te matched S and Se, and the physical and chemical properties of I matched Br and Cl. Was he justified in placing the lighter I after the heavier Te? How does this show how science is done? How can this discrepancy be explained using the modern view of the atom?

Solution: When assembling the periodic table, Mendeleev theorized that atomic weights were responsible for (or at least related to) the chemical periodicity in the elements. When he noted this discrepancy in the periodic table, he said that the physical and chemical properties of these elements dictated their placement in the table and this discrepancy showed that the masses of these two elements must have been incorrectly measured by previous chemists. The modern view is that chemical periodicity is actually related to the atomic number (Z) of an atom ($\#p^+$), not its atomic weight ($\#p^+ + \#n^0$). So, even though Te atoms have only 52 protons (compared to 53 protons in I atoms), they have an atomic weight greater than that of iodine because they have more neutrons (you will note that in the modern table, Te still weighs more than I, and that these earlier chemists did not make a mistake).

The rows in the periodic table are referred to as **periods** (numbered from 1 to 7), and the columns in the periodic table are referred to as **groups**. Atoms in each group tend to have similar physical and chemical properties (the **law of chemical periodicity**). The first two columns and the last six columns of the periodic table (groups 1A-8A) are referred to as **main group elements**. The **transition elements** are the ten columns labeled 1B-8B. The two rows of metals at the bottom of the periodic table are called the **lanthanides** and **actinides**, named after the elements preceding these rows in the periodic table (lanthanum and actinium). There are four groups that have special names—group 1A atoms are **alkali metals**, group 2A atoms are **alkaline earth metals**, group 7A atoms are **halogens**, and group 8A atoms are **noble gases**.

Chapter Review — Key Terms

The key terms that were introduced in this chapter are listed below, along with the section in which they were introduced. You should understand these terms and be able to apply them in appropriate situations.

actinides (2.9)	ion (2.1)	nucleus (2.2)
alkali metals (2.9)	isotope (2.5)	percent abundance (2.6)
alkaline earth metals (2.9)	lanthanides (2.9)	period (2.9)
atomic force microscope (p. 47)	main group elements (2.9)	periodic table (2.9)
atomic mass units (2.5)	mass (2.3)	proton (2.1)
atomic number (2.5)	mass number (2.5)	radioactivity (2.1)
atomic structure (*Introduction*)	mass spectrometer (p. 54)	scanning tunneling
atomic weight (2.6)	mass spectrum (p. 54)	microscope (p. 46)
Avogadro's number (2.7)	metric system (2.3)	significant figures (2.4)
chemical periodicity, law of (2.9)	molar mass (2.7)	SI units (2.3)
electron (2.1)	mole (2.7)	transition elements (2.9)
group (2.9)	neutron (2.2)	
halogen (2.9)	noble gases (2.9)	

Practice Test

After you have finished studying the chapter and the homework problems, the following questions can serve as a test to determine how well you have learned the chapter objectives.

1. Which of the subatomic particles (protons, neutrons, or electrons) are described in the following statements? There may be more than one right answer.
    (a)   It is found in the atom's nucleus.   (d)   It weighs about 1 amu.
    (b)   It has a negative charge.   (e)   It has no charge.
    (c)   It has a positive charge.   (f)   It is found outside the atom's nucleus.

2. Convert the numbers listed below into the specified units. List the proper number of significant figures in your response.
    (a)   486.1 nm to meters   (c)   1.5 L into gallons
    (b)   0.11 mi to inches   (d)   50 oz into kilograms

3.   Two students working together wanted to calculate the area of a crop circle. They measured the diameter of the circle to be 15.426 m. The students performed the following calculations:

$$\text{student 1:} \quad A = \pi r^2 = (3.14)\left(\frac{15.426 \text{ m}}{2}\right)^2 = 186.80 \text{ m}^2$$

$$\text{student 2:} \quad A = \pi r^2 = (3.14159)\left(\frac{15.426 \text{ m}}{2}\right)^2 = 200 \text{ m}^2$$

(a)   What did Student 1 do wrong in his/her calculation?
(b)   What did Student 2 do wrong in his/her calculation?
(c)   What should the proper answer be?

4.   For each of the following numbers, write them in scientific notation with 3 significant figures.
(a)   0.0082492   (c)   71
(b)   849,963   (d)   1.3482

5.   Complete the following chart of nuclide symbols.

nuclide symbol	#p$^+$	#n^0	A	Z
^{88}Sr	___	___	___	___
___	77	116	___	___
___	___	___	92	40

6.   When a sample of ferric bromide was decomposed into its elements, 88.1 g of iron and 378 g of bromine were collected. What is the percent iron and percent bromine in this compound?

7.   On page 56 of the textbook, you are told that copper has only two isotopes (^{63}Cu weighs 62.9296 amu and ^{65}Cu weighs 64.9278 amu).
(a)   How many neutrons are present in ^{63}Cu and ^{65}Cu?
(b)   Given the atomic weight of elemental copper is 63.55 amu, what are the percent abundances of these two isotopes of copper?

8.   Which of these samples contains the most atoms?
(a) 1.2×10^{24} molecules of C_4H_4S (c)   1,000 g of S
(b)   8.24 mol of H_2S

9.   Which of these samples weighs the most?
(a) 2.125×10^{23} atoms of ^{12}C   (c)   1.20 mol of He atoms
(b) 2×10^{22} atoms of Cs   (d)   0.025 mol of Hg atoms

10.   Use the periodic table to identify the following elements. List the element's name and symbol.
(a)   It is lightest transition element.
(b)   It is the halogen whose atomic weight is closest to 115 amu.
(c)   It is the heaviest noble gas.
(d)   It is the actinide named after the founder of the periodic table.
(e)   It is the lightest alkali metal.
(f)   It is the lanthanide whose atomic weight is closest to 150 amu.
(g)   It is the main group element that has 15 protons.
(h)   It is the first nonmetal in period 4.
(i)   It is the heaviest alkaline earth metal.
(j)   It is the metal whose atom has 25 electrons.

14

Chapter 3
Chemical Compounds

The goal of this chapter is to introduce the different types of chemical compounds normally encountered by chemists—molecular compounds, organic compounds (which are usually molecular compounds), and ionic compounds. This chapter also explains how to name these compounds, and describes some of their physical and chemical properties. The chapter also revisits molecular weights and molar masses, and uses them to convert between percent composition by mass values and empirical and molecular formulas. Finally, the chapter introduces several biologically important elements.

Interpreting Molecular, Condensed, and Structural Formulas (3.1)

Chemists use a variety of formulas to describe chemical compounds. The simplest formula, called the **molecular formula**, is simply a list of the type and number of atoms present in the compound. The molecular formula of isooctane (the chemical used to determine a gasoline's octane rating) shows that isooctane contains 8 carbon atoms and 18 hydrogen atoms, but does not show how these atoms are interconnected. While the **condensed formula** also contains subscripts showing the number of atoms present, it also shows how the atoms are connected together. The condensed formula of isooctane shows that one of the carbon atoms has three -CH_3 groups attached to it, another carbon atom has two H atoms attached to it, and a third carbon atom has one H atom and two -CH_3 groups attached to it. A **structural formula** shows the atom connections using lines between each of the atoms that are attached to each other.

In each of these formulas, there is a trade-off between ease of writing and the amount of useful information provided. Molecular formulas are used when information about structural connections is not needed since they are easy to write; when more structural information is needed, the other two formulas are used.

Naming Binary Molecular Compounds (3.2 & 3.3)

Binary molecular compounds are compounds made of two types of atoms (binary) that form molecules. In binary molecular compounds, the two types of atoms are nonmetals (or sometimes metalloids). The general rule for naming binary compounds is to list the name of the first element and then list the name of the second element, replacing its ending with the '-*ide*' suffix.

The problem with naming binary molecular compounds this way is that the same nonmetals can combine into different chemicals—for example, nitrogen and oxygen can form N_2O, NO, N_2O_3, NO_2, N_2O_4, and N_2O_5, but you can't name them all "nitrogen oxide"! To alleviate this problem, there is an additional rule for naming binary molecular compounds: Before each element name, write a Greek prefix denoting how many atoms of each element are present in the molecule. The prefixes from 1 to 10 are: '*mono-*' (1), '*di-*' (2), '*tri-*' (3), '*tetra-*' (4), '*penta-*' (5), '*hexa-*' (6), '*hepta-*' (7), '*octa-*' (8), '*nona-*' (9), and '*deca-*' (10). You should probably memorize these prefixes because you will be using them frequently.

Example: Write the names for the following compounds: (a) P_4S_7, (b) SeF_6, (c) Cl_2O

Solution:

(a) The name for P_4S_7 is tetraphosphorus heptasulfide. The '*tetra-*' tells you there are 4 P atoms, and the '*hepta-*' tells you there are 7 S atoms (sulfide is named by dropping '*-ur*' and replacing it with '*-ide*').

(b) The name for SeF_6 is selenium hexafluoride. The prefix '*mono-*' is almost always omitted, and if a prefix is missing it is assumed to be '*mono-*'. The '*hexa-*' tells you there are 6 F atoms, and the '*-ine*' ending on fluorine is changed to '*-ide*' for fluoride.

(c) The name for Cl_2O is dichlorine monoxide. The '*di-*' tells you there are 2 Cl atoms. When oxygen is the last element (replacing the '*-ygen*' with '*-ide*' to make oxide), '*mono-*' is usually used. This is one of the few examples where '*mono-*' is included. When oxygen is used and the prefix ends in 'o' or 'a', this last letter is usually dropped (i.e., monoxide instead of monooxide, pentoxide instead of pentaoxide, etc.).

There are some binary compounds that have common names that are used instead of the systematic names. The two most common examples are water (H_2O) and ammonia (NH_3).

The rules described here are not used for binary compounds of C and H. These compounds are usually referred to as **hydrocarbons** (this name simply means 'containing carbon and hydrogen'). The simplest hydrocarbons are called **alkanes** and have the general formula C_nH_{2n+2}. The rule for naming alkanes is to use a prefix to show the number of carbon atoms and an '*-ane*' ending. The prefixes for 1 to 4 C atoms are: '*meth-*' (1), '*eth-*' (2), '*prop-*' (3), and '*but-*' (4). The prefixes for 5 to 10 C atoms are the same as the Greek prefixes used in naming binary molecular compounds with the 'a' at the end of the prefixes dropped. So, CH_4 (1 C) is methane, C_3H_8 (3 C) is propane, C_5H_{12} (5 C) is pentane, C_7H_{16} (7 C) is heptane, etc.

Constitutional Isomers of Alkanes (3.4)

Isomers are two or more compounds that have the same molecular formula (same number and type of atoms) but have different structural formulas (the atoms within the molecules are bonded together in different ways). Isomers also have different physical and chemical properties. For alkanes, there are two basic forms of **constitutional** (*structural*) **isomers**. Straight-chain alkanes have the carbon atoms bonded together in a single chain, while branched alkanes do not. In *straight-chain alkanes*, you can draw a single path through all of the carbon atoms; you cannot draw a single path that goes through every carbon atom in a branched alkane. Also, *branched alkanes* contain a carbon atom bonded to three or four other carbon atoms (straight-chain alkanes have carbon atoms bonded to only one or two other carbon atoms).

Example: Write the structural formulas for the straight-chain isomer and a branched isomer of pentane.

Solution: The formula for pentane is C_5H_{12} (C_nH_{2n+2}, where $n = 5$). The straight-chain alkane has the five C atoms in a single path. There are two branched alkane isomers for pentane. One of them has a C atom with three other C atoms attached to it, and the other has a C atom with four other C atoms attached to it.

straight-chain pentane   branched pentanes

Predicting the Charges of Monatomic Ions (3.5)

Metal atoms tend to form **cations** (positively-charged ions) by losing electrons, while nonmetals tend to form **anions** (negatively-charged ions) by gaining electrons. The charge of these **monatomic ions** can be determined from the number of protons and electrons in the ion: Charge = #p$^+$ – #e$^-$. Metal cations are named using the element name plus the word 'ion'. So, Na^+ is called the sodium ion and Ca^{2+} is called the calcium ion. For main group metals (metals in groups with an 'A' in their column number), the positive charge of the metal cation is equal to the metal's column number. So, alkali metal ions (group 1A) have +1 charges (Li^+, Na^+, K^+, etc.), alkaline earth metal ions (group 2A) have +2 charges (Mg^{2+}, Ca^{2+}, Sr^{2+}, etc.), and so on. The metals at the bottom of groups 4A and 5A are an exception to this rule—they have positive charges equal to their column number minus two (Sn^{2+} and Pb^{2+} for 4A metals, and Bi^{3+} for the 5A metal).

Many of the transition elements can form more than one stable cation (e.g., copper can form Cu^+ and Cu^{2+} ions). Since you cannot name both of these ions the "copper ion", you must distinguish between these two ions by identifying the positive charge on the ion using Roman numerals and parentheses. So, Cu^+ is called the copper(I) ion and Cu^{2+} is called the copper(II) ion.

For the nonmetals, the negative charge of the nonmetal anions equals eight minus the column number. So, the halide ions (group 7A) have –1 charges (F^-, Cl^-, Br^-, etc.), group 6A ions have –2 charges (O^{2-}, S^{2-}, etc.), and so on. Nonmetal anions are named by replacing the ending of the elements name with '-*ide* ion'. So, Br^- is called the bromide ion and N^{3-} is called the nitride ion.

Common Polyatomic Ions (3.5)

Polyatomic ions are simply ions (charged objects) that have more than one atom in them (polyatomic). There are no systematic rules for naming these ions, so you will simply have to memorize the names and formulas (including charges) of the more common polyatomic ions. Table 3.7 in your textbook (p. 88) has a list of common polyatomic ions; the most commonly used ions are: the acetate ion (CH_3COO^-), the ammonium ion (NH_4^+), the carbonate ion (CO_3^{2-}), the cyanide ion (CN^-), the hydroxide ion (OH^-), the nitrate ion (NO_3^-), the phosphate ion (PO_4^{3-}), and the sulfate ion (SO_4^{2-}).

Properties of Ionic and Molecular Compounds (3.5 & 3.7)

Molecular compounds, which are made of two or more nonmetal atoms, consist of individual molecules that are weakly attracted to each other. Because of this weak attraction, these compounds tend to be gases or liquids at room temperature (low melting and boiling points), and are not particularly hard solids (easily deformed). Because these compounds are neutral and have no positive or negative ions, they are poor electrical conductors (the flow of electricity is simply the flow of charged particles, like ions or electrons). Examples include $H_2(g)$, $CO_2(g)$, $H_2O(\ell)$, $C_6H_{14}(\ell)$, and $C_{12}H_{22}O_{11}(s)$.

Ionic compounds, which are usually made from a combination of metal and nonmetal atoms, consist of cations and anions that are strongly attracted to each other through **ionic bonding** (the attraction of positive and negative ions). Unlike molecular compounds, ionic compounds do not have individual molecules in them. Instead, all of the ions in an ionic solid can be viewed as a single, incredibly large "molecule". Because ionic compounds have strong attractions holding the ions together, they tend to be solids at room temperature (high melting and boiling points) and are very hard solids. Because ionic compounds have charged particles in them, they can conduct electricity if the ions are allowed to move. That is why ionic compounds are terrible electrical conductors in their solid form (the ions can't move), but are good electrical conductors when the ions are allowed to move (e.g., in its liquid or melted state, and when dissolved in water). Examples include $NaCl(s)$, $Fe_2O_3(s)$, and $NH_4CH_3COO(s)$.

Naming Ionic Compounds (3.5 & 3.6)

Ionic compounds are named by listing the name of the cation first and the name of the anion last (omitting the word 'ion'). When naming an ionic compound, you need to determine the cation and anion present in the compound. As an example, what is the name of $BaCO_3$? You may not recognize Ba, but CO_3^{2-} is the carbonate ion (a polyatomic ion you should memorize). Since the overall charge of any compound must be zero, the Ba ion must have a +2 charge to balance the –2 charge on the carbonate ion. Ba^{2+} is the barium ion (column 2A), and its charge matches the rule (the positive charge of main group metals is equal to its column number). Since the compound is made of barium ions and carbonate ions, its name is barium carbonate (cation name first, anion name last). Similarly, an ionic compound made from the aluminum ion and the iodide ion would be called aluminum iodide. What is the chemical formula for aluminum iodide?

When writing the formula of an ionic compound, the rule is that the overall charge of the compound must be zero (neutral). Since aluminum (Al) appears in column 3A, the charge for the aluminum cation must be +3 (Al^{3+}). Since iodine (the name of the element, I) appears in column 7A, the charge for the iodide ion must be –1 (I^-; if the number is missing before a '+' or '–' sign, it is assumed to be 1). In order for aluminum iodide to be neutral, the total positive charge must be equal to the total negative charge. So, there must be 3 I^- ions for every 1 Al^{3+} ion, or AlI_3 (cations are always listed first).

Example: Write the name or formula for the following ionic compounds.
(a) $Ca_3(PO_4)_2$, (b) CrN, (c) lead(II) nitrate, (d) strontium hydride

Solution:
(a) The ions present are the calcium ion (Ca^{2+}) and the phosphate ion (PO_4^{3-}), so this is calcium phosphate. Remember that the Greek prefixes are not used in ionic compounds, only molecular compounds.
(b) The ions present are the chromium(III) ion (Cr^{3+}) and the nitride ion (the '-ide' ending tells you this is a monatomic anion of nitrogen; since N is in column 5A, its negative charge is 8 – 5 = 3 or N^{3-}), so this is chromium(III) nitride. Since chromium can have multiple charges (+2 or +3), the Roman numeral in parentheses tells you that the charge of the chromium ion in this compound is +3 (Cr^{3+}).
(c) The formula for the lead(II) ion is Pb^{2+} (the Roman numeral tells you its charge is +2), and the formula for nitrate is NO_3^-. The neutral formula is $Pb(NO_3)_2$ (1 Pb^{2+} for every 2 NO_3^-).
(d) The formula for strontium ion is Sr^{2+} (Sr is in column 2A, so its charge is +2), and the formula for hydride is H^- (the '-ide' ending tells you this is a monatomic anion of hydrogen; since H can appear in column 7A, its negative charge is 8 – 7 = 1). The neutral formula is SrH_2 (1 Sr^{2+} for every 2 H^-).

Electrolytes and Nonelectrolytes (3.7)

Electrolytes are compounds whose aqueous solutions conduct electricity. In order for the solutions to conduct electricity, they must contain charged particles (ions). So, ionic compounds dissolved in water are expected to be electrolytes because they undergo **dissociation** into ions, but molecular compounds would not because they do not produce ions. The fact that aqueous solutions of ionic compounds conduct electricity shows that the ions in these solution have the freedom to move independently—if they were stuck together, then it would still be 'neutral' and the solution would not conduct electricity. Compounds whose aqueous solutions do not conduct electricity are called **nonelectrolytes**. Compounds that do not dissolve in water, or that do not produce ions when dissolved in water, are expected to be nonelectrolytes.

The Mole Concept (3.8)

The concept of the mole was introduced in the previous chapter of the textbook (section 2.7). The mole is simply the number of atoms present in exactly 12 g of carbon-12. This definition was made so that the masses listed in the periodic table for every element are both the atomic weights (in units of amu/atom) and the molar mass (in units of g/mol). The numerical value of a mole, called *Avogadro's number*, is 6.022 × 10^{23} 'things'/mol (the 'thing' can be anything—atoms, molecules, ions, etc.).

Calculating Molar Masses (3.8)

Calculating the molar mass of a compound is done by adding up the individual molar masses of the elements present in the compound. For example, the molar mass of sucrose ($C_{12}H_{22}O_{11}$) is calculated by adding up the molar mass of 12 carbon atoms, 22 hydrogen atoms, and 11 oxygen atoms.

Example: What is the molar mass of tetraethyl lead, $Pb(C_2H_5)_4$, the additive that made gasoline 'leaded'?

Solution: The formula is $C_8H_{20}Pb$, and the molar mass would be the mass of 1 mol of $C_8H_{20}Pb$.

$$1 \text{ mol } C_8H_{20}Pb = 8 \text{ mol } C \times \frac{12.01 \text{ g C}}{\text{mol C}} + 20 \text{ mol } H \times \frac{1.008 \text{ g H}}{\text{mol H}} + 1 \text{ mol } Pb \times \frac{207.2 \text{ g Pb}}{\text{mol Pb}} = 323.4 \text{ g } C_8H_{20}Pb$$

So, the molar mass of tetraethyl lead (used as a conversion factor) is 323.4 g $C_8H_{20}Pb$ / mol $C_8H_{20}Pb$.

Converting Between Mass and Mole Values (3.8)

Molar masses of compounds, just like molar masses of elements (introduced in section 2.8 of the textbook), are mathematical equalities. Using the example for tetraethyl lead (above), you can see that the mathematical equality is 1 mol $C_8H_{20}Pb$ = 323.4 g $C_8H_{20}Pb$, and a conversion factor can be made from this molar mass value with either the mass value or the mole value on top and the other factor on bottom.

Example: You have a sample of 32.5 g of bromine pentafluoride. How many moles of bromine pentafluoride do you have? What is the mass (in g) of the same amount of disulfur tetranitride?

Solution: Before you go any further, you must determine the formulas of these two compounds. The formula for bromine pentafluoride is BrF_5, and the formula for disulfur tetranitride is S_2N_4. To answer the first question, you need to determine the molar mass of BrF_5 and use it to convert 32.5 g BrF_5 to mol BrF_5.

$$1 \text{ mol } BrF_5 = 1 \text{ mol } Br \times \frac{79.90 \text{ g Br}}{\text{mol Br}} + 5 \text{ mol } F \times \frac{19.00 \text{ g F}}{\text{mol F}} = 174.90 \text{ g } BrF_5$$

$$32.5 \text{ g } BrF_5 \times \frac{\text{mol } BrF_5}{174.90 \text{ g } BrF_5} = 0.186 \text{ mol } BrF_5$$

For the second question, determine the molar mass of S_2N_4 and use it to convert mol S_2N_4 to g S_2N_4.

$$1 \text{ mol } S_2N_4 = 2 \text{ mol } S \times \frac{32.07 \text{ g S}}{\text{mol S}} + 4 \text{ mol } N \times \frac{14.01 \text{ g N}}{\text{mol N}} = 120.18 \text{ g } S_2N_4$$

$$0.186 \text{ mol } S_2N_4 \times \frac{120.18 \text{ g } S_2N_4}{\text{mol } S_2N_4} = 22.3 \text{ g } S_2N_4$$

Hydrated Ionic Compounds (3.8)

Some ionic compounds have water molecules trapped within the solid. These solids are referred to as **ionic hydrates**, and the water molecules trapped in the solid are called **waters of hydration**. Common washing soda has two sodium ions (Na^+), one carbonate ion (CO_3^{2-}), and ten water molecules in its formula unit. Its formula is $Na_2CO_3 \cdot 10H_2O$, but could also be written $Na_2CO_3(H_2O)_{10}$. Hydrated ionic compounds are named the same way as the *anhydrous compound* (the compound without the waters of hydration), plus a Greek prefix indicating the number of waters of hydration and the word 'hydrate'. So, $Na_2CO_3 \cdot 10H_2O$ is named sodium carbonate decahydrate.

Determining Percent Composition by Mass (3.9)

Percent composition by mass is a list of the percentages (by mass) of the elements present in a compound. It is also referred to as an *elemental analysis*, and can be used to identify the molecular formula of an unknown chemical compound. To determine whether a compound has the molecular formula you think it has, you can calculate the percent composition by mass data from the proposed formula and compare these values to the actual (measured) values. The formula for percent composition by mass of element X is:

$$\% X = \frac{m_X}{m_{total}} \times 100\%$$

Example: You have collected a sample of a suspicious chemical and sent it out to a commercial laboratory for elemental analysis. The percent composition by mass data returned to you by the laboratory are 40.0% C, 6.7% H, and 53.3% O. Is this compound more likely to be glucose ($C_6H_{12}O_6$), a common sugar used by the human body, or formaldehyde (CH_2O), a known carcinogen?

Solution: You need to determine the percent composition by mass for glucose and for formaldehyde. For each sample, assume that you have exactly 1 mol of the chemical, and determine the percent composition values for each element based on those masses.

$$1 \text{ mol } C_6H_{12}O_6 = 6 \text{ mol C} \times \frac{12.01 \text{ g C}}{\text{mol C}} + 12 \text{ mol H} \times \frac{1.008 \text{ g H}}{\text{mol H}} + 6 \text{ mol O} \times \frac{16.00 \text{ g O}}{\text{mol O}} = 180.16 \text{ g } C_6H_{12}O_6$$

$$\% \text{ C} = \frac{6 \text{ mol C} \times \dfrac{12.01 \text{ g C}}{\text{mol C}}}{180.16 \text{ g } C_6H_{12}O_6} \times 100\% = 40.00\% \text{ C}$$

$$\% \text{ H} = \frac{12 \text{ mol H} \times \dfrac{1.008 \text{ g H}}{\text{mol H}}}{180.16 \text{ g } C_6H_{12}O_6} \times 100\% = 6.714\% \text{ H}$$

$$\% \text{ O} = \frac{6 \text{ mol O} \times \dfrac{16.00 \text{ g O}}{\text{mol O}}}{180.16 \text{ g } C_6H_{12}O_6} \times 100\% = 53.29\% \text{ O}$$

$$1 \text{ mol } CH_2O = 1 \text{ mol C} \times \frac{12.01 \text{ g C}}{\text{mol C}} + 2 \text{ mol H} \times \frac{1.008 \text{ g H}}{\text{mol H}} + 1 \text{ mol O} \times \frac{16.00 \text{ g O}}{\text{mol O}} = 30.03 \text{ g } CH_2O$$

$$\% \text{ C} = \frac{1 \text{ mol C} \times \dfrac{12.01 \text{ g C}}{\text{mol C}}}{30.03 \text{ g } CH_2O} \times 100\% = 40.00\% \text{ C}$$

$$\% \text{ H} = \frac{2 \text{ mol H} \times \dfrac{1.008 \text{ g H}}{\text{mol H}}}{30.03 \text{ g } CH_2O} \times 100\% = 6.714\% \text{ H}$$

$$\% \text{ O} = \frac{1 \text{ mol O} \times \dfrac{16.00 \text{ g O}}{\text{mol O}}}{30.03 \text{ g } CH_2O} \times 100\% = 53.29\% \text{ O}$$

Even though glucose and formaldehyde have different chemical formulas (as well as different chemical and physical properties), they have the same percent composition by mass values. This is because they have the same relative ratios of atoms (C:H:O in a ratio of 1:2:1). Because all chemicals with the same atom ratios have the same percent composition by mass values, the compound's molecular formula provides more information than the percent composition by mass values.

Determining Empirical and Molecular Formulas (3.10)

The percent composition by mass data measured in a chemistry laboratory consist of masses, but molecular formulas are written based on numbers of atoms (moles). In order to convert percent composition by mass data into a chemical formula (called the **empirical formula**), you need to use the molar masses of these elements to convert the mass data to mole data. The empirical formula represents the simplest ratio of atoms present in a compound. To determine the *molecular formula* (the actual chemical formula) you need to know the approximate molar mass of the compound.

Example: An ionic compound was sent for elemental analysis, and the laboratory reported that the chemical was 47.0% K, 14.5% C, and 38.5% O. What is the empirical formula of this compound? The molar mass of the compound was estimated to be about 170 g/mol. What is the molecular formula of this compound?

Solution: In order to convert the percent values into moles, assume that you have 100 g of the compound, and convert these masses to moles using each element's molar mass. This assumption is made so that the percent values are now masses in g (in 100 g of sample, there is 47.0 g K, 14.5 g C, and 38.5 g O).

$$47.0 \text{ g K} \times \frac{\text{mol K}}{39.10 \text{ g K}} = 1.20205 \text{ mol K (unrounded)}$$

$$14.5 \text{ g C} \times \frac{\text{mol C}}{12.01 \text{ g C}} = 1.20733 \text{ mol C (unrounded)}$$

$$38.5 \text{ g O} \times \frac{\text{mol O}}{16.00 \text{ g O}} = 2.40625 \text{ mol O (unrounded)}$$

The ratios of K:C:O are 1.20205:1.20733:2.40625, and you could write the formula $K_{1.20205}C_{1.20733}O_{2.40625}$. However, molecular formulas are written with whole number subscripts because you can't have fractions of K, C, or O atoms. To find the whole number ratios in the empirical formula, divide each subscript by the smallest subscript value.

$$K_{\frac{1.20205}{1.20205}}C_{\frac{1.20733}{1.20205}}O_{\frac{2.40625}{1.20205}} = K_{1.0000}C_{1.004}O_{2.002} = KCO_2$$

The empirical formula of this compound is KCO_2, but any whole-number multiple of this formula ($K_2C_2O_4$, $K_3C_3O_6$, etc.) could also be correct. To find the molecular formula (which is really a formula unit in this case, since this is an ionic compound and not a molecular compound), you need to compare the empirical mass to the molar mass of the compound, recognizing that the molar mass is some whole number times the empirical mass.

$$M \text{ molecular} = n \times M \text{ empirical}, \ n = \frac{M \text{ molecular}}{M \text{ empirical}}$$

$$1 \text{ mol } KCO_2 = 1 \text{ mol K} \times \frac{39.10 \text{ g K}}{\text{mol K}} + 2 \text{ mol C} \times \frac{12.01 \text{ g C}}{\text{mol C}} + 2 \text{ mol O} \times \frac{16.00 \text{ g O}}{\text{mol O}} = 83.11 \text{ g } KCO_2$$

$$n = \frac{M \text{ molecular}}{M \text{ empirical}} = \frac{170 \text{ g/mol}}{83.11 \text{ g/mol}} = 2.05 = 2$$

This means that the molecular formula is exactly 2 times the empirical formula: $KCO_2 \times 2 = K_2C_2O_4$. The ionic compound is actually potassium oxalate, consisting of potassium (K^+) ions and oxalate ($C_2O_4^{2-}$) ions.

For ionic compounds, the empirical formula is usually reported and a molecular formula is not calculated because the formula unit of an ionic compound is the simplest ratio of ions (although there are exceptions).

Biologically Important Elements (3.11)

Although four elements (H, O, C, and N) make up the bulk of the human body's mass, there are many other elements that are still required for human life to exist, just in smaller quantities. For example, phosphorus (P) is needed to make DNA, iron (Fe) is needed to make blood, calcium (Ca) is needed to make bones and teeth, chlorine (Cl) is needed to digest food, iodine (I) is needed to keep the thyroid gland functioning, and so on. The term **dietary mineral** is used to describe the elements (except for H, O, C, and N) needed to keep the human body functioning. Dietary minerals can be divided into two categories—**major minerals** (present in quantities greater than 0.01% of the body's total mass) and **trace elements** (present in quantities less than 0.01% of the body's total mass).

Chapter Review — Key Terms

The key terms that were introduced in this chapter are listed below, along with the section in which they were introduced. You should understand these terms and be able to apply them in appropriate situations.

alkane (3.3)
alkyl group (3.4)
anion (3.5)
binary molecular compound (3.2)
cation (3.5)
chemical bond (3.3)
condensed formula (3.1)
constitutional isomer (3.4)
Coulomb's law (3.7)
crystal lattice (3.7)
dietary mineral (3.11)
dissociation (3.7)
electrolyte (3.7)

empirical formula (3.10)
formula unit (3.7)
formula weight (3.8)
functional group (3.1)
halide ion (3.6)
hydrocarbon (3.3)
inorganic compound (3.1)
ionic bonding (3.7)
ionic compound (3.5)
ionic hydrate (3.8)
isomer (3.4)
major mineral (3.11)
molecular compound (3.1)

molecular formula (3.1)
molecular weight (3.8)
monatomic ion (3.5)
nonelectrolyte (3.7)
organic compound (3.1)
oxoanion (3.6)
percent composition by mass (3.9)
polyatomic ion (3.5)
structural formula (3.1)
trace element (3.11)
water of hydration (3.8)

Practice Test

After you have finished studying the chapter and the homework problems, the following questions can serve as a test to determine how well you have learned the chapter objectives.

1.   Complete the following chart of molecular, condensed, and structural formulas.

molecular formula	condensed formula	structural formula
_____	$NH_2CH_2CH_2NH_2$	_____
_____	$(CH_3)_2CHCH_2C{\equiv}N$	_____
_____	_____	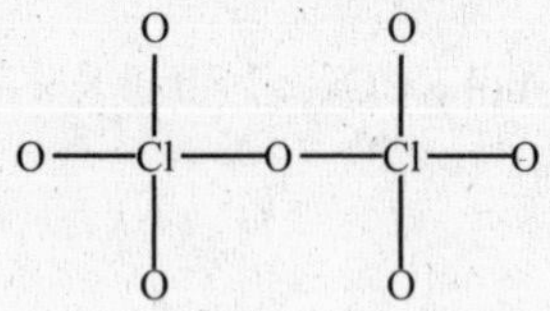 (structure shown)
_____	_____	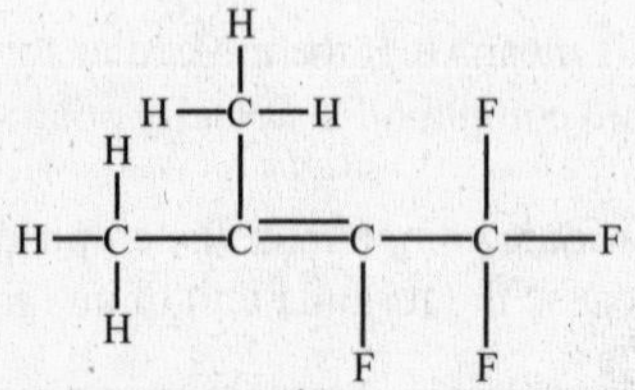 (structure shown)

2.   Consider the following structural formulas.

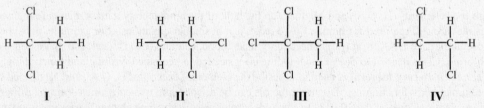

      I   II   III   IV

(a)   Can all of these structures be isomers?
(b)   Are any of these structures identical (i.e., the same compound)?
(c)   Which of these structures are constitutional isomers of each other?

3. Determine the systematic names for the following molecular compounds.
   (a) water, H_2O (d) nitric oxide, NO
   (b) ammonia, NH_3 (e) nitrous oxide, N_2O
   (c) hydrazine, N_2H_4 (f) phosphine, PH_3

4. For each of the elements listed below, predict the charge of its monatomic ion.
   (a) carbon   (d) lithium
   (b) fluorine   (e) selenium
   (c) iron   (f) strontium

5. Predict the formulas of the ionic compounds made from the following components.
   (a) bromine and calcium   (d) magnesium and oxygen
   (b) carbon and potassium   (e) zinc and the cyanide ion
   (c) chromium and the hydroxide ion   (f) sulfur and the ammonium ion

6. Complete the following chart of chemical compounds using the compound's molecular formulas.
   For the 'compound type' column, your choices are *molecular*, *organic*, *ionic*, *molecular hydrate*, and
   ionic hydrate. For the 'ions present' column, the correct answer may be *none*.

molecular formula	compound type	ions present	chemical name
B_2O_3	_____	_____	_____
C_3H_8	_____	_____	_____
$CoCl_2 \cdot 6H_2O$	_____	_____	_____
$FeCO_3$	_____	_____	_____
IF_7	_____	_____	_____
K_2Se	_____	_____	_____
S_5N_6	_____	_____	_____

7. Complete the following chart of chemical compounds using the compound's names (use the same
   choices for the 'compound type' column and the 'ions present' column as in question 6).

molecular formula	compound type	ions present	chemical name
_____	_____	_____	aluminum sulfate
_____	_____	_____	butane
_____	_____	_____	chlorine trifluoride
_____	_____	_____	iron(III) nitrate nonahydrate
_____	_____	_____	magnesium hydroxide
_____	_____	_____	nonane
_____	_____	_____	sodium nitride

23

8.   A compound of titanium and chlorine has the formula $TiCl_4$. This compound has a melting point of –25°C and is a liquid at room temperature, but does not conduct electricity.
    (a)   Based on the physical properties of this compound, is it a molecular or ionic compound?
    (b)   Given the elements in this compound, is this surprising?
    (c)   Write the name of this compound, based on its physical properties.

9.   Calcium phosphate, $Ca_3(PO_4)_2$, is a major component in fertilizers.
    (a)   What is the molar mass of calcium phosphate?
    (b)   What is the mass of 1.25 mol of calcium phosphate?
    (c)   How many moles of phosphorus atoms are present in 100.0 g of calcium phosphate?
    (d)   What is the percent phosphorus in this compound?

10.   Calculate the percent sulfur by mass for the following compounds.
    (a)   Hydrogen sulfide, H_2S (the smell of rotten eggs)
    (b)   Iron pyrite, FeS_2 (fool's gold)
    (c)   Sulfur dioxide, SO_2 (a major component of acid rain)
    (d)   Sulfuric acid, H_2SO_4 (the burn from cutting onions)

11.   The elemental analysis of a dark-green lustrous chemical was found to be 33.3% Fe, 28.6% C, and 38.1% O.
    (a)   What is the empirical formula for this compound?
    (b)   If its molar mass is approximately 500 g/mol, what is its molecular formula?

Chapter 4
Quantities of Reactants and Products

The goal of this chapter is to introduce one of the most useful tools used by chemists—the balanced chemical equation. Although balanced chemical equations are inherently symbolic (they contain atom symbols, molecular formulas, subscripts, coefficients, arrows, etc.), they also represent the macroscale interactions of reactants and products as well as the nanoscale interpretation of these interactions. This chapter also introduces the mathematical calculations known as **stoichiometry**. Stoichiometric calculations relate the quantities (amounts in grams or moles) of one chemical in the balanced chemical equation to all of the other chemicals in that equation. The skill of performing stoichiometric calculations is very important and something you will need if you take another chemistry course, whether it is second-semester general chemistry, quantitative analysis, organic chemistry, or biochemistry.

Interpreting Balanced Chemical Equations (4.1)

Balanced chemical equations tell the story of a chemical reaction as it progresses. The starting materials or *reactants* are the chemicals present before the reaction occurs, and they react with each other to form the *products*, the chemicals present after the reaction occurs. In a balanced chemical equation, the reactants are listed to the left of the arrow, the products are listed to its right, and the arrow represents the chemical reaction. As an example, calcium carbonate reacts with hydrochloric acid to form calcium chloride, water, and carbon dioxide.

$$CaCO_3(s) + 2\,HCl(aq) \rightarrow MgCl_2(aq) + H_2O(\ell) + CO_2(g)$$

The symbols in parentheses represent the state of matter of each chemical: Calcium carbonate is present as a solid (s), water is present as a liquid (ℓ), carbon dioxide is present as a gas (g), and hydrochloric acid and calcium chloride are present as aqueous solutions (aq). An **aqueous solution** is simply a sample of the chemical dissolved in water.

Balanced chemical equations contain two types of numbers. *Subscripts* are numbers that appear within a molecular formula (e.g., the 2 in H_2O, $MgCl_2$, and H_2 and the 3 in $CaCO_3$). Subscripts represent the number of atoms or ions present <u>within</u> the molecule; in water there are 2 H atoms and 1 O atom, in magnesium chloride there are 1 Mg^{2+} ion and 2 Cl^- ions, in carbon dioxide there are 1 C atom and 2 O atoms, and in calcium carbonate there are 1 Ca atom, 1 C atom, and 3 O atoms. **Coefficients** are the numbers that appear in front of the molecular formulas and tells you the total <u>number</u> of each molecule that reacts or is produced as a result of the reaction. So, one formula unit of calcium carbonate (an ionic compound) reacts with two formula units of hydrochloric acid (also ionic) to form one formula unit of magnesium chloride (also ionic), one molecule of water, and one molecule of carbon dioxide.

<u>Example</u>: What is the difference between 'N_2O_4' and '2 NO_2'? What about '3 O_2' and '2 O_3'?

<u>Solution</u>: Subscripts tell you how many atoms are in a molecule, and coefficients tell you how many of those molecules are present. So, 'N_2O_4' represents one molecule that has 2 N atoms and 4 O atoms attached together; '2 NO_2' represents two independent molecules that each have 1 N atom and 2 O atoms attached together. Similarly, '3 O_2' represents 3 molecules that each have 2 O atoms attached together (diatomic) while '2 O_3' represents 2 molecules that each have 3 O atoms attached together (triatomic). Simple drawings of these differences may help those of you who are visual learners.

General Reaction Types (4.2)

Even though there are millions of different chemical reactions, many of them fall into categories based on similarities in the ways the reactants are transformed into the products. This chapter describes five general reaction types—combustion reactions, combination reactions, decomposition reactions, displacement reactions, and exchange reactions.

Combustion reactions are just the burning of chemicals in the presence of oxygen. When a chemical is burned, most the elements in the chemical are converted to oxides of this element. For example, carbon atoms in the chemical are converted to carbon dioxide ($C \rightarrow CO_2$), hydrogen atoms are converted to water ($H \rightarrow H_2O$), sulfur atoms are converted to sulfur dioxide ($S \rightarrow SO_2$), and metals are converted to metal oxides (e.g., $Li \rightarrow Li_2O$, $Mg \rightarrow MgO$, $Fe \rightarrow Fe_2O_3$, etc.). The one exception is that nitrogen atoms are converted to nitrogen gas ($N \rightarrow N_2$). Combustion reactions can be recognized by the fact that oxygen (O_2) must be a reactant, and the products are element oxides (CO_2, H_2O, SO_2, metal oxides, etc).

Combination reactions occur when two or more chemicals react to form one product. When metal and nonmetal elements combine, they form ionic compounds; when two nonmetal elements combine they form molecular compounds. Combination reactions can be identified by the fact that there are more chemicals on the reactant side than the product side, and there is usually only one product formed in a combination reaction. **Decomposition reactions** are exact opposites of combination reactions. In a decomposition reaction, one reactant is transformed into two or more products. Decomposition reactions can be identified by the fact that there are more chemicals on the product side than the reactant side, and there is usually only one reactant in a decomposition reaction.

Displacement reactions (also called *single displacement reactions*) occur when one element reacts with a compound to produce another element. These reactions can be identified by the fact that there is an element as a reactant and a different element as a product. These reactions are usually oxidation-reduction reactions (which will be discussed in Chapter 5).

Exchange reactions (also called *double displacement* or *metathesis reactions*) usually occur when two ionic compounds react with each other. In exchange reactions, the cations of the two reactants exchange anions, resulting in two new ionic compounds. Exchange reactions will be discussed further in Chapter 5.

Example: Classify the following equations according to the reaction types discussed above.
(a) $(NH_4)_2Cr_2O_7(s) \rightarrow Cr_2O_3(s) + 4 H_2O(\ell) + N_2(g)$
(b) $BaCl_2(aq) + K_2SO_4(aq) \rightarrow BaSO_4(s) + 2 KCl(aq)$
(c) $2 Al(s) + 3 CuO(s) \rightarrow Al_2O_3(s) + 3 Cu(s)$
(d) $Fe(OH)_3(aq) + 3 HNO_3(aq) \rightarrow Fe(NO_3)_3(aq) + 3 H_2O(\ell)$
(e) $P_4(s) + 5 O_2(g) \rightarrow P_4O_{10}(s)$

Solution:
(a) This is a decomposition reaction because there is only one reactant but there are three products.
(b) This is an exchange reaction because Ba^{2+} trades the Cl^- ion for the SO_4^{2-} ion while K^+ trades the SO_4^{2-} ion for the Cl^- ion. Using the chemical names makes this exchange more obvious: barium chloride and potassium sulfate become barium sulfate and potassium chloride.
(c) This reaction is a displacement reaction because elemental Al "steals" the O atoms from the Cu-O compound, forming a Al-O compound and elemental Cu.
(d) This is an exchange reaction because Fe^{3+} trades the OH^- ion for the NO_3^- ion while H^+ trades the NO_3^- ion for the OH^- ion, making HOH or H_2O.
(c) This reaction can be classified as a combustion reaction and a combination reaction. It is a combustion reaction because oxygen, O_2, is a reactant and the product is an element oxide (tetraphosphorus decoxide); it is a combination reaction because two reactants are converted into one product.

Balancing Chemical Equations (4.3)

Balancing chemical equations is the process of making sure that there are the same numbers of each type of atom on the left (reactant) and right (product) sides. The reason these values need to be the same is based on the assumption of the modern atomic theory that atoms are indestructible. So, there must be the same number and type of atoms before and after the reaction has occurred. Most chemical equations are balanced by trial-and-error, and as long as the equation is balanced it doesn't matter how it was balanced.

There are some guidelines for balancing chemical equations. (1) Only coefficients (never subscripts) are changed when balancing the equation; (2) If an element appears more than once on either side of the equation, it is usually balanced last; (3) Balanced equations are usually written with the smallest whole-number ratios possible; and (4) Polyatomic ions that appear on both sides of the equation can usually be treated as a single object (but the equation can be balanced by looking at the atoms within these ions).

Example: When balancing the reaction of hydrogen and water, why can't you change the subscripts in water to be H_2O_2? The lines before each chemical in the equation tell you the equation is unbalanced.

$$\_ H_2(g) + \_ O_2(g) \rightarrow \_ H_2O(\ell)$$
$$H_2(g) + O_2(g) \rightarrow H_2O_2(\ell)?$$

Solution: Although the equation is balanced (2 H and 2 O atoms on each side), by changing the subscript you have changed the chemical. This is no longer a balanced equation for making water (H_2O) from its elements; it is now an equation for making hydrogen peroxide (H_2O_2) from its elements. You are allowed to change the coefficients in a chemical equations because changing the coefficients does not change the chemicals present, only how much of each chemical reacts or is produced.

Example: Write a balanced chemical reaction for the nanoscale picture below.

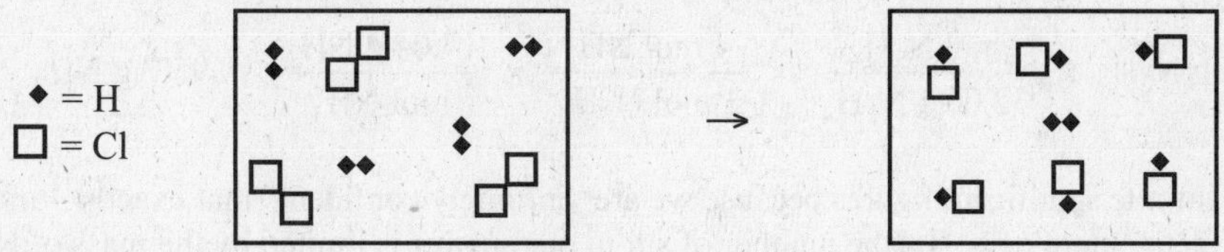

♦ = H
□ = Cl

Solution: First, you need to determine the formula and number of the chemicals present. On the reactants side, there are 4 molecules of H_2 and 3 molecules of Cl_2. For the products, there are 6 molecules of HCl (remember HCl and ClH are the same chemical) and 1 molecule of H_2 left over. The equation for the picture is written below (you could have written Cl_2 as the first reactant or H_2 as the first product, the order of the reactants and products does not matter):

$$4\,H_2 + 3\,Cl_2 \rightarrow 6\,HCl + H_2$$

When writing a balanced chemical equation, do not leave objects in the reaction that don't react. On both sides of the equation, there is an extra H_2 molecule that should be removed. This gives you the equation $3\,H_2 + 3\,Cl_2 \rightarrow 6\,HCl$, which can be reduced (divided by 3) to give the final reaction.

$$H_2 + Cl_2 \rightarrow 2\,HCl$$

Mole Ratios and Balanced Chemical Equations (4.4)

The coefficients in a balanced chemical equation show the number of reactant molecules that react and the number of product molecules that are produced. These coefficients can also be applied to molar amounts (i.e., the numbers of *moles* of reactant molecules used and the number of *moles* of product molecules produced). These coefficients can be used to create conversion factors called **mole ratios** relating the number of any two chemicals involved in the chemical reaction. While the other conversion factors you have used in the past were mathematical equalities (12 in = 1 ft, 1 kg = 1000 g, etc.), mole ratios are actually "chemical equivalents" and not mathematical equalities, even though they are used in the same ways in mathematical calculations. Consider the equation below:

$$S_8(s) + 12\,O_2(g) \rightarrow 8\,SO_3(g)$$

Mole ratios can be written with any chemical on top (with the coefficient, and units of either molecule or mol) and any other chemical on bottom. The mole ratio of '8 molecule SO_3 / 1 molecule S_8' shows the chemical equivalence (relationship) that 1 molecule of S_8 is needed to make 8 molecules of SO_3, and the mole ratio of '12 mol O_2 / 1 mol S_8' shows that 12 moles of O_2 are needed to react with 1 mole of S_8.

Combining mole ratios with conversion factors made from the molar masses of chemicals in the reaction allows you to convert masses of one chemical into masses of another chemical in the balanced equation.

Example: How many grams of ammonia can be made from 135 g of dinitrogen tetrahydride (hydrazine)?

$$\_\,N_2H_4(\ell) \rightarrow \_\,NH_3(g) + \_\,N_2(g)$$

Solution: First, you need to balance the equation. Placing a 3 in front of the N_2H_4 and a 4 in front of the NH_3 will balance H atoms (12 on each side). This leaves the N atoms to be balanced. There are 6 N atoms in 3 N_2H_4; there are 4 N atoms in 4 NH_3 and 2 N atoms in 1 N_2. So, the N atoms are balanced and so is the equation (6 N atoms and 12 H atoms on both sides).

$$3\,N_2H_4(\ell) \rightarrow 4\,NH_3(g) + N_2(g)$$

To determine the mass of NH_3, you need to: (1) Convert g N_2H_4 to mol N_2H_4 using the molar mass of N_2H_4, (2) Use a mole ratio to convert mol N_2H_4 to mol NH_3, and (3) Use the molar mass of NH_3 to convert from mol NH_3 to g NH_3.

$$135\text{ g }N_2H_4 \times \frac{\text{mol }N_2H_4}{32.06\text{ g }N_2H_4} \times \frac{4\text{ mol }NH_3}{3\text{ mol }N_2H_4} \times \frac{17.04\text{ g }NH_3}{\text{mol }NH_3} = 95.7\text{ g }NH_3$$

Mole ratios have infinite significant figures because we are "infinitely confident" that exactly 4 mol of NH_3 is formed from exactly 3 mol of N_2H_4 (the number of s.f. in our answer is limited by the mass of N_2H_4). Note also that not only do the numbers have units, but the units have units as well, and both are important.

Chemical Analysis of Mixtures (4.4)

Chemical analysis of mixtures calculations bring together two mathematical calculations that you have used in the past. The first is the mass to mole conversions introduced in Section 4.4; the second is the definition of percent (used in percent composition by mass calculations performed in Chapter 3).

Determining the Limiting Reactant (4.5)

In the last example, there was only one reactant so the amount of product made depended only on the amount of this reactant present. When you have more than one reactant, either reactant can limit the amount of chemical made because, depending on the amount of the two reactants present, either one could run out first. The chemical that runs out is called the **limiting reactant** (also called the *limiting reagent*). In most reactions performed in industry or in a chemistry laboratory, the limiting reactant is determined by chemists since they tend to use an excess of the cheaper or more plentiful reactant. The other chemicals that are not used up as part of the reaction are called *excess reactants* (or *excess reagents*).

Example: Consider the reaction and nanoscale picture below.

$$N_2(g) + 3 F_2(g) \rightarrow 2 NF_3(g)$$

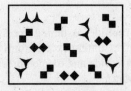

(a) How many molecules of NF_3 can be made from the N_2 molecules in the picture, assuming excess F_2?
(b) How many molecules of NF_3 can be made from the F_2 molecules in the picture, assuming excess N_2?
(c) How many molecules of NF_3 can be made from the N_2 and F_2 molecules in the picture?
(d) What is the limiting reactant, and what is the excess reactant?
(e) How many molecules of the excess reactant remain after the reaction is complete?
(f) Complete the nanoscale picture, showing the contents of the product box.

Solution:
(a) Since F_2 is in excess, N_2 must be the limiting reactant. So, calculate the number of NF_3 that can be made from the N_2 molecules present.

$$4 \text{ molecule } N_2 \times \frac{2 \text{ molecule } NF_3}{1 \text{ molecule } N_2} = 8 \text{ molecule } NF_3$$

(b) Since N_2 is now in excess, F_2 must be the limiting reactant. So, calculate the number of NF_3 that can be made from the F_2 molecules present.

$$9 \text{ molecule } F_2 \times \frac{2 \text{ molecule } NF_3}{3 \text{ molecule } F_2} = 6 \text{ molecule } NF_3$$

(c) The answer for this question is simply the smaller of the two answers above, which is 6 molecules NF_3
Steps (a)-(c) are the same as the Mass Method used on p. 139 of the textbook.
(d) The limiting reactant is F_2 and the excess reactant is N_2. There is enough N_2 and F_2 present to make up to 6 molecules of NF_3, but once 6 molecules of NF_3 are made all of the F_2 present has been used. So, F_2 limits the amount of NF_3 made, and N_2 does not (there is enough N_2 to make 8 molecules of NF_3).
(e) The way to find out how much of the excess reactant is left over is to determine how much of the excess reactant reacted with the limiting reactant, and subtract that amount from the starting amount of the excess reactant (if this number is negative, then you have chosen the wrong reactant as limiting).

$$9 \text{ molecule } F_2 \times \frac{1 \text{ molecule } N_2}{3 \text{ molecule } F_2} = 3 \text{ molecule } N_2 \text{ reacted}$$

$$4 \text{ molecule } N_2 \text{ total} - 3 \text{ molecule } N_2 \text{ reacted} = 1 \text{ molecule } N_2 \text{ left over}$$

(f) So, the product box should contain 6 molecules of NF_3 and 1 molecule of N_2. Checking to see that the atoms are balanced, we find that both boxes contain 8 atoms of N and 18 atoms of F.

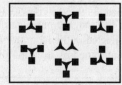

When determining the limiting reactant using the masses of products or reactants, the textbook introduces two ways to determine (calculate) which of the reactants is the limiting reactant, and you can use either method. In this study guide, the Mass Method will be used. Using this method, you calculate the mass of the product you are interested in for each of the reactants and your final answer is simply the smallest of the calculated values.

29

Example: Magnetite (Fe_3O_4) is produced by heating 632 g of iron(II) sulfide and 457 g of oxygen.

$$\_ FeS(s) + \_ O_2(g) \rightarrow \_ Fe_3O_4(s) + \_ SO_2(g)$$

(a) Balance the chemical equation.
(b) What type of reaction is this?
(c) How many grams of Fe_3O_4 can be made from the iron(II) sulfide, assuming excess oxygen?
(d) How many grams of Fe_3O_4 can be made from the oxygen, assuming excess iron(II) sulfide?
(e) How many grams of Fe_3O_4 can be made from the iron(II) sulfide and oxygen present?
(f) What is the limiting reactant, and what is the excess reactant?
(g) How many gram of the excess reactant remain after the reaction is complete?

Solution:
(a) Placing a 3 in front of the FeS will balance the Fe atoms. Then, placing a 3 in front of the SO_2 will balance the S atoms. This gives us a total of 10 O atoms on the right; in order to get 10 O atoms on the left, place a 5 in front of the O_2. The equation is now balanced (3 Fe atoms, 3 S atoms, and 10 O atoms).

$$3\ FeS(s) + 5\ O_2(g) \rightarrow Fe_3O_4(s) + 3\ SO_2(g)$$

(b) Since oxygen is one of the reactants, and the two products are element oxide compounds, this reaction is a combustion reaction.
(c) Since O_2 is in excess, Fe_3O_4 is the limiting reactant.

$$632\ g\ FeS \times \frac{mol\ FeS}{87.92\ g\ FeS} \times \frac{1\ mol\ Fe_3O_4}{3\ mol\ FeS} \times \frac{231.55\ g\ Fe_3O_4}{mol\ Fe_3O_4} = 555\ g\ Fe_3O_4$$

(d) Since FeS is now in excess, O_2 must be the limiting reactant.

$$457\ g\ O_2 \times \frac{mol\ O_2}{32.00\ g\ O_2} \times \frac{1\ mol\ Fe_3O_4}{5\ mol\ O_2} \times \frac{231.55\ g\ Fe_3O_4}{mol\ Fe_3O_4} = 661\ g\ Fe_3O_4$$

(e) The answer for this question is simply the smaller of the two answers above, or 555 g Fe_3O_4.
(f) The limiting reactant is FeS and the excess reactant is O_2. Once 555 g Fe_3O_4 is made, all of the FeS is used. So, FeS limits the amount of Fe_3O_4 made, and O_2 does not (there is enough O_2 to make 661 g Fe_3O_4).
(g) First, you need to determine how much of the excess reactant (O_2) reacted with the limiting reactant (FeS), and then subtract that amount from the starting amount of the excess reactant.

$$632\ g\ FeS \times \frac{mol\ FeS}{87.92\ g\ FeS} \times \frac{5\ mol\ O_2}{3\ mol\ FeS} \times \frac{32.00\ g\ O_2}{mol\ O_2} = 383\ g\ O_2$$
$$457\ g\ O_2\ total - 383\ g\ O_2\ reacted = 74\ g\ O_2\ left\ over$$

Actual Yield, Theoretical Yield, and Percent Yield (4.6)

Stoichiometric calculations, like the ones performed above, determine the maximum possible amount of product that can be formed. This is called the **theoretical yield**. In actual laboratories, however, the maximum theoretical yield possible may not be made due to side reactions (producing unwanted chemicals) and the fact that the purification processes often cause losses of product. The **actual yield** is what is actually collected in the laboratory, and is almost always less than the theoretical yield. The **percent yield** is determined by dividing the actual yield by the theoretical yield and multiplying by 100%.

Example: What is the percent yield if the reaction in the last example produced 514 g Fe_3O_4?

Solution: $\%\ yield = \dfrac{actual\ yield}{theoretical\ yield} \times 100\% = \dfrac{514\ g\ Fe_3O_4}{555\ g\ Fe_3O_4} \times 100\% = 92.6\%$

Determining an Empirical Formula by Combustion Analysis (4.7)

In Chapter 3, you learned how to calculate empirical formulas from percent composition by mass (elemental analysis) data. The way most chemical laboratories determine these values for you is by performing a **combustion analysis**. In a combustion analysis, the compound is burned in the presence of oxygen, and the oxide products are collected and weighed. The products of the combustion analysis not only tell you which elements were present in the original compound (CO_2 as a product tells you the original chemical has C in it, H_2O as a product means it has H in it, N_2 as a product means it has N in it, and so on), it also tells you how much of each element was present. Unfortunately, combustion analyses cannot tell you whether the original compound had oxygen in it because of the additional oxygen added as part of the combustion process. So, you will have to determine the masses of all the other elements present in the compound, and any mass unaccounted for must have been caused by oxygen in the compound.

Example: When 10.00 g of a gaseous compound (whose molar mass is approximately 95 g/mol) is burned by combustion analysis, 23.38 g of CO_2, 5.74 g of H_2O, and 2.98 g of N_2 are produced.
(a) What elements are present in this compound?
(b) Determine the number of moles of each element present.
(c) Does this compound contain oxygen? If so, how many moles of oxygen are present?
(d) What is the empirical formula for this compound?
(e) What is the molecular formula for this compound?
(f) Write a balanced chemical equation for the combustion of this compound.

Solution:
(a) Because CO_2, H_2O, and N_2 are combustion products, the compound must contain C, H, and N atoms; it may also contain O atoms.
(b) You need to convert the masses of the chemicals listed above into moles of each element.

$$23.38 \text{ g } CO_2 \times \frac{\text{mol } CO_2}{44.01 \text{ g } CO_2} \times \frac{1 \text{ mol C}}{1 \text{ mol } CO_2} = 0.53124 \text{ mol C (unrounded)}$$

$$5.74 \text{ g } H_2O \times \frac{\text{mol } H_2O}{18.016 \text{ g } H_2O} \times \frac{2 \text{ mol H}}{1 \text{ mol } H_2O} = 0.63721 \text{ mol H (unrounded)}$$

$$2.98 \text{ g } N_2 \times \frac{\text{mol } N_2}{28.02 \text{ g } N_2} \times \frac{2 \text{ mol N}}{1 \text{ mol } N_2} = 0.21271 \text{ mol N (unrounded)}$$

(c) To determine if the compound has any oxygen in it, convert the moles of elements above into masses of elements, and add them up; if they do not add to 10.00 g, then the missing mass is due to oxygen.

$$0.53124 \text{ mol C} \times \frac{12.01 \text{ g C}}{\text{mol C}} = 6.38 \text{ g C}$$

$$0.63721 \text{ mol H} \times \frac{1.008 \text{ g H}}{\text{mol H}} = 0.64 \text{ g H}$$

$$0.21271 \text{ mol N} \times \frac{14.01 \text{ g N}}{\text{mol N}} = 2.98 \text{ g N}$$

$$6.38 \text{ g C} + 0.64 \text{ g H} + 2.98 \text{ g N} = 10.00 \text{ g total}$$

Since the total mass added to 10.00 g, there is no oxygen in this compound.
(d) Using the mol data in question (b), the empirical formula is shown below.

$$C_{\frac{0.53124}{0.21271}} H_{\frac{0.63721}{0.21271}} N_{\frac{0.21271}{0.21271}} = C_{2.497}H_{2.996}N_{1.000} = C_{4.995}H_{5.991}N_{2.000} = C_5H_6N_2$$

(e) The empirical mass of $C_5H_6N_2$ is 94.118 g/mol; since this value matches the molar mass given in the question, the formula for the compound must also be $C_5H_6N_2$.
(f) The reactants are $C_5H_6N_2$ and O_2; the products are CO_2, H_2O, and N_2 (all gases).

$$2 C_5H_6N_2(g) + 13 O_2(g) \rightarrow 10 CO_2(g) + 6 H_2O(g) + 2 N_2(g)$$

You can also calculate the empirical formula of a compound if you are given mass data instead of percent composition by mass values. In this case, you do not have to assume there is 100 g of the compound present since you are already given mass data.

Example: When 5.00 g of solid copper is reacted with excess chlorine gas, 10.57 g of a solid copper-chlorine compound is made.
(a) Determine the empirical formula for this compound.
(b) Why is it unnecessary to determine the molecular formula?
(c) What is the name for this compound?
(d) Write a balanced equation for this reaction. What kind of reaction is this?

Solution:
(a) The new compound is made of 5.00 g Cu and 10.57 g – 5.00 g = 5.57 g Cl. First determine the number of moles of each element, then determine the empirical formula.

$$5.00 \text{ g Cu} \times \frac{\text{mol Cu}}{63.55 \text{ g Cu}} = 0.078678 \text{ mol Cu (unrounded)}$$

$$5.57 \text{ g Cl} \times \frac{\text{mol Cl}}{35.45 \text{ g Cl}} = 0.15712 \text{ mol Cl (unrounded)}$$

$$Cu_{\frac{0.078678}{0.078678}}Cl_{\frac{0.15712}{0.078678}} = Cu_{1.000}Cl_{1.997} = CuCl_2$$

(b) Since this is an ionic compound (metal and nonmetal), the formula unit is reported as the simplest ratio of ions (the empirical formula), so there is no need to determine a molecular formula.
(c) The name is copper(II) chloride. Because copper can form Cu^+ or Cu^{2+} ions, the Roman numeral and parentheses are needed to show that this compound contains Cu^{2+} ions along with the Cl^- ions.
(d) This reaction is a combination reaction because elemental copper and chlorine react to form $CuCl_2$.

$$Cu(s) + Cl_2(g) \rightarrow CuCl_2(s)$$

Chapter Review — Key Terms

The key terms that were introduced in this chapter are listed below, along with the section in which they were introduced. You should understand these terms and be able to apply them in appropriate situations.

actual yield (4.6)	combustion analysis (4.7)	mole ratio (4.4)
aqueous solution (4.1)	combustion reaction (4.1)	percent yield (4.6)
atom economy (4.6)	decomposition reaction (4.2)	stoichiometric coefficient (4.1)
balanced chemical equation (4.1)	displacement reaction (4.2)	stoichiometry (4.1)
coefficient (4.1)	exchange reaction (4.2)	theoretical yield (4.6)
combination reaction (4.2)	limiting reactant (4.5)	

Practice Test

After you have finished studying the chapter and the homework problems, the following questions can serve as a test to determine how well you have learned the chapter objectives.

1. When solid elemental sulfur (S_8) is heated, it forms gaseous diatomic molecules.
   (a) Write the balanced chemical equation for this reaction.
   (b) What kind of reaction is this?
   (c) What information are subscripts and coefficients telling you about this reaction?
   (d) Draw a nanoscale picture of this reaction, showing the difference between subscripts and coefficients.

2. The following reactions occur by two successive reactions. Balance each chemical equation and classify it according to the five reaction types.

    (a) Potassium carbonate and hydrobromic acid react to form potassium bromide and carbonic acid, H_2CO_3. Carbonic acid reacts further to produce water and carbon dioxide.

    $$\_\ K_2CO_3(s) \ + \ \_\ HBr(aq) \ \rightarrow \ \_\ KBr(aq) \ + \ \_\ H_2CO_3(aq)$$
    $$\_\ H_2CO_3(aq) \ \rightarrow \ \_\ H_2O(\ell) \ + \ \_\ CO_2(g)$$

    (b) Copper(II) oxide reacts with sulfuric acid to produce copper(II) sulfate and water. Copper(II) sulfate then reacts with titanium to make copper and titanium(IV) sulfate.

    $$\_\ CuO(s) \ + \ \_\ H_2SO_4(aq) \ \rightarrow \ \_\ CuSO_4(aq) \ + \ \_\ H_2O(\ell)$$
    $$\_\ CuSO_4(aq) \ + \ \_\ Ti(s) \ \rightarrow \ \_\ Cu(s) \ + \ \_\ Ti(SO_4)_2(aq)$$

3. Write the balanced chemical equation for the nanoscale picture below. What kind of reaction is this?

     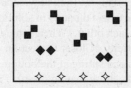 →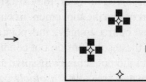

4. When a reddish-purple sample of solid cobalt(II) chloride hexahydrate is heated, the waters of hydration evaporate, leaving behind the anhydrous cobalt(II) chloride as a light blue solid.

    (a) Write a balanced chemical equation for this reaction.
    (b) What kind of reaction is this?
    (c) If you start with 5.000 g of the hydrated compound, how many grams of the anhydrous compound can be made?

5. Modern (post-1982) pennies are made of a copper-zinc mixture. You can separate the two metals by reacting the zinc with sulfuric acid, dissolving it (copper does not react with the acid). If a 1998 penny weighing 2.52 g is allowed to react with excess sulfuric acid, 6.00 g of $ZnSO_4$ is collected. The reaction for zinc and sulfuric acid is: $Zn(s) + H_2SO_4(aq) \rightarrow ZnSO_4(aq) + H_2(g)$

    (a) What kind of reaction is this?
    (b) How many grams of zinc are in the penny? How many grams of copper are in the penny?
    (c) What is the percent composition by mass of the penny?

6. Calcium phosphate can be made from the reaction of calcium cyanide and sodium phosphate.

    $$\_\ Ca(CN)_2(aq) \ + \ \_\ Na_3PO_4(aq) \ \rightarrow \ \_\ Ca_3(PO_4)_2(s) \ + \ \_\ NaCN(aq)$$

    (a) Write the balanced chemical equation for this reaction.
    (b) What kind of reaction is this?
    (c) If 16.4 g of $Ca(CN)_2$ and 25.3 g Na_3PO_4 react, how much calcium phosphate can be made? What is the limiting reactant?
    (d) If 8.25 g of calcium phosphate was produced, what is the percent yield?

7. When 50.0 g of a liquid organic compound (whose molar mass is approximately 90 g/mol) is burned by combustion analysis, 99.9 g of CO_2 and 40.9 g of H_2O are collected.

    (a) What is the empirical formula for this compound?
    (b) What is the molecular formula for this compound?
    (c) Write a balanced chemical equation for the combustion of this compound.

8. When a 3.000 g sample of steel wool is burned, the mass of the iron-oxygen compound is 4.279 g.

    (a) What are the percent composition by mass values for this compound?
    (b) What is the empirical formula of this compound?
    (c) Write the balanced chemical equation for this reaction.
    (d) What kind of reaction is this?

33

Chapter 5
Chemical Reactions

The goal of this chapter is to introduce specific types of reactions that occur in aqueous solutions and to relate them back to the types of reactions discussed in the previous chapter. Precipitation, acid-base, and gas forming reactions are examples of exchange reactions. Oxidation-reduction reactions, on the other hand, can be combination reactions, combustion reactions, decomposition reactions, and displacement reactions. This chapter also describes stoichiometric calculations based on aqueous solutions by introducing the concept of molar concentration and molarity.

Predicting Products for Precipitation, Acid-Base, and Gas-Forming Reactions (5.1 & 5.2)

When ionic compounds are dissolved in water, they are **strong electrolytes** (i.e., they conduct electricity). Because a flow of electricity requires the movement of charged particles, the ions in solution from the aqueous ionic compounds must be free to move independently of each other. When some ionic compounds are placed in water, however, these solutions do not conduct electricity. Closer inspection of the container shows that the ionic compound did not actually dissolve and is instead sitting at the bottom of the container. Solids that do not dissolve in water are called *insoluble compounds*; those that do dissolve in water are called *soluble compounds*. Table 5.1 in the textbook (p. 163) contains a list of *solubility rules* used for determining whether an ionic compound will dissolve in water. These rules were determined empirically (i.e., scientists tested the solubility of many compounds in the laboratory and proposed these rules to explain the observed solubilities).

Example: Use the solubility rules to predict the solubility of the following solid ionic compounds. Write a balanced chemical equation showing what happens when these solids are placed in water.
(a) silver phosphate, Ag_3PO_4, (b) iron(II) chloride $FeCl_2$, (c) ammonium carbonate, $(NH_4)_2CO_3$.

Solution:
(a) The solubility rules say that phosphates are generally insoluble. There are exceptions (alkali metals, NH_4^+ ion) but the Ag^+ ion is not one of them, so this compound is insoluble and will not dissolve in water: $Ag_3PO_4(s) \rightarrow Ag_3PO_4(s)$. You can also write this as: $Ag_3PO_4(s) \rightarrow$ N.R., where 'N.R.' stands for 'no reaction'.
(b) According to the solubility rules chlorides are generally soluble, and the Fe^{2+} ion is not an exception. So, this compound is soluble in water and will dissolve: $FeCl_2(s) \rightarrow Fe^{2+}(aq) + 2 Cl^-(aq)$. Note that the 2 Cl^- ions are written separately (using coefficients not subscripts). These ions do <u>not</u> exist as "$(Cl^-)_2$" or "Cl_2^{2-}" clusters because these ions have the same charge, and would repel each other and try to get as far apart from each other as possible.
(c) Carbonates compounds are generally insoluble, but ammonium is an exception, so this compound is soluble in water. Also, the rule for ammonium compounds says that they are always soluble. When placed in water, ammonium ions and carbonate ions will be produced: $(NH_4)_2CO_3(s) \rightarrow 2 NH_4^+(aq) + CO_3^{2-}(aq)$.

The solubility rules can also be used to predict whether an insoluble solid (called a **precipitate**) will form when two or more aqueous solutions of ionic compounds are mixed together. When aqueous solutions of two ionic compounds are mixed, an exchange reaction can occur in which the cation of one solution reacts with the anion of the other solution and vice versa. If either of these exchange products is insoluble in water, a precipitate (solid) will form; if they are both soluble in water, then nothing happens.

Example: Write the exchange reactions for the following reactants. Will a precipitate form?
(a) Li_3PO_4 and $NiBr_2$, (b) $NaCH_3COO$ and $(NH_4)_2SO_4$

Solution:

(a) First, identify the ions present in the mixture. Li_3PO_4 contains Li^+ and PO_4^{3-} ions; $NiBr_2$ contains Ni^{2+} and Br^- ions. The exchange products will be ionic compounds containing Li^+ and Br^- ions and Ni^{2+} and PO_4^{3-} ions. The formulas for these compounds are $LiBr$ and $Ni_3(PO_4)_2$. Looking at the solubility rules, Br^- are generally soluble and Ni^{2+} and Li^+ are not exceptions, so $NiBr_2$ and $LiBr$ are soluble (aqueous). Phosphate compounds are generally insoluble, but Li^+ is an exception. So, Li_3PO_4 is soluble, but $Ni_3(PO_4)_2$ is insoluble and a precipitate of $Ni_3(PO_4)_2$ will form. The balanced equation is:

$$2\,Li_3PO_4(aq)\ +\ 3\,NiBr_2(aq)\ \rightarrow\ Ni_3(PO_4)_2(s)\ +\ 6\,LiBr(aq)$$

Remember that ionic compounds are written as the simplest ratio of ions, so "Li_6Br_6" would be wrong.

(b) The ions present in $NaCH_3COO$ are Na^+ and CH_3COO^- ions, and the ions present in $(NH_4)_2SO_4$ are NH_4^+ and SO_4^{2-} ions. The exchange products would be Na_2SO_4 (the ionic compound made from Na^+ and SO_4^{2-} ions) and NH_4CH_3COO (the ionic compound made from NH_4^+ and CH_3COO^- ions). Acetate ions are always soluble, and sulfates are generally soluble. Also, Na^+ ions and NH_4^+ ions are always soluble, so all four compounds are soluble and no precipitate forms.

$$2\,NaCH_3COO(aq)\ +\ (NH_4)_2SO_4(aq)\ \rightarrow\ Na_2SO_4(aq)\ +\ 2\,NH_4CH_3COO(aq)$$

Remember that you can only change coefficients in the balanced equation and not the subscripts.

Acid-base reactions are another example of exchange reactions. **Acids** are chemicals that produce $H^+(aq)$ ions when placed in water. They also have several similar chemical and physical properties (caused by the H^+ ion)—they taste sour (although you should never taste any chemical in the laboratory), turn blue litmus paper red, make carbonate solutions bubble, corrode certain metals, etc.

Whenever H^+ ions are present in water, they attach themselves to water molecules making $H_3O^+(aq)$, **hydronium ions**. **Bases** are chemicals that produce $OH^-(aq)$, **hydroxide ions**, when placed in water. Bases have common characteristics based on the hydroxide ion: They taste bitter, feel slippery to the touch, turn red litmus paper blue, and produce precipitates with many metal cations.

Acids and bases that form strong electrolyte solutions are called **strong acids** and **strong bases**. Strong acids and bases *dissociate* (fall apart into ions) completely; acids and bases that do not dissociate 100% are called **weak acids** and **weak bases**.

Example: Hydrochloric acid (HCl) is a strong acid, but hydrofluoric acid (HF) is a weak acid. How do these samples behave differently in water? Draw a nanoscale picture showing this difference.

Solution: HCl dissociates completely in water, while HF does not, as shown in the nanoscale pictures to the right. This means that if there is the same total number of acid molecules in each container, there will be more $H^+(aq)$ ions in the HCl container because of its complete dissociation. This makes HCl behave more like an acid (or like a "better" acid) than HF, dissolving metals better, making carbonate solutions bubble better, etc.

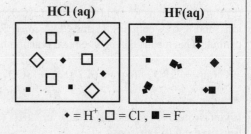

$\blacklozenge = H^+$, $\square = Cl^-$, $\blacksquare = F^-$

Some exchange reactions result in the formation of gases that bubble out of the solution. The most common gases formed are carbon dioxide (CO_2), sulfur dioxide (SO_2), ammonia (NH_3), and dihydrogen sulfide (H_2S). These gas-forming reactions are summarized in the table on the next page. Many of these gases are at least slightly soluble in water and can stay as aqueous solutions, so you may not see rapid bubbling when these solutions are mixed.

gas formed	reacting ions	equation
CO_2	H_3O^+ and CO_3^{2-}	$2\,H_3O^+(aq) + CO_3^{2-}(aq) \rightarrow 3\,H_2O(\ell) + CO_2(g)$
	H_3O^+ and HCO_3^-	$H_3O^+(aq) + HCO_3^-(aq) \rightarrow 3\,H_2O(\ell) + CO_2(g)$
SO_2	H_3O^+ and SO_3^{2-}	$2\,H_3O^+(aq) + SO_3^{2-}(aq) \rightarrow 3\,H_2O(\ell) + SO_2(g)$
NH_3	NH_4^+ and OH^-	$NH_4^+(aq) + OH^-(aq) \rightarrow H_2O(\ell) + NH_3(g)$
H_2S	H_3O^+ and S^{2-}	$2\,H_3O^+(aq) + S^{2-}(aq) \rightarrow 2\,H_2O(\ell) + H_2S(g)$

Writing Net Ionic Equations (5.1)

One of the characteristics of an acid is that it causes carbonate salts to bubble. This is a general reaction, and it works for all kinds of acids (no matter what the anion of the acid is) and all kinds of carbonate salts (no matter what the metal cation is). Chemists have defined **net ionic equations** to describe these general types of reactions. When ionic compounds dissolve in water, they produce two kinds of ions (cations and anions). Once these ions are floating free in the aqueous solution, they can react independently of the other ion. Net ionic equations are determined by first writing an *ionic equation*, in which the soluble <u>aqueous</u> solutions are written as independent ions. The ions that are identical (same formula, coefficient, and state of matter) are canceled and the chemicals remaining represent the net ionic equation. If two solutions are mixed and nothing happens, all of the ions will cancel and the net ionic equation will be nothing.

<u>Example</u>: Write the ionic equation and the net ionic equation for the following reactions.
(a) $Pb(NO_3)_2(aq) + H_2SO_4(aq) \rightarrow PbSO_4(s) + 2\,HNO_3(aq)$
(b) $2\,K_3PO_4(aq) + 3\,Li_2CO_3(aq) \rightarrow 2\,Li_3PO_4(aq) + 3\,K_2CO_3(aq)$
(c) $Al_2(SO_3)_3(s) + 6\,HCl(aq) \rightarrow 2\,AlCl_3(aq) + 3\,SO_2(g) + 3\,H_2O(\ell)$

<u>Solution</u>:
(a) Only ionic compounds that are aqueous are written as ions, so the solid $PbSO_4$ should be left as it is. $Pb(NO_3)_2$ contains Pb^{2+} and NO_3^- ions, H_2SO_4 contains H^+ and SO_4^{2-} ions, and HNO_3 contains H^+ and NO_3^- ions. The net ionic equation is determined by canceling the ions that are identical on both sides of the equation [$2\,NO_3^-(aq)$ and $2\,H^+(aq)$], leaving the net ionic equation (the objects listed in boldface). If you wrote the H^+ ions as "H_2^{2+}" or the NO_3^- ions as "$(NO_3)_2^{2-}$" (which is wrong), they would not cancel with the ions on the other side of the equation since $2\,H^+$ and "H_2^{2+}" are not the same thing.

$$\mathbf{Pb^{2+}(aq)} + 2\,NO_3^-(aq) + 2\,H^+(aq) + \mathbf{SO_4^{2-}(aq)} \rightarrow \mathbf{PbSO_4(s)} + 2\,H^+(aq) + 2\,NO_3^-(aq)$$

$$Pb^{2+}(aq) + SO_4^{2-}(aq) \rightarrow PbSO_4(s)$$

(b) K_3PO_4 contains K^+ and PO_4^{3-} ions, Li_2CO_3 contains Li^+ and CO_3^{2-} ions, Li_3PO_4 contains Li^+ and PO_4^{3-} ions, and K_2CO_3 contains K^+ and CO_3^{2-} ions. For the net ionic equation, every ion cancels leaving nothing behind. This means that no chemical reaction has occurred, and can be written as 'N.R.'.

$$6K^+(aq) + 2PO_4^{3-}(aq) + 6Li^+(aq) + 3CO_3^{2-}(aq) \rightarrow 6Li^+(aq) + 2PO_4^{3-}(aq) + 6K^+(aq) + 3CO_3^{2-}(aq)$$

<u>net ionic equation</u>: N.R.

If you wrote $6\,K^+(aq)$ as "$3K_2^{2+}(aq)$" or "$2K_3^{3+}(aq)$" (which is wrong), they would not cancel. The same is true for the $6\,Li^+(aq)$ ions.
(c) The solid $Al_2(SO_3)_3$, liquid water, and gaseous SO_2 are left alone since only aqueous ionic compounds are written as ions. HCl contains H_3O^+ and Cl^- ions (once the H^+ ion reacts with water) and $AlCl_3$ contains Al^{3+} and Cl^- ions. Only the Cl^- ions are canceled for the net ionic equation, leaving everything else (in boldface).

$$\mathbf{Al_2(SO_3)_3(s)} + \mathbf{6\,H_3O^+(aq)} + 6\,Cl^-(aq) \rightarrow \mathbf{2\,Al^{3+}(aq)} + 6\,Cl^-(aq) + \mathbf{3\,SO_2(g)} + \mathbf{9\,H_2O(\ell)}$$

$$Al_2(SO_3)_3(s) + 6\,H_3O^+(aq) \rightarrow 2\,Al^{3+}(aq) + 3\,SO_2(g) + 9\,H_2O(\ell)$$

Net ionic equations are more general equations because they can describe several reactions. The net ionic equation for the first example above shows that any solution containing Pb^{2+} ions will react with any solution containing SO_4^{2-} ions to produce a precipitate of $PbSO_4$, regardless of the anion accompanying Pb^{2+} or the cation accompanying the SO_4^{2-} cation. So, if you wanted to make $PbSO_4$, and you didn't have $Pb(NO_3)_2$, then $Pb(CH_3COO)_2$, $Pb(ClO_4)_2$, or any other soluble lead(II) salt could be substituted. Similarly, you could substitute H_2SO_4 for any other soluble sulfate salt (K_2SO_4, Na_2SO_4, $CuSO_4$, etc.).

The ions that do not react in an exchange reaction are called **spectator ions**. These ions are not needed for the reaction to occur, but are still important. They are present as charge balancers because we cannot have aqueous solutions with just cations or just anions in them.

Common Acids and Bases and Neutralization Reactions (5.2)

Acids and bases are often considered to be chemical opposites of each other. Their effect on litmus paper is a good example of their opposite behavior. As another example of this opposite behavior, when a base is added to a metal ion and a precipitate forms, adding acid to this mixture will cause the precipitate to disappear. This is because acids and bases react with each other, producing water. The net ionic equation for the *neutralization reaction* of acids and bases is: $H_3O^+(aq) + OH^-(aq) \rightarrow 2\,H_2O(\ell)$.

Most acids are soluble in water, and most of them are weak acids. The strong acids commonly encountered in the chemistry laboratory are hydrochloric acid (HCl), nitric acid (HNO_3), and sulfuric acid (H_2SO_4). The other strong acids (hydrobromic acid, HBr; hydroiodic acid, HI; perchloric acid, $HClO_4$) are not used as much in lab due to their expensive (HBr, HI) or explosive ($HClO_4$) nature. Common weak acids include acetic acid (CH_3COOH), formic acid (HCOOH), phosphoric acid (H_3PO_4), and carbonic acid (H_2CO_3).

Strong bases are soluble metal hydroxide salts. Most hydroxide salts (except the alkali metal and the heavier alkaline earth metals) are insoluble in water; these exceptions are the strong bases commonly used in the laboratory. Sodium hydroxide (NaOH) is by far the most commonly used strong base, followed by potassium hydroxide (KOH). The other strong bases (lithium hydroxide, LiOH; calcium hydroxide $Ca(OH)_2$; barium hydroxide, $Ba(OH)_2$; and strontium hydroxide, $Sr(OH)_2$) are not used as much because they are more expensive, less soluble, or have cations that are toxic to humans (e.g., Ba^{2+} ions). Common weak bases include nitrogen-containing compounds like ammonia (NH_3) or methylamine (CH_3NH_2).

Whenever acids and bases are mixed together, they will react. In fact, the addition of an acid to an insoluble metal hydroxide (base) will cause this compound to dissolve as the H^+ and OH^- ions react.

$$Fe(OH)_3(s) + 3\,HNO_3(aq) \rightarrow Fe(NO_3)_3(aq) + 3\,H_2O(\ell)$$

Salts Made from Acid-Base Reactions (5.2)

When chemists want to make an ionic compound, they will often use an acid-base reaction. This is because the only other product (water) can be easily removed by filtration if the salt is insoluble in water or by boiling away the water if the salt is soluble in water. To make this salt, the chemist simply needs to use the acid containing the anion of interest and the base (hydroxide compound) containing the cation of interest.

Example: What chemicals would a chemist mix to make the copper(II) sulfate from an acid-base reaction? Write the balanced chemical equation for this reaction.

Solution: The anion is SO_4^{2-}, so the acid used is sulfuric acid, H_2SO_4; the cation is Cu^{2+}, so the base used is copper(II) hydroxide, $Cu(OH)_2$. Both H_2SO_4 and $CuSO_4$ are soluble (aqueous), but $Cu(OH)_2$ is not (solid). Water will be in the liquid state.

$$H_2SO_4(aq) + Cu(OH)_2(s) \rightarrow CuSO_4(aq) + 2\,H_2O(\ell)$$

Oxidation-Reduction Reactions and Oxidizing and Reducing Agents (5.3)

Oxidation-reduction reactions (also called **redox reactions**) involve the transfer of electrons from one reactant to another. The reactant that loses the electron is said to be **oxidized** (undergoes **oxidation**) while the reactant that gains the electron is said to be **reduced** (undergoes **reduction**). Sometimes chemists write two *half-reactions* (one for oxidation, one for reduction) to show which reactant has gained electrons and which has lost them.

<u>Example</u>: Write the net ionic equation for the displacement reaction shown below, then write the two half-reactions for this equation. Which reactant is oxidized and which is reduced?

$$3\ AgNO_3(aq) + Fe(s) \rightarrow Fe(NO_3)_3(aq) + 3\ Ag(s)$$

<u>Solution</u>: Only ionic compounds that are aqueous are written as ions, so the solid Fe and Ag should be left alone. $AgNO_3$ contains Ag^+ and NO_3^- ions and $Fe(NO_3)_3$ contains Fe^{3+} and NO_3^- ions. The NO_3^- ion is a spectator ion, and is canceled to give the net ionic equation.

$$\mathbf{3\ Ag^+(aq) + 3\ NO_3^-\ (aq) + Fe(s) \rightarrow Fe^{3+}(aq) + 3\ NO_3^-\ (aq) + 3\ Ag(s)}$$

$$3\ Ag^+(aq) + Fe(s) \rightarrow Fe^{3+}(aq) + 3\ Ag(s)$$

To write the two half-reactions, split the equation into two parts by the elements (Ag in one equation and Fe in the other). The atoms are already balanced, but the charge is not. To balance charge, add electrons (e^-) to each of the equations:

$$3\ Ag^+(aq) + 3\ e^- \rightarrow 3\ Ag(s)$$

$$Fe(s) \rightarrow Fe^{3+}(aq) + 3\ e^-$$

The top equation shows that Ag^+ ions gain electrons in order to make $Ag(s)$, so the Ag^+ ion in $AgNO_3$ is reduced. The bottom equation shows that Fe loses electrons to make Fe^{2+} ions, so the Fe metal is oxidized.

A reduction cannot occur without a subsequent oxidation because a chemical cannot gain electrons unless there is another chemical present that will release them. Half-reactions (like the two written above) cannot happen alone, but chemists write these separate equations to help them identify the oxidation and reduction processes. A chemical that undergoes reduction is called an **oxidizing agent** because it allows another chemical to be oxidized, and a chemical that undergoes oxidation is called a **reducing agent** because it allows another chemical to be reduced. Although this may seem confusing, the analogy of a travel agent may help clarify these terms. A travel agent is not the person who travels; instead, the travel agent is the person who allows <u>others</u> to travel. Common oxidizing agents include nonmetal elements at the top right of the periodic table (O_2, F_2, Cl_2, Br_2, I_2) or chemicals containing oxygen like hydrogen peroxide (H_2O_2) or ozone (O_3). Common reducing agents include metals (especially reactive metals like the alkali metals, the alkaline earth metals, and aluminum), carbon (C), and hydrogen (H_2).

Assigning and Using Oxidation Numbers (5.4)

Exchange reactions do not involve redox reactions, and combustion and displacement reactions are always redox reactions. Combination and decomposition reactions, on the other hand, may or may not be redox reactions. How can you determine whether they are redox reactions? One way is to calculate the oxidation numbers for each atom in the reaction. **Oxidation numbers** (also called *oxidation states*) are a method of counting electrons within a compound.

In neutral atoms and monatomic ions, the oxidation number is the charge on the atom. In molecules or polyatomic ions, the oxidation number is an estimate of the charges that could exist on each of the atoms. Although the textbook lists four rules (p. 183) for assigning oxidation numbers of elements (*O.N.*), only two of the rules are actually needed (Rules 3 and 4).

Rule 4 states that the sum of the *O.N.* values in a molecule or polyatomic ion equals the overall charge on the molecule (which must be zero) or the ion. This rule can be applied to neutral atoms (all elements have *O.N.* = 0, Rule 1) or monatomic ions (the oxidation number of a monatomic ion equals its charge, Rule 2). Rule 4 is used when you know the oxidation number of every element in a molecule or polyatomic ion except one. Rule 3 is a series of guidelines for assigning oxidation number based on observed trends (this order is different than the order used in the textbook):

(a) In compounds, F always has *O.N.* = –1 and Cl, Br, and I have *O.N.* = –1 unless bonded to oxygen or a halogen above it in the periodic table.

(b) In compounds, alkali metals have *O.N.* = +1, alkaline earth metals have *O.N.* = +2, and Al has *O.N.* = +3.

(c) In compounds, H has *O.N.* = +1 except when bonded to metals.

(d) In compounds, O has *O.N.* = –2 unless bonded to fluorine or another oxygen.

The assignment rules described above are *hierarchical*, which means that if the rules give different oxidation numbers for the same atom, the rule that is higher on the list is the correct one (and if any of the four guidelines from Rule 3 give different oxidation numbers than Rule 4, then Rule 4 is right).

Example: Determine the oxidation number for each atom in the following chemicals.
(a) Mg, (b) Mg^{2+}, (c) Cl_2, (d) F_2O, (e) ClO_3^-, (f) $H_2O_2(s)$, (g) MgH_2, (h) SO_4^{2-}, (i) $FeCO_3$

Solution:
(a) Rule 4 says the sum of the oxidation numbers equals zero because the atom is neutral, so $O.N._{Mg} = 0$. Rule 3b says that Mg (an alkaline earth metal) should have an oxidation number of +2, but when there is a conflict, Rule 4 wins.
(b) Rule 4 tells you that $O.N._{Mg} = +2$, the charge of the monatomic ion. Note that atoms can have different oxidation numbers depending on the formula of the chemical it is in (e.g., Mg versus Mg^{2+}).
(c) Rule 4 says $2 \times O.N._{Cl} = 0$, dividing by 2 gives $O.N._{Cl} = 0$. In general, all elements have *O.N.* = 0.
(d) Using Rule 4, $2 \times O.N._F + O.N._O = 0$. Once you know the value of one of the oxidation numbers, you can determine the other. Rule 3a tells you that $O.N._F = –1$, and $O.N._O = 0 – 2 \times O.N._F = 0 – 2(–1) = +2$. Note that this does not follow Rule 3d for oxygen, but Rules 4 and 3a take priority over Rule 3d.
(e) Rule 4 tells you $O.N._{Cl} + 3 \times O.N._O = –1$. Rule 3a says Cl should have an oxidation number of –1, except when bonded to O (like it is here), so skip Rule 3a and use Rule 3d to assign $O.N._O = –2$. Solving for $O.N._{Cl}$ gives: $O.N._{Cl} = –1 – 3 \times O.N._O = –1 – 3(–2) = –1 +6 = +5$.
(f) Rule 4 tells you that $2 \times O.N._H + 2 \times O.N._O = 0$. Using Rule 3c, you find $O.N._H = +1$, and $2 \times O.N._O = 0 – 2 \times O.N._H = 0 – 2 \times (+1) = –2$. Dividing by 2 gives $O.N._O = –1$. This does not follow Rule 3d, but H_2O_2 (H-O-O-H) has an O atom bonded to another O atom, which is mentioned as a reason for this exception.
(g) Using Rule 4, $O.N._{Mg} + 2 \times O.N._H = 0$. Using Rule 3b, you find that $O.N._{Mg} = +2$, and solving for hydrogen gives you $2 \times O.N._H = 0 – O.N._{Mg} = 0 – 2 = –2$. Dividing by 2 gives $O.N._H = –1$, which is the common oxidation number of hydrogen when bonded to metals.
(h) Rule 4 tells you that $O.N._S + 4 \times O.N._O = –2$. Rule 3b says $O.N._O = –2$, and $O.N._S = –2 – 4 \times O.N._O = –2 –4(–2) = –2 + 8 = +6$.
(i) Using Rule 4, $O.N._{Fe} + O.N._C + 3 \times O.N._O = 0$. Rule 3d tells $O.N._O = –2$, but you still don't have values for Fe or C. Based on these rules, you can't solve for these values unless you remember that $FeCO_3$ is an ionic compound made of Fe^{2+} and CO_3^{2-} ions. Now, you can determine $O.N._{Fe} = +2$ from Rule 4 and the monatomic Fe^{2+} ion. You can solve for $O.N._C$ using the equation above, or using the CO_3^{2-} ion. From the equation, $O.N._C = 0 – O.N._{Fe} – 3 \times O.N._O = 0 – 2 – 3(–2) = –2 +6 = +4$.

When an element is oxidized in a redox equation, its oxidation number will become more positive (less negative), and when an element is reduced, its oxidation number will become less positive (more negative). If none of the oxidation numbers change from the reactant to product side, then no oxidation or reduction has occurred and this reaction is <u>not</u> a redox reaction.

39

Example: Determine the oxidation number for each atom in the following combination reactions. Are these redox reactions? If they are, which atom is oxidized and which atom is reduced?
(a) $CaS(s) + O_3(g) \rightarrow CaSO_3(s)$, (b) $CaO(s) + SO_2(g) \rightarrow CaSO_3(s)$

Solution:
(a) First, identify the oxidation number of the atoms in each compound. In CaS the oxidation numbers are +2 for Ca (Rule 3b) and –2 for S (Rule 4), in O_3 the oxidation number of O is 0 (Rule 4), and in $CaSO_3$, the oxidation numbers are +2 for Ca (Rule 3b), –2 for O (Rule 3d), and +4 for S (Rule 4). For S, the oxidation number changes from –2 in CaS to +4 in $CaSO_3$, so the S atom is oxidized. Since the oxidation number for O changed from 0 in O_3 to –2 in $CaSO_3$, the O atoms were reduced. So, this is a redox reaction.
(b) In CaO the oxidation numbers are +2 for Ca (Rule 3b) and –2 for O (Rule 4), and in SO_2 the oxidation number of O is –2 (Rule 3d) and +4 for S (Rule 4). You have already determined the oxidation numbers for $CaSO_3$ in part (a). Since none of the oxidation numbers changed for any of the atoms, this must not be a redox equation.

The Activity Series and Predicting Displacement Products (5.5)

When a sample of one metal is placed in contact with an aqueous solution containing an ion of another metal, sometimes the metal atoms will trade places in a displacement reaction (i.e., the metal atoms become aqueous ions and the aqueous ions become metal atoms). Chemists created a **metal activity series**, which is a ranking of the relative reactivities of these metals, based on experiments performed in the laboratory. A copy of the activity series appears in Table 5.5 (p. 187) of the textbook.

At one end of the series are the reactive metals that prefer to exist in their ion form. These metals include the alkali metals and alkaline earth metals. At the other end of the series are the unreactive metals that prefer to exist in their metal form. These metals include copper, silver, platinum, and gold; these metals are sometimes referred to as *noble metals* (reflecting their nonreactive nature, just like the noble gases) or *coinage* or *jewelry metals* (reflecting their uses as coins and jewelry due to their nonreactive nature).

Unfortunately, not all textbooks draw this chart the same way so sometimes the reactive metals are at the top (or left) and sometimes they are at the bottom (or right). To use the activity series, determine the end with the reactive metals (alkali, alkaline earth metals). If a metal closer to the reactive metals is mixed with a metal ion that is farther from them, a displacement reaction will occur; if a metal farther from the reactive metals is mixed with a metal ion that is closer to them, no reaction will occur.

Example: Predict whether the following chemicals will react. Write balanced equations for the reactions.
(a) $Al(s)$ and $Sn^{2+}(aq)$, (b) $Fe(s)$ and $Li^+(aq)$, (c) Mg and $H^+(aq)$, (d) $Pt(s)$ and $Ag^+(aq)$, (e) $K(s)$ and $Al^{3+}(aq)$

Solution:
(a) Since Al metal is closer to the reactive metals than Sn, a displacement reaction will occur. So, $Sn^{2+}(aq)$ becomes $Sn(s)$, and $Al(s)$ becomes $Al^{3+}(aq)$ because it is in column 3A and a main group metal cation's charge is equal to its column number: $2\ Al(s) + 3\ Sn^{2+}(aq) \rightarrow 2\ Al^{3+}(aq) + 3\ Sn(s)$.
(b) Since Fe metal is farther away from the reactive metals than Li, no displacement reaction will occur: $Fe(s) + Li^+(aq) \rightarrow N.R.$
(c) Since Mg metal is closer to the reactive metals than H_2 gas (the element form of the H^+ ion), a displacement reaction will occur. So, $H^+(aq)$ becomes $H_2(g)$, and $Mg(s)$ becomes $Mg^{2+}(aq)$ because it is in column 2A (alkaline earth metal): $Mg(s) + 2\ H^+(aq) \rightarrow Mg^{2+}(aq) + H_2(g)$.
(d) Since Pt metal is farther away from the reactive metals than Cu, no displacement reaction will occur: $Pt(s) + Cu^{2+}(aq) \rightarrow N.R.$
(e) Since K metal is closer to the reactive metals than Al, a displacement reaction will occur. So, $K(s)$ becomes $K^+(aq)$ and $Al^{3+}(aq)$ ion becomes Al: $3\ K(s) + Al^{3+}(aq) \rightarrow 3\ K^+(aq) + Al(s)$.

Molarity and Molar Concentration Calculations (5.6)

Many chemical reactions occur from reactants in aqueous solutions. In order to perform stoichiometric calculations, chemists could weigh these aqueous solutions. Unfortunately, aqueous solutions are mostly water and weighing the solution does not tell you the weight of the compound dissolved in the water. So, chemists defined the concept of **concentration** to aid in stoichiometric calculations involving aqueous solutions. Concentration is related to density—while density is the 'crowdedness' of matter (in units of g/mL), concentration is the 'crowdedness' of molecules (in units of mol/L). In fact, density is often thought of as a *mass concentration* and concentration is often thought of as a *molar density*.

In mathematical terms, the concentration of an aqueous solution is the number of moles of **solute**, the aqueous compound dissolved in water, per liter of the resulting solution: $[X] = n_{solute} / V_{soln}$. Note that the volume used is the volume of solution and not the volume of the **solvent** (the liquid in which the compound is dissolved; in this case, water). The symbol for concentration is the chemical formula of the compound placed in square brackets. The concentration of a solution is also called its **molarity**, and the units of 'mol/L' are also referred to as *molar* 'M'.

Example: A solution is made by dissolving 10.0 g of potassium phosphate (K_3PO_4) and 15.0 g of potassium monohydrogen phosphate (K_2HPO_4) in 750.0 mL water; the final volume is 754.6 mL of solution. What are the values of $[K_3PO_4]$, $[K_2HPO_4]$, and $[K^+]$ in this solution?

Solution: You need to determine the number of moles of K_3PO_4, K_2HPO_4, and K^+ ions. The first two are determined using their molar masses; the moles of K^+ ions are determined by the molecular formulas.

$$10.0 \text{ g } K_3PO_4 \times \frac{\text{mol } K_3PO_4}{212.27 \text{ g } K_3PO_4} = 0.047110 \text{ mol } K_3PO_4 \text{ (unrounded)}$$

$$15.0 \text{ g } K_2HPO_4 \times \frac{\text{mol } K_2HPO_4}{174.18 \text{ g } K_2HPO_4} = 0.086119 \text{ mol } K_2HPO_4 \text{ (unrounded)}$$

$$0.047110 \text{ mol } K_3PO_4 \times \frac{3 \text{ mol } K^+}{1 \text{ mol } K_3PO_4} + 0.086119 \text{ mol } K_2HPO_4 \times \frac{2 \text{ mol } K^+}{1 \text{ mol } K_2HPO_4} = 0.31357 \text{ mol } K^+$$

The concentrations of these chemicals are determined by dividing by the *solution* volume (754.6 mL, or 0.7546 L) not the volume of water actually used.

$$[K_3PO_4] = \frac{0.047110 \text{ mol } K_3PO_4}{0.7546 \text{ L solution}} = 0.0624 \text{ M}$$

$$[K_2HPO_4] = \frac{0.086119 \text{ mol } K_2HPO_4}{0.7546 \text{ L solution}} = 0.114 \text{ M}$$

$$[K^+] = \frac{0.31357 \text{ mol } K^+}{0.7546 \text{ L solution}} = 0.416 \text{ M}$$

Preparing Solutions from Pure Solutes and More Concentrated Solutions (5.6)

Aqueous solutions are usually made using one of two methods. If the solute is a solid or liquid, then it is weighed on a balance and dissolved in enough water to reach the target volume. If the solute is a gas, the aqueous solution is usually made by diluting a more concentrated solution of the solute with water. The exact amount of solid to be used (for the first method) or the exact volume of concentrated solution to be used (for the second method) can be determined by stoichiometric calculations.

Example: Describe how you would make 2.50 L of an aqueous solution of 0.136 M $MgCl_2$ from solid magnesium chloride and water.

41

Solution: The general method is to dissolve a certain amount of solid magnesium chloride in a little bit of water (about 0.5 L would work), and after the solid has dissolved add enough water to get a solution volume of 2.50 L. The only question is how much solid $MgCl_2$ to use.

$$2.50 \text{ L solution} \times \frac{0.136 \text{ mol } MgCl_2}{\text{L solution}} \times \frac{95.21 \text{ g } MgCl_2}{\text{mol } MgCl_2} = 32.4 \text{ g } MgCl_2$$

Although this answer has only 3 significant figures, you should try to weigh out the sample as close to the calculated value (32.3714 g) as your balance will allow.

Example: Describe how you would make 125 mL of an aqueous solution of 4.16 M ammonia (NH_3) from a concentrated solution of 14.8 M ammonia and water.

Solution: The general method is to dilute a certain amount of concentrated ammonia with enough water to get a solution volume of 125 mL (0.125 L). The only question is how much concentrated NH_3 to use.

$$0.125 \text{ L dil. solution} \times \frac{4.16 \text{ mol } NH_3}{\text{L dil. solution}} \times \frac{\text{L conc. solution}}{14.8 \text{ mol } NH_3} = 0.0351 \text{ L conc. solution}$$

So, you should use 0.0351 L (35.1 mL) of the concentrated ammonia and add enough water to get 125 mL.

Solving Stoichiometry Problems Using Molarity (5.7)

Just as molar masses can be viewed as conversion factors between mass and mole values for a chemical, concentrations (molarities) can be viewed as conversion factors between the volume of an aqueous solution and the number of moles of the solute.

Example: How many grams of hydrogen gas can be made from excess aluminum metal and 8.75 L of a 0.256 M hydrobromic acid solution? What is the resulting concentration of aluminum bromide?

$$2 \text{ Al(s)} + 6 \text{ HBr(aq)} \rightarrow 2 \text{ AlBr}_3\text{(aq)} + 3 \text{ H}_2\text{(g)}$$

Solution: For the first question, convert the solution volume to moles using the molarity, then use a mole ratio from the balanced equation and the molar mass of hydrogen to solve for mass of H_2. For the second question, solve for moles of $AlBr_3$ using the volume and concentration of HBr and the balanced equation, and then divide the moles of $AlBr_3$ by the solution volume (which is still 8.75 L).

$$8.75 \text{ L solution} \times \frac{0.256 \text{ mol HBr}}{\text{L solution}} \times \frac{3 \text{ mol } H_2}{6 \text{ mol HBr}} \times \frac{2.016 \text{ g } H_2}{\text{mol } H_2} = 2.26 \text{ g } H_2$$

$$8.75 \text{ L solution} \times \frac{0.256 \text{ mol HBr}}{\text{L solution}} \times \frac{2 \text{ mol } AlBr_3}{6 \text{ mol HBr}} = 0.74667 \text{ mol } AlBr_3 \text{ (unrounded)}$$

$$[AlBr_3] = \frac{0.74667 \text{ mol } AlBr_3}{8.75 \text{ L solution}} = 0.0853 \text{ M}$$

You may choose to use the $M_{conc.}V_{conc.} = M_{dil.}V_{dil.}$ equation introduced in the textbook (p. 193) for **dilution** problems like the second problem in the previous section, but you should never use it when you are solving solution stoichiometry problems or titration problems (like the problem solved in this section) because it can give incorrect answers.

Solving Stoichiometry Problems Involving Titrations (5.8)

Chemists often perform **titrations** to determine the concentration of an acid or base solution. In a titration experiment, a solution whose concentration is known (this is called a **standard solution**) is used to determine the concentration of a solution whose concentration is unknown. Unfortunately, when acids and bases react with each other, no noticeable changes occur in either solution (color change, precipitate, etc.).

Chemists tend to use *acid-base indicators* to signal when there are equal amounts of acid and base present, which is also called the **equivalence point**. Acid-base indicators are compounds that change characteristics (usually color) in the presence of acids or bases. For example, phenolphthalein is an indicator that is colorless in acidic and neutral solutions but turns bright pink in bases. The equivalence point of an acid-base reaction is estimated to occur when characteristic of the indicator (color) is halfway between its acid characteristic and base characteristic. So, for phenolphthalein the color at the equivalence point would be faint pink because that is the color that is halfway between colorless and bright pink.

After performing a titration experiment, you should know the volume of both the acid and base solutions and the concentration of one of those solutions. You can use this information and the balanced equation for the acid-base reaction to perform a stoichiometric calculation to determine the concentration of the unknown solution.

Example: When a 100.0 mL sample of nitric acid (HNO_3) was titrated with a solution of 0.497 M barium hydroxide, $Ba(OH)_2$, in the presence of phenolphthalein, a total of 250.0 mL was added. The solution started out colorless, turned faint pink after 205.8 mL of $Ba(OH)_2$ was added, and ended bright pink.
(a) Write the balanced equation for this reaction.
(b) What was the concentration of the nitric acid solution?

Solution:
(a) This is an acid-base (exchange) reaction, so the products are barium nitrate, $Ba(NO_3)_2$, and water. Both of the nitrate salts are soluble (nitrates are always soluble), barium hydroxide is soluble (hydroxides are generally insoluble, but Ba^{2+} is an exception), and water is in its liquid form:

$$Ba(OH)_2(aq) + 2\ HNO_3(aq) \rightarrow Ba(NO_3)_2(aq) + 2\ H_2O(\ell)$$

(b) Since you know the concentration and volume of the barium hydroxide solution, you can determine the moles of $Ba(OH)_2$ present, and using a mole ratio from the balanced equation will give you the moles of HNO_3 present. Then you can use the moles of HNO_3 and the volume of HNO_3 to determine the molarity (concentration) of HNO_3.

$$0.2058\ L\ solution \times \frac{0.497\ mol\ Ba(OH)_2}{L\ solution} \times \frac{2\ mol\ HNO_3}{1\ mol\ Ba(OH)_2} = 0.20457\ mol\ HNO_3\ (unrounded)$$

$$[HNO_3] = \frac{0.20457\ mol\ HNO_3}{0.1000\ L\ solution} = 2.05\ M$$

Remember that you should use the volume at the equivalence point (where the number of H^+ and OH^- ions are equal) when performing titrations.

Chapter Review — Key Terms

The key terms that were introduced in this chapter are listed below, along with the section in which they were introduced. You should understand these terms and be able to apply them in appropriate situations.

acid (5.2)	dilution (5.6)	hydroxide ion (5.2)
base (5.2)	equivalence point (5.8)	metal activity series (5.5)
concentration (5.6)	hydronium ion (5.2)	molarity (5.6)

net ionic equation (5.1)
oxidation (5.3)
oxidation number (5.4)
oxidation-reduction reaction (5.3)
oxidized (5.3)
oxidizing agent (5.3)
precipitate (5.1)
redox reaction (5.3)

reduced (5.3)
reducing agent (5.3)
reduction (5.3)
salt (5.2)
solute (5.6)
solvent (5.6)
spectator ion (5.1)
standard solution (5.8)

strong acid (5.2)
strong base (5.2)
strong electrolyte (5.1)
titration (5.8)
weak acid (5.2)
weak base (5.2)
weak electrolyte (5.2)

Practice Test

After you have finished studying the chapter and the homework problems, the following questions can serve as a test to determine how well you have learned the chapter objectives.

1. What is wrong with the following solubility and exchange reactions? Write the correct equation for each of these reactions.

   (a) $ZnCl_2(s) \rightarrow Zn^{2+}(aq) + (Cl^-)_2(aq)$

   (b) $Ag_2CO_3(s) \rightarrow 2\,Ag^+(aq) + CO_3^{2-}(aq)$

   (c) $NaNO_3(aq) + HCl(aq) \rightarrow NaH(aq) + ClNO_3(aq)$

   (d) $K_3PO_4(aq) + H_2SO_4(aq) \rightarrow K_3SO_4(aq) + H_2PO_4(aq)$

2. Predict the products and write the equations for the following exchange reactions. Classify the reactions as precipitation, acid-base, gas-forming, or none of these.

   (a) $Co(OH)_3$ and HBr
   (b) HNO_3 and Al_2S_3
   (c) $Ba(OH)_2$ and $LiCH_3COO$
   (d) $AgNO_3$ and KI
   (e) H_2SO_4 and LiOH
   (f) $Fe(NO_3)_2$ and K_3PO_4
   (g) $(NH_4)_2SO_4$ and KOH

3. In a laboratory experiment, two students were given four unlabeled aqueous solutions containing the following compounds: potassium acetate, sodium carbonate, sulfuric acid (H_2SO_4), and zinc chloride. The students mixed these each sample together, and their results appear in this table (the abbreviation 'ppt' is commonly used in chemistry laboratories for 'precipitate').

	Unknown A	Unknown B	Unknown C	Unknown D
Unknown A	—	bubbles formed	white ppt formed	N.R.
Unknown B	bubbles formed	—	N.R.	N.R.
Unknown C	white ppt formed	N.R.	—	N.R.
Unknown D	N.R.	N.R.	N.R.	—

   (a) Why was it unnecessary for the students to mix Unknown A with Unknown A (Unknown B and Unknown B, etc.)? Did the students need to mix Unknown A with Unknown B after mixing Unknown B with Unknown A?
   (b) Write the balanced equations for the six different reactions.
   (c) Identify the compounds in each of the unknown solutions.

4. Write the ionic equation and the net ionic equation for the following reactions.

   (a) $2\,FeBr_3(aq) + 3\,Na_2S(aq) \rightarrow Fe_2S_3(s) + 6\,NaBr(aq)$

   (b) $Ag_2SO_4(aq) + Cu(ClO_4)_2(aq) \rightarrow 2\,AgClO_4(aq) + CuSO_4(aq)$

   (c) $2\,Al(s) + 3\,MnCl_2(aq) \rightarrow 2\,AlCl_3(aq) + 3\,Mn(s)$

   (d) $CaC_2(s) + 2\,HNO_3(aq) \rightarrow Ca(NO_3)_2(aq) + C_2H_2(g)$

   (e) $Ag_2SO_4(aq) + Ba(OH)_2(aq) \rightarrow 2\,AgOH(s) + BaSO_4(s)$

5.   How would you make the following compounds from acid-base reactions?
     (a) $Cr(ClO_3)_3$ (c) Na_3PO_4
     (b) $Mg(NO_3)_2$ (d) $SrSO_4$

6.   Write the two-half reactions for the following equations, and identify the chemicals that have been oxidized and reduced.
     (a) $Cl_2(aq) + 2\,I^-(aq) \rightarrow 2\,Cl^-(aq) + I_2(aq)$
     (b) $4\,Fe(s) + 3\,O_2(g) \rightarrow 2\,Fe_2O_3(s)$

7.   Determine the oxidation number for each atom in the following reactions. Which of these reactions represent redox reactions? For these reactions, which atom is oxidized and which atom is reduced?
     (a) $CH_4(g) + 2\,O_2(g) \rightarrow CO_2(g) + 2\,H_2O(g)$
     (b) $2\,CaO(s) + CF_4(g) \rightarrow 2\,CaF_2(s) + CO_2(g)$
     (c) $Na_2S(s) + 2\,HBr(g) \rightarrow 2\,NaBr(s) + H_2S(g)$
     (d) $(NH_4)_2Cr_2O_7(s) \rightarrow Cr_2O_3(s) + N_2(g) + 4\,H_2O(g)$
     (e) $3\,N_2H_4(g) \rightarrow 4\,NH_3(g) + N_2(g)$

8.   Gallium metal reacts with Fe^{2+} ions to make Ga^{3+} ions and iron metal, but gallium metal does not react with Zn^{2+} ions.
     (a)   Write balanced equations for these reactions.
     (b)   Where does Ga belong in the metal activity series?
     (c)   Would gallium metal react with Al^{3+} ions? Explain. Write this balanced equation.
     (d)   Would Ga^{3+} ions react with potassium metal? Explain. Write this balanced equation.

9.   A student in a chemistry lab has 25.0 L of water and 582.5 g of potassium permanganate ($KMnO_4$).
     (a)   How many liters of a 0.235 M solution can the student make? What is the limiting reactant?
     (b)   What would the concentration of the solution be if the student mixed the two components?

10. If you have 415.2 mL of concentrated (12.6 M) hydrochloric acid (HCl), can you make 35.0 L of 0.160 M HCl?

11. Magnesium hydroxide can be made from the reaction of magnesium bromide and lithium hydroxide. If 768.0 mL of 0.215 M $Mg(NO_3)_2$ and 242.0 mL of 0.594 M LiOH are mixed, how much magnesium hydroxide can be made? What is the limiting reactant?

$$Mg(NO_3)_2(aq) + 2\,LiOH(aq) \rightarrow Mg(OH)_2(s) + 2\,LiNO_3(aq)$$

12. When a 200.0 mL sample of aqueous calcium hydroxide is titrated with 0.0746 M HNO_3, it takes 33.5 mL of HNO_3 to reach the equivalence point. The indicator used in this experiment was methyl red, which appears red in the acidic solutions and yellow in basic solutions.
     (a)   Write a balanced equation for this reaction.
     (b)   How would you know that you are at the equivalence point?
     (c)   What is the concentration of the calcium hydroxide solution?
     (d)   What is the concentration of the hydroxide ions in the solution?

Chapter 6
Energy and Chemical Reactions

The goal of this chapter is to introduce energy concepts, and how they are related to chemical reactions. The relationship between chemistry and energy is often referred to as *thermochemistry* (the chemistry of heat) or **thermodynamics** (the science of heat transfer). Chemists became interested in the energy changes of chemical reactions for both practical and theoretical reasons. One practical use of knowing which chemical reactions give off or absorb heat is in making hot or cold packs for medical injuries or sprains, or in treating patients suffering from heat stroke or frostbite. However, early chemists learned that heat changes in chemical reactions can be used to predict whether a reaction will actually occur or not (called the *spontaneity* of the reaction). In general, reactions that release heat are spontaneous and happen automatically, while reactions that absorb heat are nonspontaneous and will not happen without some sort of outside intervention.

This chapter introduces three important chemical concepts. The first concept is the measurement of heat changes using calorimetric techniques. These methods are how chemists derive thermochemical data that are placed in reference books. The second concept is the path independence of state functions, which allows chemists to perform theoretical calculations using calorimetry data and Hess's law. Through these calculations, chemists can determine energy changes for reactions that do not occur (nonspontaneous reactions) or are difficult to control. Finally, this chapter introduces standard molar enthalpies of formation, which allow chemists to calculate energy changes for any reaction based solely on tabulated data collected in reference books.

Kinetic and Potential Energy (6.1)

Kinetic energy is the energy of motion, and *energy* is usually defined as the ability to do work (to do something of use like lighting a light bulb, moving a heavy object, etc.). Kinetic energy at the macroscale (the motion of big objects) is referred to as *mechanical energy*, while kinetic energy at the nanoscale (the motion of atoms, molecules, and ions) is referred to as *thermal energy*. The kinetic energy of moving electrons (electricity) is called *electrical energy*. Kinetic energy can be calculated as $KE = 1/2\ mv^2$, where m is the mass of the moving object and v is its velocity. This leads to an important rule of kinetic energy: The faster an object is moving, the more kinetic energy the object has. At the nanoscale level, the kinetic-molecular theory also relates temperature to kinetic energy and particle motion: The higher the temperature of a particle is, the more kinetic energy it has and the faster it is moving.

Potential energy is energy at rest, or stored energy. The most common macroscale form of potential energy is *gravitational energy*, which measures the ability (potential) of an object to fall and release its potential energy as the kinetic energy of movement as it falls. At the nanoscale level, potential energy is described as the energy stored in ionic bonds (*electrostatic energy*) or molecular bonds (*chemical potential energy*).

Converting Between Energy Units (6.1)

The SI (metric) unit for energy is the joule (J), which can be derived from the other SI units of kilogram, meter, and second: $1\ J = 1\ kg\ m^2/s^2$. The energy unit in the English system is the calorie (cal). The definition of a calorie is based on empirical (measured) data: It takes exactly 1 cal to raise the temperature of 1 g of water by 1.0°C (from 14.5°C to 15.5°C). The scientific calorie (cal, lower case c) and the nutritional Calorie (Cal, capital c) are not the same. In fact, the nutritional Calorie is actually equal to 1,000 scientific calories or 1 scientific kilocalorie (kcal).

Example: A 1.3-ounce Cherry Nutri-Grain bar contains 130 Cal. What is this energy in calories, joules, and kilojoules?

Law of Conservation of Energy and Energy Transfer (6.2)

The **law of conservation of energy** (also called the **first law of thermodynamics**) states that energy cannot be created or destroyed, it can only be transferred from one object to another or from one form of energy to another. It is difficult to measure the total energy of an object, and chemists are usually more interested in <u>changes</u> in energy as a process (physical or chemical) occurs. The change in energy is calculated as $\Delta E = E_{final} - E_{initial} = E_f - E_i$. The triangle is a capital Greek letter delta (d), and changes (ΔT, Δm, ΔV, etc.) are always calculated as 'final minus initial'. Energy transfers (changes) occur in one of two ways—by doing work or having work done (w) and by adding or removing heat (q): $\Delta E = q + w$.

When an object exerts energy to do **work** (like moving another object), the value of w is negative because the final energy of the object is less than the initial energy—energy was transferred away from the object when it did work. When an object has work done on it, the value of w is positive because the final energy of the object is more than the initial energy—energy was given to the object from the work done on it.

Heat is usually referred to as 'energy <u>in</u> motion' because heat is measured when one object transfers energy to another object. This is different from the definition of kinetic energy as the 'energy <u>of</u> motion'. When an object gives energy to something else, the value of q is negative because the object giving away heat has less energy at the end, so ($E_f < E_i$, and $\Delta E < 0$). When an object gains energy from something else, the value of q is positive because the object has gained energy ($E_f > E_i$, and $\Delta E > 0$). The changes in heat are usually measured by the object's *temperature* because objects that gain heat tend to get hotter (temperature increase) while objects that lose heat tend to get colder (temperature decrease).

Example: Nonscientists tend to use the terms 'heat' and 'temperature' interchangeably. Are they the same? Is it possible to change the heat of an object without changing its temperature?

Solution: No, they are not the same. The easiest way to convince yourself that they are different is that they have different units. Heat values have energy units (J, kJ, cal, etc.), but temperature values do not (°C, °F, K, etc.). Heat values represent energy (kinetic or potential) transferred from one object to another. At the nanoscale, temperature is a measure of the average kinetic energy of the particles in an object.

If you add heat to an object but do not change the particles' average kinetic energy, then its temperature will not change. One way this can happen is if the added (or removed) energy changes the potential energy of the object but not its kinetic energy. An example of this would be phase changes (boiling, melting, etc.). Energy added to liquid water molecules at 100°C goes to boiling the water (converting molecules in the liquid phase into gas molecules, a potential energy change) but doesn't change their average speeds (no change in kinetic energy) so the temperature does not increase above 100°C even though energy is being added to the sample of water.

In general, chemists tend to focus on heat changes (q) and like to ignore any work changes (w). Physicists, on the other hand, like to ignore heat changes and focus on changes in work. So, when looking at a simple circuit containing a light bulb and a battery, chemists will ignore the fact that the battery is causing the bulb to light and will focus on the fact that the bulb gets hot, while physicists will ignore the fact that the bulb gets hot and focus on what the battery is doing to make the bulb light.

Using Thermodynamic Terms (6.2 & 6.4)

When looking at changes in heat, chemists classify the environment where this change takes place into two categories—the system and the surroundings. The **system** is the part of the environment that the chemist is concerned about. All other parts of the environment are referred to as the **surroundings**. The medium where the reaction occurs (e.g., the water in an aqueous solution or the reaction container) is usually considered to be part of the system but sometimes it's part of the surroundings. When performing energy calculations, chemists must consider whether heat is being transferred between system and surroundings.

When an object gains heat (and q is positive), this process is referred to as **endothermic** and when an object loses heat (and q is negative), this process is referred to as **exothermic**. The '-*thermic*' part of these terms means that this is a heat change (just like a thermometer measures heat changes). The prefix '*endo-*' means 'in' and the prefix '*exo-*' means 'out'. These prefixes are used in biology when talking about an animal with its skeleton on the inside (endoskeleton) or with its skeleton on the outside (exoskeleton).

Using Heat Capacities (6.3)

The **heat capacity** (C) of an object is the amount of energy needed to increase its temperature by 1°C, and heat capacities have units of J/°C or kJ/°C (energy per degree Celsius). The heat capacity is a property of a specific object (an iron pot, a ceramic coffee mug, a sample of water, etc.), and is usually measured only for specific objects that will be used over and over again in *calorimetry* (heat change) experiments. The heat change of an object is related to its heat capacity and its temperature change: $q = C\Delta T$.

The **specific heat capacity**, also called the *specific heat* (c) of an object, is the amount of energy needed to increase one gram of the substance by 1°C, and the units of specific heat capacities are J/g °C or kJ/g °C. The specific heat capacity is a property of a chemical substance (iron, ceramic, water, etc.). Specific heat capacities for some common elements and compounds appear in Table 6.1 of the textbook (p. 222). The heat change of an object is related to the specific heat capacity, mass, and temperature change of the object: $q = mc\Delta T$. The heat capacity of an object is also related to the object's specific heat capacity and its mass: $C = mc$.

The **molar heat capacity** (c_m) of an object is the amount of energy needed to increase one mole of the substance by 1°C, and the units of molar heat capacities are J/mol °C or kJ/mol °C. Just like the specific heat capacity, the molar heat capacity is a property of a chemical substance. The heat change of an object is related to its molar heat capacity, number of moles (n), and temperature change: $q = nc_m\Delta T$. The molar heat capacity of an object is also related to its heat capacity and specific heat capacity: $C = nc_m$, and $c_m = cM$, where M is the molar mass of the substance.

<u>Example:</u> A sample of 325.0 mL of water is placed in a ceramic mug. The specific heat capacity of water is 4.184 J/g °C. What is the heat capacity and molar heat capacity of this sample of water?

<u>Solution:</u> For the heat capacity of the water sample, use the specific heat capacity of water and the mass of the water sample; for the molar heat capacity, use the specific heat capacity and the molar mass of water.

$$C = mc = \left(325.0 \text{ mL} \times 1.00 \; \frac{g}{mL}\right)\left(4.184 \; \frac{J}{g \; C}\right) = 1{,}360. \; \frac{J}{C}$$

$$c_m = cM = \left(4.184 \; \frac{J}{g \; C}\right)\left(18.02 \; \frac{g}{mol}\right) = 75.40 \; \frac{J}{mol \; C}$$

<u>Example:</u> How much energy is absorbed by the ceramic mug and the 325.0 mL of water at 17.5°C when they are heated in a microwave oven to 45.2°C? *C(mug)* = 475 J/°C, and *c(water)* = 4.184 J/g °C

48

48

Changes in Internal Energy and Enthalpy (6.4)

In looking at energy changes, chemists tend to ignore changes in work and focus on heat changes. When a change occurs in a sealed container (so the volume is constant and $\Delta V = 0$), then there is no work being done on or by the system ($w = 0$) and chemists are justified in ignoring the work terms. In this case, the energy change (called the **internal energy**) is equal to the heat change: $\Delta E = q_V$ (the subscript V tells you that this is the heat change at constant volume).

However, chemists are often more interested in doing reactions at constant pressure (i.e., open to the atmospheric pressure) instead of at constant volume. In this case, chemists cannot ignore the work term in the internal energy formula because it is not zero. So, chemists have defined a new unit called **enthalpy change** (ΔH) which represents the heat change under constant pressure conditions: $\Delta H = q_P$.

The relationship between the internal energy and enthalpy is $\Delta H = \Delta E + P\Delta V$, where P is the pressure of the system and ΔV is the volume change. If the volume of a reaction does not change (which is true when the number of gas particles at the beginning and end are the same), then ΔE and ΔH are about the same.

Thermochemical Expressions and Thermostoichiometric Factors (6.5 & 6.6)

Most chemical reactions either require heat to occur (endothermic, $\Delta H > 0$) or give off heat once they occur (exothermic, $\Delta H < 0$). To indicate the enthalpy change of a chemical reaction, the balanced chemical equation is written with a ΔH value. This combination is called a **thermochemical expression**.

$$N_2O(g) + NO_2(g) \rightarrow 3\;NO(g) \qquad\qquad \Delta H° = 155.52 \; kJ$$

The little zero indicates that this reaction was performed under standard conditions (25°C and a pressure of 1 bar), and $\Delta H°$ is referred to as the **standard enthalpy change**. The value of 155.52 kJ refers only to the reaction of 1 mole of N_2O and 1 mole of NO_2 to form 3 moles of NO. Since $\Delta H°$ is tied to the equation, conversion factors (moles ratios) can also be made using $\Delta H°$ as well: 155.52 kJ/mol NO_2' tells you that 155.52 kJ are absorbed when 1 mole of NO_2 reacts and '3 mol NO/155.52 kJ' tells you that 3 moles of NO are made when 155.52 kJ are absorbed. Mole ratios that use $\Delta H°$ are called *thermostoichiometric factors*.

Calculating Enthalpy Changes Using Thermostoichiometric Factors (6.6)

Thermostoichiometric factors can be used (just like mole ratios) to perform stoichiometric calculations relating the amount of energy absorbed (if $\Delta H°$ is positive) or released (if $\Delta H°$ is negative) to the amount of chemicals reacting or being produced (in moles, grams, or volumes of aqueous solutions).

Example: How many grams of boron trichloride will be formed if 500.0 kJ of energy is released from the system? How many grams of chlorine gas will be consumed?

$$2\;B(s) + 3\;Cl_2(g) \rightarrow 2\;BCl_3(g) \qquad\qquad \Delta H° = -807.6 \; kJ$$

Solution: Since this reaction is exothermic ($\Delta H°$ negative), energy is released when the reaction occurs. To determine the amount of reactants and products involved, you need to use thermostoichiometric factors.

$$-500.0 \text{ kJ} \times \frac{2 \text{ mol BCl}_3}{-807.6 \text{ kJ}} \times \frac{117.16 \text{ g BCl}_3}{\text{mol BCl}_3} = 145.1 \text{ g BCl}_3$$

$$-500.0 \text{ kJ} \times \frac{3 \text{ mol Cl}_2}{-807.6 \text{ kJ}} \times \frac{70.90 \text{ g Cl}_2}{\text{mol Cl}_2} = 131.7 \text{ g Cl}_2$$

Enthalpy Changes and Bond Enthalpies (6.7)

Chemists (and biologists) often talk of energy being stored in chemical bonds. Unfortunately, this choice of wording sometimes causes students to think of bonded atoms as being forced together and when the bond breaks, the energy is released. This would imply that breaking a chemical bond is an exothermic process (ΔH° negative). However, breaking bonds is always an endothermic process because it requires added energy to pull apart two atoms that are attracted to each other:

$$H_2(g) \rightarrow 2 \text{ H}^\cdot(g) \qquad\qquad \Delta H^\circ = 436 \text{ kJ}$$

So, how is energy stored in chemical bonds, and how is it released? It would be better for chemists and biologists to say that energy is stored in weak bonds. It takes a smaller amount of energy to break weak bonds into their individual atoms. If stronger bonds are formed between the atoms, then more energy will be released than was used to break the weak bonds. The net result is a loss of 'stored' energy and an exothermic reaction. Figure 6.15 in the textbook (p. 237) shows this concept in graphical form. **Bond enthalpies** (the energy required to break a chemical bond) will be discussed in Chapter 8.

Calorimetry Calculations (6.8)

The ΔH° values listed in thermochemical expressions are often determined in the chemistry laboratory using calorimetry calculations. A **calorimeter** is a device used in chemistry laboratories to measure heat changes. It is usually either a steel container that can withstand large pressures without changing its volume (a *bomb calorimeter*) immersed in water inside a container made of a thermal insulator (an object that will not absorb or release heat energy, like a Styrofoam cup). When a bomb calorimeter is used, the volume of the system is a constant and ΔE° values are determined; when a calorimeter open to the air (constant pressure) is used, ΔH° values are determined.

Calorimetry calculations relate the amount of energy absorbed or released by each object in the universe. The law of conservation of energy says that there is no change in the total energy of the universe (system plus surroundings). This is usually expressed mathematically as $\sum \Delta E = 0$ (the sum of all energies is zero), or as $\sum q = 0$ since chemists tend to ignore work changes. Some of these q values will be positive (heat gain) and some will be negative (heat loss), but the total change over all of the q values must be zero.

Each of these q values can be calculated using several different equations. The first three, which have already been described, involve the heat capacity ($q = C\Delta T$), the specific heat capacity ($q = mc\Delta T$), or the molar heat capacity ($q = nc_m\Delta T$). The next two equations involve the thermostoichiometric factors derived from thermochemical expressions. At constant volume, $q = n\Delta E^\circ$, where n is the number of moles of reacting chemical and ΔE° is a thermostoichiometric ratio from the balanced equation (with units of J/mol or kJ/mol). At constant pressure, $q = n\Delta H^\circ$, where ΔH° is also a thermostoichiometric ratio from the balanced equation with units of J/mol or kJ/mol. If you are given a **fuel value**, then the formula to use is $q = mFV$, where m is the mass of the object burned, and FV is the fuel value in units of J/g or kJ/g. Which equation is used for each object depends on what sort of information you have about each object.

Example: If you have a ceramic mug filled with 325.0 mL of coffee at 71.6°C, how much cream (at 8.1°C) must be added to the coffee to cool it down to 70.0°C? Assume the coffee and cream are mostly water. C(mug) = 263 J/°C, and c(water) = 4.184 J/g °C.

Solution: First, you must sum all of the q values to zero. There are three things in this experiment that can gain or lose heat—the mug, the coffee, and the cream.

$$\sum q = q_{mug} + q_{coffee} + q_{cream} = 0$$

Since you have a heat capacity for the mug, it makes sense to use the equation $q_{mug} = C_{mug}\Delta T_{mug}$.

$$q_{mug} = C_{mug}\Delta T_{mug} = \left(263\ \frac{J}{C}\right)(70.0\ C - 71.6\ C) = -420.8\ J$$

Remember that ΔT is always $T_f - T_i$. This q value is negative because the mug lost heat as it cooled. Since you have the masses (from the volumes, assuming a density of 1.00 g/mL) and specific heats for the coffee and the cream (water), you should use the formula $q = mc\Delta T$ for both of them. Since you do not know the mass of the cream, leave it as m in the equation.

$$q_{coffee} = m_{coffee}c_{coffee}\Delta T_{coffee} = (325.0\ g)\left(4.184\ \frac{J}{g\ C}\right)(70.0\ C - 71.6\ C) = -2,175.68\ J$$

$$q_{cream} = m_{cream}c_{cream}\Delta T_{cream} = (m_{cream})\left(4.184\ \frac{J}{g\ C}\right)(70.0\ C - 8.1\ C) = 258.99\ \frac{J}{g} \times m_{cream}$$

Plugging back into the first equation, and solving for m_{cream} gives:

$$(-420.8\ J) + (-2,175.68\ J) + 258.99\ \frac{J}{g} \times m_{cream} = 0$$

$$258.99\ \frac{J}{g} \times m_{cream} = 2,596.48\ J,\ m_{cream} = \frac{2,596.48\ J}{258.99\ J/g} = 10.0\ g = 10.0\ mL$$

Example: You have a Styrofoam cup with 125.0 mL of water at 23.4°C in it. When you dissolve 10.0 g of KOH, the final temperature is 41.6°C. What is $\Delta H°$ (in kJ/mol) for this process? $c(water) = 4.184$ J/g °C.

Solution: You must sum all of the q values to zero. Since Styrofoam is a good thermal insulator, you may assume $q_{cup} = 0$. Use $q = mc\Delta T$ for the solution (using the total mass of the solution) and $q = n\Delta H°$ for the dissolution reaction (rxn).

$$\sum q = q_{solution} + q_{rxn} = 0$$

$$q_{solution} = m_{solution}c_{solution}\Delta T_{solution} = (135.0\ g)\left(4.184\ \frac{J}{g\ C}\right)(41.6\ C - 23.4\ C) = 10,280.1\ J$$

$$q_{rxn} = n_{rxn}\Delta H_{rxn} = \left(10.0\ g\ KOH \times \frac{mol\ KOH}{56.11\ g\ KOH}\right)\Delta H_{rxn} = 0.178228\ mol\ KOH \times \Delta H_{rxn}$$

$$(10,280.1\ J) + 0.178228\ mol \times \Delta H_{rxn} = 0$$

$$\Delta H_{rxn} = \frac{-10,280.1\ J}{0.178228\ mol} = -57,700\ \frac{J}{mol} = -57.7\ \frac{kJ}{mol}$$

$\Delta H°_{rxn}$ is negative because this is an exothermic reaction that released heat to the water solution.

Applying Hess's Law (6.9 & 6.10)

Because enthalpy and internal energy are **state functions**, their values are the same no matter how those values are determined. This means that if these values are determined indirectly, they will be the same as if they were actually measured using calorimetric methods. **Hess's law** takes advantage of this fact. Hess's law states that individual thermochemical expressions (balanced chemical equations and $\Delta H°$ values) can be added together to get new balanced chemical equations and the $\Delta H°$ values for those reactions. Hess's law provides chemists with a way of determining the enthalpy of reactions that do not actually happen on their own (nonspontaneous reactions).

There are three rules for adding equations together using Hess's law: (1) If you multiply a balanced chemical equation by a number, then you must multiply its $\Delta H°$ value by the same number; (2) If you write a balanced chemical equation backwards (products as reactants and vice versa), then the value of $\Delta H°$ is the same but its sign must be changed; and (3) If you add two balanced chemical equations together (canceling chemicals that are identical on both sides of the equation), then you must add their $\Delta H°$ values together.

Example: You want to know the $\Delta H°$ value for the following reaction, but it cannot be measured directly. Use the thermochemical expressions below to determine $\Delta H°$ for this reaction.

$$2 \, CH_4(g) + O_2(g) \rightarrow 2 \, CH_3OH(\ell) \qquad\qquad \Delta H° = ?$$

$CH_4(g) + 2 \, O_2(g) \rightarrow CO_2(g) + 2 \, H_2O(g)$	$\Delta H_1° = -802.33$ kJ
$2 \, H_2(g) + O_2(g) \rightarrow 2 \, H_2O(g)$	$\Delta H_2° = -483.64$ kJ
$CO(g) + 3 \, H_2(g) \rightarrow CH_4(g) + H_2O(g)$	$\Delta H_3° = -206.10$ kJ
$2 \, H_2(g) + CO(g) \rightarrow CH_3OH(\ell)$	$\Delta H_4° = -128.14$ kJ

Solution: When solving Hess's law problems, you should look at the target equation (the one whose $\Delta H°$ value you want to determine), and add the other equations together to get this formula. The chemicals that will appear in the final equation will be put in boldface, and all others should be canceled from both sides of the equations. First, you need 2 $CH_4(g)$ molecules listed as reactants; since CH_4 appears in both equation 1 and equation 3, save it for later. Since $O_2(g)$ also appears in two equations (1 and 2), save it for later as well. To get 2 molecules of $CH_3OH(\ell)$, double equation 4 and its $\Delta H°$ value.

$$4 \, H_2(g) + 2 \, CO(g) \rightarrow \textbf{2 } \textbf{CH}_3\textbf{OH}(\ell) \qquad\qquad 2 \times \Delta H_4° = -256.28 \text{ kJ}$$

This equation also brings 4 $H_2(g)$ and 2 $CO(g)$, which you do not want and must be canceled before you are finished. In order to get the 2 CH_4, you can use equation 1 or 3. Since equation 1 has $CO_2(g)$ and equation 3 has $CO(g)$, you should use equation 3 (the CO_2 in equation 1, which you do not want, cannot be canceled because no other equation contains CO_2, so you should not use this equation). To get 2 CH_4 from equation 3, double it and write it backwards. This will double the $\Delta H°$ value and change its sign.

$$\textbf{2 } \textbf{CH}_4\textbf{(g)} + 2 \, H_2O(g) \rightarrow 2 \, CO(g) + 6 \, H_2(g) \qquad\qquad -2 \times \Delta H_3° = +412.20 \text{ kJ}$$

Note that the 2 CO and 4 of the H_2 could be canceled from the equations at this point. The only thing missing from the equation is $O_2(g)$ and there is still $H_2(g)$ and $H_2O(g)$ to get rid of, so it makes sense to use equation 2 because it contains all three of these chemicals. Using Equation 2 as written will provide the O_2 and cancel the remaining H_2 and H_2O.

$$2 \, H_2(g) + \textbf{O}_2\textbf{(g)} \rightarrow 2 \, H_2O(g) \qquad\qquad \Delta H_2° = -483.64 \text{ kJ}$$

To calculate the value for $\Delta H°$, add up the three individual $\Delta H°$ values.
$$\Delta H° = 2 \times \Delta H_4° + -2 \times \Delta H_3° + \Delta H_2° = -256.28 \text{ kJ} + 412.20 \text{ kJ} + -483.64 \text{ kJ} = -327.72 \text{ kJ}$$

Using Standard Molar Enthalpies of Formation (6.10)

Using Hess's law, chemists can calculate $\Delta H°$ for just about any reaction, but sometimes chemists would like to determine the $\Delta H°$ value for a reaction without doing any actual experiments. If $\Delta H°$ values could be tabulated in a book, this would allow for strictly theoretical calculations. But, there are so many possible reactions to tabulate that these books would be huge. Instead, chemists invented a way to tabulate enthalpy values for individual chemical substances. **Standard molar enthalpy of formation** $(\Delta H°_f)$ values represent enthalpy changes that occur when chemicals are made from elements in their **standard states** (normal state of matter at 25°C and 1 bar). The standard state of most elements are solid (Hg and Br_2 are liquids, and H_2, N_2, O_2, F_2, Cl_2, and the noble gases are gases); most elements are monatomic, except the diatomic elements listed above, $I_2(s)$, $P_4(s)$, and $S_8(s)$. Thermochemical expressions written for standard molar enthalpy of formation values that are tabulated in the textbook (Appendix J) require that there be only one product, the product's coefficient must be 1, and the reactants are elements in their standard states.

Example: Write the thermochemical expressions for the standard molar enthalpy of formation for the following chemicals: (a) $PF_5(g)$, (b) $H_2SO_4(aq)$, (c) $CH_3OH(aq)$, $Cl_2(g)$

Solution:
(a) The standard states for elemental phosphorus and fluorine are $P_4(s)$ and $F_2(g)$; these are the reactants. The product must be $PF_5(g)$ with a coefficient of 1, so the coefficients are 1/4 for P_4 (1 P atom) and 5/2 for F_2 (5 F atoms). Fractional coefficients are *very* common in thermochemical expressions.

$$\frac{1}{4}\,P_4(s) \;+\; \frac{5}{2}\,F_2(g) \;\rightarrow\; PF_5(g) \qquad\qquad \Delta H°_f\{PF_5(g)\}$$

(b) The standard states for elemental hydrogen, sulfur, and oxygen are $H_2(g)$, $S_8(s)$, and $O_2(g)$. The product must be $H_2SO_4(aq)$ with a coefficient of 1, so the coefficients are 1 for H_2 (2 H atoms), 1/8 for S_8 (1 S atom), and 2 for O_2 (4 O atoms).

$$H_2(g) \;+\; \frac{1}{8}\,S_8(s) \;+\; 2\,O_2(g) \;\rightarrow\; H_2SO_4(aq) \qquad\qquad \Delta H°_f\{H_2SO_4(aq)\}$$

(c) The standard states for elemental carbon, hydrogen, and oxygen are $C(s)$, $H_2(g)$, and $O_2(g)$. The product must be $CH_3OH(aq)$ with a coefficient of 1, so the coefficients are 2 for H_2 (4 H atoms), 1 for C (1 C atoms), and 1/2 for O_2 (1 O atom).

$$C(s) \;+\; 2\,H_2(g) \;+\; \frac{1}{2}\,O_2(g) \;\rightarrow\; CH_3OH(aq) \qquad\qquad \Delta H°_f\{CH_3OH(aq)\}$$

(d) The standard state for elemental nitrogen is $Cl_2(g)$, so $Cl_2(g)$ is the reactant. The product is also $Cl_2(g)$ with a coefficient of 1, and the equation is:

$$Cl_2(g) \;\rightarrow\; Cl_2(g) \qquad\qquad \Delta H°_f\{Cl_2(g)\}$$

Note that for any element, $\Delta H°_f$ is no reaction (N.R.), and since it takes no energy to make the element from itself under standard conditions, the $\Delta H°_f$ values for all elements are defined to be zero.

The following equation is used to calculate $\Delta H°$ values for a chemical reaction based on the $\Delta H°_f$ values of the reactants and products. The n values are the coefficients from the balanced chemical equation. The units of the $\Delta H°_f$ values are kJ/mol and the units of n are mol, so the final answer has units of kJ.

$$\Delta H \;=\; \sum \{n_{product}\Delta H_f\,(product)\} \;-\; \sum \{n_{reactant}\Delta H_f\,(reactant)\}$$

Example: Calculate $\Delta H°$ for the following reaction: $8\,NH_3(g) + 6\,NO_2(g) \rightarrow 7\,N_2(g) + 12\,H_2O(g)$. $\Delta H°_f$ values (in kJ/mol): $H_2O(g) = -241.82$, $N_2(g) = 0.00$, $NH_3(g) = -46.11$, $NO_2(g) = 33.18$.

Solution: $\Delta H \;=\; \left\{ \left(7\text{ mol }N_2 \times 0.00\,\dfrac{kJ}{mol\ N_2}\right) + \left(12\text{ mol }H_2O \times -241.82\,\dfrac{kJ}{mol\ H_2O}\right) \right\} -$

$\qquad\qquad \left\{ \left(8\text{ mol }NH_3 \times -46.11\,\dfrac{kJ}{mol\ NH_3}\right) + \left(6\text{ mol }NO_2 \times 33.18\,\dfrac{kJ}{mol\ NO_2}\right) \right\} =$

53

$$\left\{0.00 \text{ kJ} + (-2{,}901.84 \text{ kJ})\right\} - \left\{-368.88 \text{ kJ} + 199.08 \text{ kJ}\right\} = -2{,}901.84 \text{ kJ} - (-169.80 \text{ kJ}) = -2{,}732.04 \text{ kJ}$$

Chemical Fuels as Sources of Energy (6.11)

Chemical fuels are chemicals that react with oxygen exothermically (i.e., when they burn, or undergo combustion, they release energy in the form of heat). In general, chemical fuels have weak chemical bonds in them, and the products of combustion have strong chemical bonds , which explains the loss of heat when these fuels are burned. Common chemical fuels include hydrocarbons (natural gas, gasoline, diesel, coal, kerosene, etc.) containing C and H atoms; biomass (wood, ethanol from corn, etc.) containing C, H, and O atoms; and hydrogen gas (containing only H atoms).

Caloric Values of Foods (6.12)

Foods are basically chemical fuels for living creatures. The energy stored in food is usually measured in terms of its **caloric value** (the amount of nutritional Calories in the food). In general, carbohydrates and proteins have caloric values of 4 Cal/g while fats have caloric values of 9 Cal/g. The body stores excess energy beyond what it needs to function in the form of fats.

Chapter Review — Key Terms

The key terms that were introduced in this chapter are listed below, along with the section in which they were introduced. You should understand these terms and be able to apply them in appropriate situations.

basal metabolic rate (6.12)	exothermic (6.4)	standard enthalpy change (6.5)
bond enthalpy (bond energy) (6.7)	first law of thermodynamics (6.2)	standard molar enthalpy of
caloric value (6.12)	fuel value (6.11)	formation (6.10)
calorimeter (6.8)	heat/heating (6.2)	standard state (6.10)
change of state (6.4)	heat capacity (6.3)	state function (6.4)
chemical fuel (6.11)	Hess's law (6.9)	surroundings (6.2)
conservation of energy, law of (6.2)	internal energy (6.2)	system (6.2)
endothermic (6.4)	kinetic energy (6.1)	thermal equilibrium (6.2)
energy density (6.11)	molar heat capacity (6.3)	thermochemical expression (6.5)
enthalpy change (6.4)	phase change (6.4)	thermodynamics (*Introduction*)
enthalpy of fusion (6.4)	potential energy (6.1)	work/working (6.2)
enthalpy of vaporization (6.4)	specific heat capacity (6.3)	

Practice Test

After you have finished studying the chapter and the homework problems, the following questions can serve as a test to determine how well you have learned the chapter objectives.

1. When a 263 g ball is dropped from the top of a 593 m cliff, it hits the ground after 11 seconds.
   (a) What is the potential energy of the ball before it is dropped? $PE = mgh$ ($g = 9.81 \text{ m/s}^2$)
   (b) What is the velocity of the ball when it hits the ground?

2. Eggs have a caloric value of 5.86 kJ/g. How many kilojoules are present in an egg weighing 45 g? How many joules is this? How many Calories is this?

3. A 785 g glass coffee mug containing 163 g of water and a 218.9 g iron spoon is cooled from 76.5°C to 15.2°C. Which object has released the most heat? $C(mug) = 417 \text{ J/°C}$, $c(water) = 4.184 \text{ J/g °C}$, and $c_m(iron) = 25.2 \text{ J/mol °C}$.

4. How are endothermic reactions different from exothermic reactions based on the following ideas?
    (a)   The change in energy of the system.
    (b)   The signs of the energy terms (q, $\Delta E°$, and $\Delta H°$).
    (c)   The temperature change of the surroundings.
    (d)   The spontaneity of a reaction (will the reaction happen on its own).

5. When acids and bicarbonates (HCO_3^-) react, heat is absorbed. Consider the following reaction:

$$2\ NaHCO_3(s) + H_2SO_4(aq) \rightarrow Na_2SO_4(aq) + 2\ H_2O(\ell) + 2\ CO_2(g) \qquad \Delta H° = +62.8\ kJ$$

    (a)   Is this reaction endothermic or exothermic? How do you know?
    (b)   If 25.0 g of solid sodium bicarbonate reacts with 200.0 mL of 0.749 M H_2SO_4, how much energy (in kJ) is absorbed? (*Hint:* This is a limiting reactant problem).
    (c)   What was the percent yield of this reaction if the temperature of the solution decreased by 9.5°C? Assume that the solution has the same density and specific heat capacity as water.

6. In order to determine the heat capacity of the steel container in a bomb calorimeter, a sample of benzoic acid (whose $\Delta E°$ is known) is burned. When the steel container with 3.05 g of benzoic acid is placed in 2.50 L of water, the temperature of the system increases from 22.7°C to 30.2°C. What is the heat capacity of the steel container (in J/°C), assuming $\Delta E° = \Delta H°$? $c(water) = 4.184$ J/g °C.

$$2\ C_6H_5COOH(s) + 15\ O_2(g) \rightarrow 14\ CO_2(g) + 6\ H_2O(\ell) \qquad \Delta H° = -6{,}453.7\ kJ$$

7. When a 48.2 g sample of an unknown metal is heated in boiling water until its temperature is 97.3°C and then dropped into a Styrofoam cup holding 100.0 mL of water at 24.7°C, the final temperature of the metal and water is 27.5°C. What is the specific heat capacity (in J/g °C) for the metal? Is this metal more likely to be aluminum or zinc? c values (in J/g °C): Al = 0.90, H_2O = 4.184, Zn = 0.38.

8. Determine $\Delta H°$ for the following reaction using Hess's law. Is it endothermic or exothermic?

$$TiCl_4(g) + 2\ COCl_2(g) \rightarrow TiO_2(s) + 2\ CCl_4(\ell) \qquad \Delta H° = ?$$
$$2\ COCl_2(g) \rightarrow CO_2(g) + CCl_4(\ell) \qquad \Delta H_1° = -91.3\ kJ$$
$$COCl_2(g) + H_2O(g) \rightarrow CO_2(g) + 2\ HCl(g) \qquad \Delta H_2° = -117.5\ kJ$$
$$Ti(s) + 4\ HCl(g) \rightarrow TiCl_4(g) + 2\ H_2(g) \qquad \Delta H_3° = -435.0\ kJ$$
$$Ti(s) + 2\ H_2O(g) \rightarrow TiO_2(s) + 2\ H_2(g) \qquad \Delta H_4° = -456.1\ kJ$$

9. Calculate $\Delta H°$ for the following reaction: $2\ NH_3(aq) + 7\ H_2O_2(aq) \rightarrow N_2O_4(g) + 10\ H_2O(\ell)$.
    $\Delta H°_f$ values (in kJ/mol): $N_2O_4(g) = 9.16$, $NH_3(aq) = -80.29$, $H_2O_2(aq) = -191.17$, $H_2O(\ell) = -285.83$.

10. For this question, you will use the thermochemical expression in question 6 above to determine $\Delta H°_f$ for benzoic acid, $C_6H_5COOH(s)$. $\Delta H°_f$ values (in kJ/mol): $CO_2(g) = -393.5$, $H_2O(\ell) = -285.8$.
    (a)   Write the balanced chemical equation for $\Delta H°_f\{C_6H_5COOH(s)\}$.
    (b)   You were not given $\Delta H°_f\{O_2(g)\}$. Why is it unnecessary to list this value?
    (c)   What is the value for $\Delta H°_f\{C_6H_5COOH(s)\}$?

11. You decide to eat 236 g of white grapes, which have a caloric value of 2.9 kJ/g, and then go swimming to use up these Calories at a rate of 9.0 Cal/min.
    (a)   How many kilojoules and nutritional Calories are there in 236 g of grapes?
    (b)   How long would you have to swim to use up these Calories?

Chapter 7
Electron Configurations and the Periodic Table

The goal of this chapter is to introduce the electronic structure of the atom (i.e., how the electrons are arranged around the nucleus of an atom). The reason chemists are interested in an atom's electronic structure is that this structure ultimately determines the atom's chemical and physical properties. When Mendeleev organized the atoms in the periodic table, he placed elements that had similar chemical reactivities in columns. At the time he could not explain why these elements behaved similarly, but with the assistance of the modern atomic theory, this chapter explains this *chemical periodicity* in terms of the relative positions of electrons around the atom's nucleus. The modern explanation of the atom's electronic structure comes from laboratory experiments on the interaction of matter (atoms) with light.

Frequencies and Wavelengths of Electromagnetic Radiation (7.1)

Visible light is a form of **electromagnetic radiation**, which are linked electric and magnetic fields. Other examples of electromagnetic radiation include infrared light, microwaves, radio waves, ultraviolet light, and X-rays. The interaction of matter and electromagnetic radiation is also an issue of energy: When matter releases electromagnetic radiation it loses energy, and when matter absorbs electromagnetic radiation it gains energy. So, electromagnetic radiation can be viewed as a form of energy that can be transferred from one object to another. Although all forms of electromagnetic radiation are made of waves that travel at the speed of light (2.998×10^8 m/s), they have different values for certain properties including wavelengths, frequencies, and energies.

All waves (electromagnetic radiation, waves in the oceans, the "wave" that happens at football games, etc.) have wavelengths and frequencies. The **wavelength** of a wave, symbolized by the Greek letter lambda (λ, 'l' for 'length'), is the shortest repeat distance within the wave that completely describes the wave. The entire wave can be viewed as several wavelengths placed one after the other (much like an entire wall can be wallpapered by placing one roll of wallpaper after another around the room). Wavelengths have units of length (m, km, in, ft, mi, etc.), and can be measured from trough to trough (the bottom part of the wave) or from crest to crest (the top part of the wave).

The **frequency** of the wave, symbolized by the Greek letter nu (ν, 'n' for 'number'), is the number of waves passing by a particular point in a given timeframe. You can measure the frequency of other things, like how often you breathe (10 times per minute), how often you eat meals (3 times per day), how often you go to chemistry laboratory (1 time per week). In each case, the units for frequency are 'per time' or '1 divided by time': 10 1/min, 3 1/d, 1 1/wk, etc. These units can also be written with an exponent of -1: 10 min^{-1}, 3 d^{-1}, 1 wk^{-1}, etc. The unit $1/s = s^{-1}$ is also called a hertz (Hz).

Example: Which of these two waves (A or B) has the longer wavelength? Which of the two waves has the larger frequency?

wave A:

wave B:

λ_A _____ _____ λ_B _____ _____

Solution:
The wavelength of the waves is the smallest repeat distance (the first line is drawn from crest to crest; the second line is drawn trough to trough). As you can see, the size of the wavelength is the same whether you measure from crest to crest or trough to trough and λ_A is larger than λ_B. The frequency of the waves is the number of waves passing by a point at any given time. If you assume that the two wave fragments above represent the same timeframe, then Wave A has three complete waves passing by (3 1/time) but Wave B has five complete waves passing by (5 1/time). So, ν_B is larger than ν_A.

The mathematical relationship between frequency and wavelength is: $c = \lambda v$, where c is the speed of the wave (the speed of light for electromagnetic radiation). Since all electromagnetic radiation travels at the same speed, waves that are rather long (larger λ) will have fewer waves passing by any point in the same timeframe (smaller v), and vice versa.

Example: The longest wavelength of light the human eye can see is 750 nm (red). The largest frequency of light the human eye can see is 7.5×10^{14} s^{-1} (purple). What are the wavelength and frequency regions for visible light?

Solution: To determine the wavelength region for visible light, you need to convert the frequency of purple light into a wavelength value in nm.

$$c = \lambda v, \quad \lambda = \frac{c}{v} = \frac{2.998 \times 10^8 \text{ m/s}}{7.5 \times 10^{14} \text{ s}^{-1}} = 4.0 \times 10^{-7} \text{ m} \times \frac{1 \text{ nm}}{10^{-9} \text{ m}} = 400 \text{ nm}$$

To determine the frequency region for visible light, you need to convert the wavelength of red light into a frequency value in s^{-1}.

$$c = \lambda v, \quad v = \frac{c}{\lambda} = \frac{2.998 \times 10^8 \text{ m/s}}{750 \text{ nm}} \times \frac{1 \text{ nm}}{10^{-9} \text{ m}} = 4.0 \times 10^{14} \text{ s}^{-1}$$

So, human eyes can detect electromagnetic radiation with a wavelength from 400-750 nm and with a frequency from 4.0-7.5 $\times 10^{14}$ s^{-1}.

Planck's Quantum Theory and the Photoelectric Effect (7.2)

Before Dalton proposed his atomic theory that matter was made of indivisible particles called atoms, chemists believed that matter was *continuous*—that matter was made of one big 'blob' instead of lots of tiny particles. For noncontinuous substances, the smallest quantity is called a **quantum**, so the quantum of matter is the atom. Likewise, Planck proposed that electromagnetic radiation emitted from a quantum of matter (atom) produces a quantum of light (called a **photon**) containing a quantum of energy. Planck noted a relationship between the quantum of energy and the frequency or wavelength of light emitted by the atom: $E_{quantum} = hv = hc/\lambda$, where h is called **Planck's constant** and equals 6.626×10^{-34} J $\times$ s.

Example: What are the energy values for red light (750 nm) and purple light (7.5×10^{14} s^{-1})?

Solution: To determine the energy of red light, you must convert wavelength in energy.

$$E_{quantum} = \frac{hc}{\lambda} = \frac{(6.626 \times 10^{-34} \text{ J} \times \text{s})(2.998 \times 10^8 \text{ m/s})}{750 \text{ nm}} \times \frac{1 \text{ nm}}{10^{-9} \text{ m}} = 2.6 \times 10^{-19} \text{ J}$$

To determine the energy of purple light, convert frequency into energy.

$$E_{quantum} = hv = (6.626 \times 10^{-34} \text{ J} \times \text{s})(7.5 \times 10^{14} \text{ s}^{-1}) = 5.0 \times 10^{-19} \text{ J}$$

Planck's **quantum theory** was not widely accepted until Einstein showed it could explain a phenomenon called the **photoelectric effect**. Some metals emit electrons when illuminated by certain types of light, and Einstein suggested that the quantum of light must have some minimum energy to cause this effect to occur.

Line Emission Spectra and the Bohr Model of the Atom (7.3)

When light from the sun or an incandescent bulb is sent through a prism, the light is refracted into a rainbow of colored light with no missing spots (a **continuous spectrum**). However, when light emitted from pure elements is refracted through a prism, a **line emission spectrum** containing thin lines of colored light and lots of empty spaces (missing colors) is obtained. These line emission spectra are unique for different elements, and can be used to identify the presence of an element; in fact, helium was first identified by looking at a line emission spectrum of the sun.

The phenomenon of thin colored lines in the line emission spectra of elements intrigued chemists, but they could not explain why it happened when it was first observed. Rydberg noticed that for the line emission spectrum of hydrogen, each line could be predicted by the following formula:

$$\frac{1}{\lambda} = R_H \left(\frac{1}{n_1^2} - \frac{1}{n_2^2} \right)$$

This formula is called the *Rydberg equation*, where R_H is the Rydberg constant for hydrogen (1.097×10^7 m^{-1}), n_1 and n_2 are whole numbers (integers), and $n_1 < n_2$. Although Rydberg's equation could be used to calculate the correct wavelengths of known lines (and predict new ones that were subsequently found in the infrared and ultraviolet regions), Rydberg could not explain why this formula worked, only that it did.

A few decades later, Bohr proposed a theory of the atom that could explain Rydberg's equation. Bohr based his model on the solar system—the nucleus acts like the sun and the electrons orbit around the nucleus like planets orbit the sun. Just like planets, Bohr theorized that electrons could only appear at certain distances from the nucleus, but not in between these distances. He assumed that these values were limited to whole numbers times the shortest distance ($1r_0$, $2r_0$, $3r_0$, etc., but not $1.3r_0$ or $4.9r_0$). From these assumptions, he could calculate the energy of an electron. For hydrogen, the energy is $E = -b/n^2$, where b is Bohr's constant for hydrogen (2.179×10^{-18} J), and n is the whole number distance from the nucleus.

Bohr further theorized that the electricity from the lamp caused an electron to move from a lower-energy orbit (smaller n value, closer to the nucleus) to a higher-energy orbit (larger n value, farther from the nucleus), and that the light emitted by the atom comes from an electron moving from a higher-energy orbit to a lower-energy orbit.

$$\left| \Delta E \right| = \left| E_f - E_i \right| = \left| \frac{-b}{n_f^2} - \frac{-b}{n_i^2} \right| = b \left| \frac{1}{n_f^2} - \frac{1}{n_i^2} \right|$$

The absolute value signs are needed because an electron moving from a higher orbit to a lower one is exothermic (ΔE negative), but wavelengths must be positive. Since $E = hc/\lambda$, solving for $1/\lambda$ gives $1/\lambda = E/hc$. If you take Bohr's equation and divide it by hc, then you get $b/hc = 1.097 \times 10^7$ m^{-1} = R_H. Note also that in Bohr's equation the final orbit was smaller than the initial orbit ($n_f < n_i$), which also matches the requirement of Rydberg's equation that $n_1 < n_2$.

Calculations Using Bohr's Equation (7.3)

You can use Bohr's or Rydberg's equations to calculate changes in energy, wavelength, or frequency when an electron changes orbits in a hydrogen atom. When an electron moves closer to the nucleus, its n value decreases and so does its energy (ΔE, which is the same as $E_{quantum}$, is negative); this is an exothermic process and the energy is released (emitted) from the atom in the form of light. When an electron moves farther from the nucleus, its n value increases and so does its energy (ΔE and $E_{quantum}$ are positive); this is an endothermic process and the atom must absorb energy from its surroundings (as light, heat, or electricity).

Example: An electron in hydrogen moves from the $n = 1$ to the $n = 4$ orbit.
(a) Is this process endothermic or exothermic? Is ΔE positive or negative?
(b) Calculate the energy change (in J) for this electron.
(c) Calculate the wavelength (in m) for this change. In what region of the electromagnetic spectrum does this photon appear?

Solution:
(a) Because the electron is moving farther from the nucleus, this process is endothermic and ΔE is positive.
(b) The energy change can be calculated from Bohr's equation (note that this value should be positive).

$$|\Delta E| = b \left| \frac{1}{n_f^2} - \frac{1}{n_i^2} \right| = 2.179 \times 10^{-18} \text{ J} \times \left| \frac{1}{1^2} - \frac{1}{4^2} \right| = 2.043 \times 10^{-18} \text{ J}$$

(c) When calculating wavelength from energy, use $E_{quantum} = hc/\lambda$, remembering that $\Delta E = E_{quantum}$.

$$E_{quantum} = \frac{hc}{\lambda}, \lambda = \frac{hc}{E_{quantum}} = \frac{(6.626 \times 10^{-34} \text{ J} \times \text{s})(2.998 \times 10^8 \text{ m/s})}{2.043 \times 10^{-18} \text{ J}} = 9.723 \times 10^{-8} \text{ m}$$

A wavelength of 9.723×10^{-8} m ($\approx 10^{-9}$ m) falls in the X-ray range in Figure 7.1 of the textbook (p. 273).

Quantum Mechanical Model of the Atom (7.4)

Although Rydberg's equation and Bohr's theory were successful in explaining the line emission spectrum of hydrogen, they do not work as well for the other hundred atoms in the periodic table. Although they can predict lines that actually exist, there are many more lines that these formulas cannot predict or explain. Chemists realized that there must be more than one number that has restricted values; these numbers, which are called **quantum numbers**, will be described in the next section.

Chemists also changed their view of how the electrons exist around an atom's nucleus. While the *planetary model of the atom* worked for explaining hydrogen's line emission spectrum, it was not an accurate description of the atom. The biggest difference is that the path of planets around the sun is very orderly and predictable, but the paths of electrons are not. They 'buzz about' very randomly (more like a swarm of bees than planets), and it is impossible to predict exactly where an electron will be at any given time. Instead, chemists talk about *probabilities* of finding electrons in certain regions around the nucleus. The three-dimensional spaces that electrons occupy are referred to as **orbitals** (not orbits).

The First Three Quantum Numbers (7.5)

Chemists realized that to explain the line emission spectra of all elements, four quantum numbers were required. The first quantum number, called the **principal quantum number** (n), is the whole number value used in Rydberg's and Bohr's equations. The possible values for n are restricted to whole numbers greater than zero: $n = 1, 2, 3$, etc. The value of n tells you the *distance* the electron is from the nucleus (as n increases, the electron gets farther from the nucleus), the *size* of the electron's orbital (as n increases, the orbital gets bigger), and the *energy* of the electron (as n increases, the energy of the electron increases).

The values for the *angular momentum quantum number* (ℓ) are restricted by the n value of the electron: $\ell = 0, 1, 2, \ldots (n-1)$. The value of ℓ must be a positive whole number including zero that is smaller than n. When writing symbols for atomic orbitals, chemists use the n value and a letter symbol for ℓ. The letter symbols are: s ($\ell = 0$), p ($\ell = 1$), d ($\ell = 2$), f ($\ell = 3$), etc. From $\ell = 3$ on, the symbols are used in alphabetical order (i.e., the next is g, then h, etc.). The value of ℓ tells you the *shape* of the orbital (3-D space occupied by the electrons around the nucleus) and the *relative energy* of electrons with the same n value (when n is the same, the energy of the electron increases as the ℓ value increases).

The values for the *magnetic quantum number* (m_ℓ) are restricted by the value of ℓ for the electron: $m_\ell = -\ell, \ldots 0, \ldots +\ell$. The m_ℓ value must be a whole number that can be as low as $-\ell$ and as high as $+\ell$. The m_ℓ values determine the *3-D orientation* of the orbital in space. While the n and ℓ values affect an electron's energy, as long as n and ℓ stay the same, changing m_ℓ doesn't change the electron's energy.

Example: Suppose you have an electron in the $n = 4$ shell.
(a) What are the possible values of ℓ for this electron? Write the orbital label for each of these.
(b) What are the possible values of m_ℓ for an electron in the maximum ℓ value orbital?

Solution:
(a) When $n = 4$, ℓ can be 0, 1, 2, or 3. The label for $n = 4$ and $\ell = 0$ is $4s$, the label for $n = 4$ and $\ell = 1$ is $4p$, the label for $n = 4$ and $\ell = 2$ is $4d$, and the label for $n = 4$ and $\ell = 3$ is $4f$.
(b) The maximum ℓ value is 3. When $\ell = 3$, m_ℓ can be $-3, -2, -1, 0, +1, +2,$ or $+3$.

Electron Spin Quantum Numbers (7.5)

When a charged object (like an electron) is spinning, it creates a magnetic field. Chemists realized that electrons within an atom could produce magnetic fields in two different directions. They interpreted this as coming from the electron spinning in one of two directions (clockwise and counterclockwise). The *electron spin quantum number* (m_s) is restricted to two values: $m_s = -1/2$ or $+1/2$ ('spin up' and 'spin down'). Chemists don't tend to focus on which direction the electron is spinning; instead, they focus on how many electrons are spinning in one direction and how many are spinning in the other direction.

The **Pauli exclusion principle** states that no two electrons in the same atom can have the same four quantum numbers. Since any specific orbital is defined by its n, ℓ, and m_ℓ values, this rules tells you that each orbital can hold only two electrons, and they must have opposite spins ($m_s = -1/2$ and $+1/2$).

The Relationship Between Shells, Subshells, and Orbitals (7.5)

All electrons in an atom with the same n value are said to be in the same **shell**, all electrons in an atom with the same n and ℓ values are said to be in the same **subshell**, and all electrons in an atom with the same n, ℓ, and m_ℓ values are said to be in the same **orbital**. In other words, a shell is defined by the n value, a subshell is defined by the n and ℓ values ($1s$, $3p$, $4d$, etc.), and an orbital is defined by the n, ℓ, and m_ℓ values ($1s$, $3p_x$, $4d_{yz}$, etc.).

Shapes of Atomic Orbitals (7.6)

The angular momentum quantum number (ℓ) determines the shapes of the atomic orbitals. When $\ell = 0$ (s orbitals), the atomic orbitals are spherical. There is only one orientation of s orbitals for each n value (when $\ell = 0$, $m_\ell = 0$). When $\ell = 1$ (p orbitals), the atomic orbitals are 'dumbbell' shaped. There are three orientations of p orbitals for each n value (when $\ell = 1$, $m_\ell = -1, 0, +1$). These dumbbell-shaped orbitals lay on one of the Cartesian axes (x, y, or z) and are labeled by the axis on which they lie (i.e., the p_x orbital lies on the x-axis, etc.). When $\ell = 2$ (the d orbitals), the atomic orbitals are 'cloverleaf' shaped (except one that looks like a p orbital with a donut around its middle). There are five orientations of d orbitals for each n value (when $\ell = 2$, $m_\ell = -2, -1, 0, +1, +2$). Three of these cloverleaf-shaped orbitals lay between two of the Cartesian axes and are labeled by the axes they lie between (i.e., the d_{xy} orbital lies between the x- and y-axes, etc.). The other cloverleaf-shaped orbital (the $d_{x^2-y^2}$ orbital) lies on the x- and y- axes (like the d_{xy} orbital, but rotated 45°), and the d_{z^2} orbital lies with its lobes on the z-axis and the donut in the xy plane.

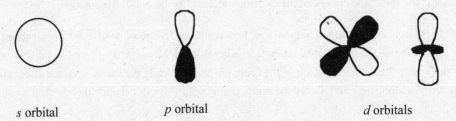

s orbital p orbital d orbitals

Writing Electron Configurations of Atoms and Ions (7.7 & 7.8)

Chemists use **electron configurations** to describe how the electrons occupy the three-dimensional space around the nucleus of the atom. In general, the electrons completely fill the lowest-energy orbitals first before filling higher-energy orbitals (the *aufbau principle*). Using the aufbau principle, the Pauli exclusion principle (only two electrons with opposite spins per orbital), **Hund's rule** (when placing electrons in a set of subshells, each orbital is occupied by a single electron with parallel spin before the electrons are paired up), and the periodic table, you can write the electron configuration of any atom by starting at hydrogen and adding electrons until you reach the target atom. Electron configurations are often written using lines or boxes representing the atom's orbitals. This configuration is called an *orbital box diagram*.

Example: What is the electron configuration (orbital box diagram) for the manganese atom ($Z = 25$)?

Solution: Since manganese has 25 protons (Z), a neutral manganese atom also has 25 electrons. Figure 7.17 in the textbook (p. 297) shows the orbitals used by the electrons in an atom. The first two electrons (H and He) are placed in the $1s$ orbital with opposite spins, and the next two electrons are placed in the $2s$ orbital (Li and Be) with opposite spins. The first three electrons added to the $2p$ orbitals are each added to a different p orbital with parallel spin (B to N); the next three electrons are added with opposite spins to fill the three $2p$ orbitals (O to Ne). The next eight electrons are added to the $3s$ orbital (Na and Mg) and the three $3p$ orbitals (Al to Ar). The next two electrons are added to the $4s$ orbital (K and Ca). So far, 20 electrons have been added and there are 5 electrons remaining to be added to the five $3d$ orbitals. These electrons are added to a different $3d$ orbital with parallel spins, so all five $3d$ orbitals are half-filled:

$1s$	$2s$	$2p$	$3s$	$3p$	$4s$	$3d$
$\uparrow\downarrow$	$\uparrow\downarrow$	$\uparrow\downarrow$ $\uparrow\downarrow$ $\uparrow\downarrow$	$\uparrow\downarrow$	$\uparrow\downarrow$ $\uparrow\downarrow$ $\uparrow\downarrow$	$\uparrow\downarrow$	$\uparrow$ $\uparrow$ $\uparrow$ $\uparrow$ $\uparrow$

Certain regions of the periodic table are often referred to by the type of orbital in which the outermost electron is placed. So, the first two columns (1A-2A) are referred to as the **s-block elements**, the last six columns (3A-8A) are called the **p-block elements**, the **transition elements** (columns 3B-2B) are referred to as the *d-block elements*, and the lanthanides and actinides are called the *f-block elements*.

Although there are ways of remembering the relative energies of the orbitals, most chemists simply use the periodic table to determine the order in which the orbitals are filled. For elements with many electrons, the orbital box diagrams are rather long. So, chemists use an abbreviated electron configuration (called the *condensed method*) in which orbital designations (number for n and letter for ℓ) are used with a superscript showing how many electrons are in the set of orbitals. For the vanadium atom, the condensed electron configuration is: $1s^2 2s^2 2p^6 3s^2 3p^6 4s^2 3d^3$. Notice that the $4s$ orbital is filled before the $3d$ orbitals even though it has a larger n value. Sometimes, condensed electron configurations are written with the orbitals of the same n value placed together (i.e., $1s^2 2s^2 2p^6 3s^2 3p^6 3d^3 4s^2$). Either form is usually acceptable.

For atoms with a very large number of electrons, even the condensed method gets to be pretty long. So, chemists use another abbreviation method called the **noble gas notation**. In the noble gas notation, the electron configuration of the largest noble gas within the atom is replaced by its chemical symbol in square brackets. In the vanadium atom, the largest noble gas within the atom is argon ($1s^2 2s^2 2p^6 3s^2 3p^6$), so the noble gas notation of vanadium would be [Ar] $3d^3 4s^2$ or [Ar] $4s^2 3d^3$.

Electron configurations for ions in the *s*- and *p*-blocks are determined by adding or subtracting electrons (as appropriate) from the electron configuration of the neutral atom. So, the electron configuration of the Al^{3+} ion is found by looking at the electron configuration of the Al atom ($1s^2 2s^2 2p^6 3s^2 3p^1$) and removing three electrons to give a +3 charge ($1s^2 2s^2 2p^6$). Similarly, the electron configuration of the oxide ion is found from the electron configuration of the oxygen atom ($1s^2 2s^2 2p^4$) and adding two electrons ($1s^2 2s^2 2p^6$). Atoms or ions with the same number of electrons (and the same electron configurations, like Al^{3+} and O^{2-}) are said to be **isoelectronic**. All stable monatomic *s*- and *p*-block ions are isoelectronic with noble gases.

Unexpected Electron Configurations of Transition Elements and Ions (7.8)

Electrons in orbitals with the largest n value in an atom are referred to as **valence** (or *outer*) **electrons**, and electrons in orbitals with lower n values are referred to as **core** (or *inner*) **electrons**. Transition elements have a filled ns orbital but the $(n–1)d$ orbitals are only partially filled. For these elements, electrons in the ns and $(n–1)d$ orbitals are considered to be valence electrons. These two sets of orbitals are very close in energy, and this leads to some unexpected electron configurations in transition element atoms and ions.

For example, Cr is expected to have the configuration [Ar] $3d^4 4s^2$ and Cu's configuration is expected to be [Ar] $3d^9 4s^2$. However, these atoms actually have electron configurations of [Ar] $3d^5 4s^1$ and [Ar] $3d^{10} 4s^1$, respectively. Chemists believe that Cr and Cu have these unexpected configurations because subshells that are empty (0%), half-filled (50%), or filled (100%) are more stable than partially-filled orbitals.

Another set of anomalous electron configurations appear when looking at transition element cations. For example, you might predict that the Fe^{3+} ion would have an electron configuration of [Ar] $3d^3 4s^2$, but instead it has a configuration of [Ar] $3d^5$. In general, when transition elements make cations, electrons are removed from the filled s orbital before removing electrons from the partially-filled d orbitals. Even though the s orbital was lower in energy before electrons were added to the atom (so it was filled before the d orbitals), once the atom has electrons in the s and d orbitals, the d orbitals drop in energy so that the s orbital (which is filled) is now higher in energy. This is why electrons are removed from the s orbital in transition elements when making cations.

Trends in Atomic Radii (7.9)

Electron configurations can be used to explain physical and chemical properties of the elements. One of these properties is an element's **atomic radius**. In general, as you look down a column in the periodic table, atomic radii increase and as you look across a row in the periodic table, atomic radii decrease.

To explain the trend in the columns of the periodic table, look at the electron configurations and the atomic radii (p. 307 in the textbook) of the first four atoms in column 6A of the periodic table—O: [He] $2s^2 2p^4$ (74 pm), S: [Ne] $3s^2 3p^4$ (104 pm), Se: [Ar] $4s^2 3d^{10} 4p^4$ (117 pm), Te: [Ar] $5s^2 4d^{10} 5p^4$ (137 pm). The reason each of these atoms is larger than the previous atom is that the outer (valence) electrons are being placed in orbitals with a larger n value. The n value determines the size of the atomic orbital (orbital sizes increase as n increases). So, atoms get larger as you look down a column of the periodic table because the outer electrons are being placed in larger orbitals.

Effective Nuclear Charge and Atomic Radii (7.9)

The trend in atomic radii down a column in the periodic table suggests that adding electrons to an atom makes the atomic radii bigger. However, the trend across a row in the periodic table shows that these atoms get smaller when electrons are added. To explain this trend, look at the electron configurations and the atomic radii for the first four atoms in the second row of the periodic table—Li: [He] $2s^1$ (157 pm), Be: [He] $2s^2$ (112 pm), B: [He] $2s^2 2p^1$ (86 pm), C: [He] $2s^2 2p^2$ (77 pm). Since the outer electrons are being added to orbitals with the same n value, it makes sense that the sizes aren't getting any bigger (and you might predict they would have the same size), but it doesn't explain why they are getting smaller.

The radius of an atom is determined by the attractions of the outer (valence) electrons by the protons in the nucleus, so both the number of electrons and the number of protons affect the atomic radius. These four atoms are adding electrons to orbitals with the same n value, the numbers of protons are also changing. So, the electrons in Li are attracted to the 3 protons in the nucleus (3+ charge) while the electrons in C are attracted to the 6 protons in its nucleus (6+ charge). The increased attraction of the electrons to a stronger positive charge in the nucleus pulls the electrons closer to the nucleus and the atomic radius decreases.

This trend can also be explained using the concept of **effective nuclear charge**, which explains that since the inner (core) electrons are between the outer electrons and the nucleus, they shield the outer electrons (called a **screening effect**) from the full positive charge of the nucleus.

Trends in Ionic Radii (7.10)

Electron configurations can also be used to explain trends in the **ionic radii** of cations and anions made from neutral atoms. Cations always have smaller radii than their neutral atoms, and anions always have larger radii than their neutral atoms. When a cation is made from the neutral atom, the #p^+ remains the same but the #e^- around the nucleus is smaller. With fewer electrons, the electron-electron repulsions are weaker and the electrons are pulled closer to the nucleus by their attractions to the protons in the nucleus. The explanation for why anions are larger than their neutral atoms is the reverse argument for the cations— anions have the same #p^+ but the #e^- around the nucleus is larger, so the electron-electron repulsions are greater and this forces the electrons away from each other and from the nucleus, resulting in a larger radius.

This change in radius is often dramatic for cations, and this can be explained by looking at the electron configurations. For example, Mg ($1s^2 2s^2 2p^6 3s^2$) has an atomic radius of 160 pm, but Mg^{2+} ($1s^2 2s^2 2p^6$) has an ionic radius of 86 pm. For Mg, the outer electrons are in the $n = 3$ shell but those electrons are lost when Mg^{2+} is formed and now its outer electrons are in the $n = 2$ shell, which is smaller than the $n = 3$ shell.

Example: When calcium and sulfur atoms react, they make Ca^{2+} and S^{2-} ions. Which of the reacting atoms has the larger atomic radius? Which of the product ions has the larger ionic radius?

Solution: Looking at the electron configurations, Ca ([Ar] $4s^2$) has its outer electron in the $n = 4$ shell, but S ([Ne] $3s^2 3p^4$) has its outer electrons in the $n = 3$ shell, so S should have a smaller atomic radius than Ca. The ions Ca^{2+} ([Ar]) and S^{2-} ([Ne] $3s^2 3p^6$ = [Ar]) are isoelectronic. You might assume that they have the same size, but since Ca^{2+} has 18 p^+ attracting the electrons and S^{2-} has only 16 p^+, Ca^{2+} is smaller than S^{2-}.

Trends in Ionization Energies (7.11)

Ionization energy (*IE*) is the energy required to remove the outermost electron from a gaseous atom:

$$X(g) \rightarrow X^+(g) + e^- \qquad \qquad \Delta H = IE$$

The trends for ionization energies are that *IE* values decrease as you look down a column and increase as you look across a row. These trends can be explained by atomic radii—the smaller the atom, the closer the outer electrons are to the nucleus and the stronger the attraction between the outer electrons and the nucleus. So, it takes more energy (*IE*) to remove an electron closer to the nucleus.

Example: Looking at Figure 7.26 in the textbook (p. 312), the trend across rows in the periodic table is not perfect. Why are the *IE* values lower for B than Be? Why are the *IE* values lower for O than N?

Solution: In general, IE increases as the atomic radius decreases. The atomic radii for Be (112 pm), B (86 pm), N (74 pm), and O (73 pm) show that this trend holds, so there must be some other explanation. Looking at the electron configurations (orbital box diagrams) may help explain these anomalies.

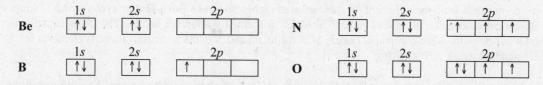

The electron removed from Be comes from the $2s$ orbital, but the electron removed from B comes from the $2p$ orbital. Since $2p$ orbitals are slightly higher in energy than $2s$ orbitals (same n value, larger ℓ value) it takes less energy to remove this outer electron.

63

> The electrons removed from N and O are both in $2p$ orbitals, so there must be another reason for this anomaly. The electron removed from O comes from the filled orbital, while the electron removed from N comes from a half-filled orbital. When two electrons share the same orbital in the same 3-D space, they repel each other and this raises the electron's energy. So, removing an electron from a filled orbital takes less energy than removing an electron in a half-filled orbital due to electron-electron repulsions.

Since it is possible to remove more than one electron from an atom, you can measure successive ionization energies for the removal of each electron. The energy needed to remove the first electron (IE_1) is always the lowest, and each successive ionization energy is larger than the previous one ($IE_1 < IE_2 < IE_3$, etc.). A table of successive ionization energies (Table 7.7 on p. 313 of the textbook) shows that each IE value is about two to three times as large as the previous IE. However, there are places when the jump in IE is much larger—about five to ten times as large as the previous IE. These values occur when the electron to be removed is from a lower shell (lower n value). Since these orbitals are much lower in energy, it takes much more energy (IE) to remove these electrons.

Trends in Electron Affinity (7.12)

Electron affinity (EA) is the energy change when an additional electron is added to a gaseous atom:

$$X(g) + e^- \rightarrow X^-(g) \qquad\qquad \Delta H = EA$$

Most electron affinities are negative (exothermic), showing that the added electron is attracted to the atom's nucleus. The trends for electron affinities are not as clear those for IE, but in general EA values become less negative as you look down a column and more negative as you look across a row. These trends can also be explained by atomic radii—the smaller the atom, the closer the new electron is to the nucleus and the stronger the attraction between the new electron and the nucleus. So, more energy (EA) is released when an electron is added closer to the nucleus.

Ionization Energies, Electron Affinities, and the Formation of Ions (7.13)

Metals (especially alkali and alkaline earth metals) have the lowest ionization energies in the periodic table; so, these elements are the most likely to lose electrons and become positive (cations). Likewise, the large negative electron affinities for nonmetals (especially the halogens) imply that nonmetals are the most likely to gain electrons and become negative (anions). However, the lowest energy needed to make a metal cation ($Cs \rightarrow Cs^+$, $IE = 376$ kJ/mol) is larger in magnitude than the most negative electron affinity ($F \rightarrow F^-$, $EA = -328$ kJ/mol), so this process would be endothermic. Yet, you know that metals and nonmetals will react exothermically to make ionic compounds. Clearly, there are other energy factors to consider.

$$Cs(g) + F(g) \rightarrow Cs^+(g) + F^-(g) \qquad \Delta H = IE(Cs) + EA(F) = +48 \text{ kJ}$$

Energy Changes in the Formation of Ionic Compounds (7.13)

The problem with the calculation above is that ionic compounds are not typically made from gaseous elements, and ionic compounds are not gases. In order to calculate the energy change for the formation of ionic compounds from their elements at standard states, chemists use a **Born-Haber cycle**, which is simply a Hess's law calculation using data that can be found in chemistry reference books. The target equation is the formation of the ionic compound (where M is the metal and X is the nonmetal) from its elements in their standard states, $\Delta H°_f\{MX(s)\}$.

The information used to calculate $\Delta H°_f\{MX(s)\}$ are: (1) The energy needed to convert the solid metal into a gas, $\Delta H°_{sublimation}$ or $\Delta H°_f\{M(g)\}$; (2) The energy needed to convert the nonmetal into a monatomic gas atom, $\Delta H°_f\{X(g)\}$—this value involves the energy needed to break the X-X bonds in the element; (3) The energy needed to make the gaseous cation from the gaseous metal atom, successive IE values (IE_1 for M^+, $IE_1 + IE_2$ for M^{2+}, etc.); (4) The energy needed to make the gaseous anion from the gaseous nonmetal atom, successive *EA* values (EA_1 for X^-, $EA_1 + EA_2$ for X^{2-}, etc.); and (5) The energy needed to convert the gaseous ions into the solid ionic compound, called the **lattice energy**.

Calculating Lattice Energies from a Born-Haber Cycle (7.13)

Although lattice energies appear in reference books, they are usually calculated from a Born-Haber cycle. You can also use the lattice energy to determine other values within the Born-Haber cycle that are harder to measure or that are not tabulated in the reference books.

<u>Example</u>: When solid sodium and solid sulfur react, they make solid sodium sulfide (Na_2S). Use the lattice energy for Na_2S (–2,230 kJ/mol) and the information below to determine the second electron affinity (in kJ/mol) for sulfur. This is the electron affinity for the S^- ion.

$$2\,Na^+(g) + S^{2-}(g) \rightarrow Na_2S(s) \qquad\qquad \Delta H_{LE} = -2{,}203 \text{ kJ}$$

$$2\,Na(s) + \frac{1}{8}\,S_8(s) \rightarrow Na_2S(s) \qquad\qquad \Delta H°_f\{Na_2S(s)\} = -373 \text{ kJ}$$

$$Na(s) \rightarrow Na(g) \qquad\qquad \Delta H°_f\{Na(g)\} = +108 \text{ kJ}$$

$$\frac{1}{8}\,S_8(s) \rightarrow S(g) \qquad\qquad \Delta H°_f\{S(g)\} = +279 \text{ kJ}$$

$$Na(g) \rightarrow Na^+(g) + e^- \qquad\qquad IE_1(Na) = +496 \text{ kJ}$$

$$S(g) + e^- \rightarrow S^-(g) \qquad\qquad EA_1(S) = -200 \text{ kJ}$$

$$S^-(g) + e^- \rightarrow S^{2-}(g) \qquad\qquad EA_2(S) = \quad ?$$

<u>Solution</u>: Use Hess's law to solve for the $EA_2(S)$ equation. To get S^-(aq) as a reactant write equation 6 backwards, and to get S^{2-}(g) as a product, write equation 1 backwards. To cancel $Na_2S(s)$, used equation 2 as written. Use equation 4 backwards to cancel 1/8 S8(s) and S(g). To cancel 2 Na^+(g) use equation 5 backwards and doubled; to cancel the 2 Na(s) and the 2 Na(g) use equation 3 backwards and doubled.

$$\mathbf{S^-(g)} \rightarrow S(g) + e^- \qquad\qquad -EA_1(S) = +200 \text{ kJ}$$

$$Na_2S(s) \rightarrow 2\,Na^+(g) + \mathbf{S^{2-}(g)} \qquad\qquad -\Delta H_{LE} = +2{,}203 \text{ kJ}$$

$$2\,Na(s) + \frac{1}{8}\,S_8(s) \rightarrow Na_2S(s) \qquad\qquad \Delta H°_f\{Na_2S(s)\} = -373 \text{ kJ}$$

$$S(g) \rightarrow \frac{1}{8}\,S_8(s) \qquad\qquad -\Delta H°_f\{S(g)\} = -279 \text{ kJ}$$

$$2\,Na^+(g) + e^- + \mathbf{e^-} \rightarrow 2\,Na(g) \qquad\qquad -2 \times IE_1(Na) = -992 \text{ kJ}$$

$$2\,Na(g) \rightarrow 2\,Na(s) \qquad\qquad -2 \times \Delta H°_f\{Na(g)\} = -216 \text{ kJ}$$

The final energy is: $\Delta H = (200 \text{ kJ} + 2{,}203 \text{ kJ} - 373 \text{ kJ} - 279 \text{ kJ} - 992 \text{ kJ} - 216 \text{ kJ}) = 543 \text{ kJ}$, and $EA_2(S) = 543 \text{ kJ}/1 \text{ mol } S^- = 543 \text{ kJ/mol}$.

Chapter Review — Key Terms

The key terms that were introduced in this chapter are listed below, along with the section in which they were introduced. You should understand these terms and be able to apply them in appropriate situations.

atomic radius (7.9)
Born-Haber cycle (7.13)
boundary surface (7.4)
continuous spectrum (7.3)
core electrons (7.7)

diamagnetic (7.8)
effective nuclear charge (7.9)
electromagnetic radiation (7.1)
electron affinity (7.12)
electron configuration (7.7)

electron density (7.4)
excited state (7.3)
ferromagnetic (7.8)
frequency (7.1)
ground state (7.3)

Hund's rule (7.7)   paramagnetic (7.8)   radial distribution plot (7.6)
ionic radius (7.10)   Pauli exclusion principle (7.5) s-block elements (7.7)
ionization energy (7.11)   p-block elements (7.7)   screening effect (7.9)
isoelectronic (7.8)   photoelectric effect (7.2)   shell (7.5)
lattice energy (7.13)   photons (7.2)   spectrum (7.1)
Lewis dot symbol (7.7)   Planck's constant (7.2)   subshell (7.5)
line emission spectrum (7.3) principal energy level (7.5) transition elements (7.7)
momentum (7.4)   principal quantum number (7.3) uncertainty principle (7.4)
noble gas notation (7.7)   quantum (7.2)   valence electrons (7.7)
orbital (7.4)   quantum number (7.5)   wavelength (7.1)
orbital shape (7.6)   quantum theory (7.2)

Practice Test

After you have finished studying the chapter and the homework problems, the following questions can serve as a test to determine how well you have learned the chapter objectives.

1. When ozone molecules absorb light at a wavelength of 260 nm, an O–O bond is broken.
    (a) What is the wavelength (in m) of this light? What kind of light is this?
    (b) What is the frequency (in s^{-1}) of light with this energy?
    (c) What is the energy (in J) needed to break the bond in one molecule?
    (d) What is this energy in kJ/mol?

2. Consider what happens when an electron in hydrogen moves from the $n = 6$ to the $n = 2$ orbit.
    (a) Does the electron move closer to or farther from the nucleus?
    (b) Does the electron gain or lose energy? Is ΔE positive or negative?
    (c) What is the wavelength (in nm) for this change? What kind of light is this?

3. List one advantage and one disadvantage associated with Rydberg's equation.

4. List one advantage and one disadvantage associated with Bohr's planetary model of the atom.

5. Which quantum number (n, ℓ, m_ℓ, and/or m_s) tells you about the following properties of an orbital?
    (a) The distance of an electron in the orbital from the nucleus.
    (b) The shape of the orbital.
    (c) The energy of an electron in the orbital.
    (d) The region of space the electron in the orbital occupies.

6. Are the following set of quantum numbers possible? If not, explain why. What is the orbital label ($1s$, $3p$, etc.) for each set of quantum numbers?
    (a) $n = 2$, $\ell = 1$, $m_\ell = 0$, $m_s = -1/2$ (c) $n = 4$, $\ell = 4$, $m_\ell = +3$, $m_s = -1/2$
    (b) $n = 3$, $\ell = 2$, $m_\ell = -3$, $m_s = +1/2$ (d) $n = 5$, $\ell = 3$, $m_\ell = -2$, $m_s = +1/2$

7. Identify the following orbitals by their boundary-surface diagrams.

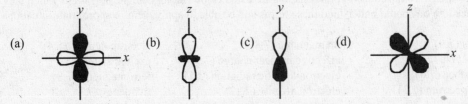

8.   Write the electron configurations for the following atoms or ions.
     (a)   N (orbital box diagram)   (d) Co^{2+} (condensed notation)
     (b) Si^- (orbital box diagram)   (e)   Pt (noble gas notation)
     (c)   I (condensed notation)   (f) Se^{2-} (noble gas notation)

9.   Consider the following orbital box diagrams for the same neutral element.

     I
     1s   2s   2p
     [↑↓] [↑↓] [↑↓][ ][ ]

     III
     1s   2s   2p
     [↑↓] [↑↑] [↑][↑][ ]

     II
     1s   2s   2p
     [↑↓] [↑↓] [↑][↑][ ]

     IV
     1s   2s   2p
     [↑↓] [↑] [↑][↑][↑]

     (a)   What is the identity of this element? Explain your answer.
     (b)   Which orbital box diagram represents the correct electron configuration for this atom?
     (c)   What is wrong with the other three orbital box diagrams?

10. Use electron configurations to rank the following atoms or ions in order of increasing radii.
     (a)   Br, Cl, F, I   (c)   Fe, Fe^-, Fe^{2+}, Fe^{3+}
     (b)   Ar, Mg, S, Si   (d) Cl^-, K^+, P^{3-}, Sc^{3+}

11. Even though nitrogen and sodium both have fewer electrons than phosphorus, P (110 pm) is larger
     than N (74 pm) but smaller than Na (191 pm). Explain this observation.

12. Which of the following atoms or ions have the higher ionization energy. Explain your answer.
     (a)   Al or B   (c)   Cl or Cl^-
     (b)   C or F   (d) Mg^{2+} or Na^+

13. Use the electron affinity of chlorine ($EA = -349$ kJ/mol) to determine the ionization energy of the
     chloride ion.

14. Consider the graph of successive ionization energies for
     nitrogen to the right.
     (a)   What is the electron configuration (condensed
           notation) for nitrogen?
     (b)   Why is there a dramatic increase in *IE* for the
           sixth electron?
     (c)   Why does the curve end at the seventh electron?

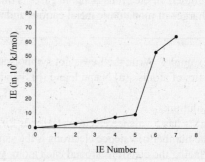

Chapter 8
Covalent Bonding

The goal of this chapter is to introduce the concept of covalent bonds, which are chemical bonds made between two nonmetal atoms using their valence electrons. The properties of bonds (including bond length, bond enthalpy, bond polarity, and electronegativity) are also described in this chapter. The number and types of bonds formed in nonmetal (molecular) compounds are usually described in terms of Lewis structures. This chapter also describes two important concepts in Lewis structures—formal charges and resonance structures. In addition to these topics, this chapter also introduces several new categories of organic compounds including alkenes, alkynes, and aromatic compounds.

Covalent Bonding (8.1-8.3)

When metals and nonmetals combine together to make ionic compounds, the metal gives up one or more electrons and the nonmetal accepts them (electron transfer). When two nonmetals combine together to make molecular compounds, neither atom is willing to give up its electrons. So, the only way to make stable compounds between two nonmetals is for the two atoms to share electrons, forming **covalent bonds**. Each covalent bond consists of two nuclei sharing two electrons. The forces holding a covalent bond together are the attractions between the positively-charged nuclei and the negatively-charged electrons, balanced by the repulsions of the atomic nuclei and the repulsions of the two electrons.

You can use Lewis dot symbols (introduced in the Chapter 7 of the textbook) and the octet rule to predict the number of bonds each nonmetal will make. *Lewis dot symbols* contain the chemical symbol of the atom (representing the atom's nucleus and its core electrons) and a dot to represent each valence electron. The dots are placed around the atom following a variation of Hund's rule—place the first four dots around the atom as unpaired electrons (1 e⁻ for the *s* and the three *p* orbitals), then pair any additional electrons with the unpaired electrons. The **octet rule** states that atoms will make covalent bonds until they have eight electrons surrounding them (the electron configuration of a noble gas). The fact that atoms want the same numbers of electrons as noble gases is the same rule used in Chapter 3 that allows you to predict the charges of monatomic metal cations and nonmetal anions.

Example: Write the Lewis dot symbols and predict the molecular compounds made from the following pairs of atoms. (a) N and I, and (b) S and H.

Solution:
(a) N has 5 valence electrons ([He] $2s^2 2p^3$), and I has 7 ([Kr] $5s^2 5p^5$). In general, the number of valence electrons is the same as the atom's column number. Placing the electrons around each atom, you see that N has 3 unpaired electrons and 1 **lone pair** (two paired electrons that appear on a single atom) while I has 1 unpaired electron and 3 lone pairs. It does not matter where the single or paired atoms are drawn in Lewis dot symbols—all that matters is how many pairs and how many single electrons each atom has. Since N is 3 e⁻ short of an octet, it will make 3 bonds with other atoms in order to get an octet; similarly, I is 1 e⁻ short of an octet, so it will make 1 bond in order to get an octet. So, 1 N atom will bond with 3 I atoms (NI_3).
(b) S has 6 valence electrons (column 6A) and H has 1 (column 1A). The Lewis dot symbols show that S has 2 unpaired electrons and 2 lone pairs and H has 1 unpaired electron. To reach an octet, S must make 2 bonds, but H is 7 electrons away from an octet. H is an exception to the octet rule, and instead follows a *duet rule* (H wants only 2 e⁻ because its nearest noble gas—He = $1s^2$—has only 2 e⁻). Since H needs to form only one bond, 2 H atoms will bond with 1 S atom (H_2S).

Molecular structures made from Lewis dot symbols are called **Lewis structures** (or *Lewis dot structures*). The pairs of electrons on a single atom are lone pair electrons, and the line between atoms represents two electrons being shared by the two atoms (**bonding electrons**). In the structures above, you can see that each atom (except H) has 8 e⁻ around it from a combination of lone pairs and bonding pairs of electrons.

Example: Use the Lewis dot symbols for C and H to predict the formulas of C-H compounds containing one carbon, two carbons, and three carbons.

Solution: The Lewis dot symbols for C shows that C will make 4 bonds (4 e⁻ short of an octet), and H will make one bond (1 e⁻ short of a duet). The simplest formula will be when 1 C atom bonds to 4 H atoms (CH_4). When the compound has 2 C atoms, the C atoms must be bonded to each other, leaving each C atom needing to make 3 more bonds, so this formula is C_2H_6. When 3 C atoms are bonded to each other, the middle C needs 2 more bonds and the outer two C atoms need 3 more bonds, so this compound needs eight H atoms for a formula of C_3H_8.

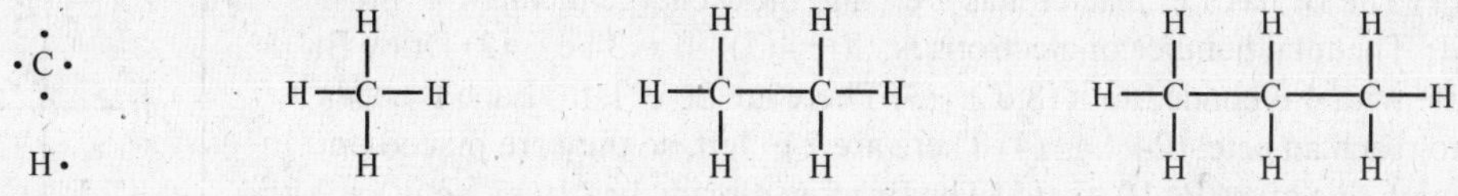

Although constructing Lewis structures this way works for many molecules, it is not always successful at predicting actual chemical compounds. For example, when you predict the structures of carbon-oxygen compounds, you will see that C makes 4 bonds and O makes 2 bonds. From these rules, you can predict structures of CO_2 (which exists) and C_2O_2 (which does not exist and decomposes into CO). However, you cannot predict the structure of CO (a molecule known to exist) from these rules.

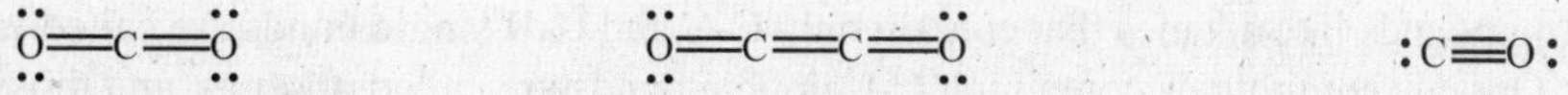

predicted, and exists   predicted, but doesn't exist   can't predict, but exists

Writing Lewis Structures for Molecules and Polyatomic Ions (8.1-8.3)

Since Lewis dot symbols cannot be used to draw Lewis structures for every molecule, chemists came up with a series of rules for drawing Lewis structures: (1) Count the total number of electrons coming from each atom within the molecule or ion and the overall charge of the molecule or ion; (2) Draw a skeleton structure using Lewis dot symbols (element symbols) and a single line representing two bonding electrons between each pair of bonded atoms; (3) Place lone pairs on the outer atoms to satisfy the octet rule for these atoms; (4) Place any additional electrons in pairs on the central atom; (5) If the central atom has fewer than 8 e⁻, share one or more of the lone pairs on an outer atom with the central atom, forming multiple bonds.

Example: Draw the Lewis structure for bromomethane, CH_3Br.

Solution: (1) The C has 4 e⁻, each H has 1 e⁻, and the Br has 7 e⁻. So, the total number of electrons is $4 + 3(1) + 7 = 14$ e⁻. (2) Generally, the central atom is written first. So, draw C in the middle with 3 H and 1 Br bonded to it (8 e⁻). (3) There are 6 e⁻ left. Remembering that H only wants 2 e⁻, the H atoms do not need lone pairs but the Br needs 3 lone pairs (6 e⁻) to reach an octet. (4) There are no electrons left. (5) The C atom has an octet from the four bonds, so no multiple bonds are needed.

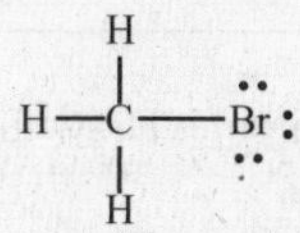

Example: Draw the Lewis structure for hydrogen cyanide, HCN.

Solution: (1) The C has 4 e⁻, the H has 1 e⁻, and the N has 5 e⁻. So, the total number of electrons is $4 + 1 + 5 = 10$ e⁻. (2) When there are three atoms bonded together, the formula is often written so that the middle atom (in this case, C) is the central atom. So, draw C in the middle with H and N bonded to it (4 e⁻). (3) There are 6 e⁻ left. Since H only wants 2 e⁻, it does not need lone pairs; N needs 3 lone pairs (6 e⁻) to reach an octet. (4) There are no electrons left. (5) The C atom has only 4 e⁻ (4 e⁻ short of an octet) from the two bonds, so multiple bonds need to be made. Since N is the only outer atom with lone pairs, N shares two lone pairs with C, making a C-N triple bond.

Example: Draw the Lewis structure for the BrF_4^+ ion.

Solution: (1) The Br has 7 e⁻, each F has 7 e⁻, and the + charge means 1 e⁻ must be removed. The total number of electrons is: $7 + 4(7) - 1 = 34$ e⁻. (2) Draw Br in the middle with 4 F bonded to it (8 e⁻). (3) There are 26 e⁻ left. Each F needs 3 lone pairs to reach an octet (24 e⁻). (4) There are 2 e⁻ left, so they are placed on Br even though this gives Br 10 e⁻. (5) The Br atom already has 10 e⁻, so it doesn't need more, and no multiple bonds are made. The Lewis structure of an ion is usually placed in square brackets with the ion charge on the outside.

Multiple Bonds in Alkenes and Alkynes (8.4)

Alkanes (described in Chapter 3 of the textbook) are organic compounds containing only C-C and C-H single bonds. Compounds like alkanes that contain only C-C and C-H single bonds are called **saturated hydrocarbons**. Organic compounds containing C-C double bonds are called **alkenes**, and those containing C-C triple bonds are called **alkynes**. Hydrocarbons with C-C double or triple bonds are called **unsaturated hydrocarbons**. **Double bonds** occur when two atoms share two pairs of electrons (4 e⁻) between them; similarly, **triple bonds** occur when two atoms share three pairs of electrons (6 e⁻) between them. Together, double bonds and triple bonds are referred to as **multiple covalent bonds**.

Recognizing *cis-trans* Isomerism (8.5)

In alkenes, the carbon atoms cannot rotate about the C-C double bond. This restricted rotation leads to two types of isomers that differ only in their arrangement of the bonds around the C-C double bond. These isomers can only occur when both C atoms have two different substituents bound to them and there is at least one substituent that is the same on both C atoms (the general formulas are $XYC=CXY$ or $XYC=CXZ$, where X, Y, and Z are the substituents on the C atoms). The *cis* **isomer** occurs when the common substituent is on the same side of the double bond; the *trans* **isomer** occurs when the common substituent appears on the opposite side of the double bond.

Example: Draw the *cis* isomer and the *trans* isomer for 1-fluoropropene, $FHC=CHCH_3$.

Solution: The four substituents on the C atoms of the double bonds are F, H, H, and CH_3. The H atoms are on the same side of the double bond in the *cis* isomer and on opposites sides in the *trans* isomer.

Predicting Bond Lengths from Atomic Radii (8.6)

Bond lengths are the distances between the nuclei of two atoms bonded to each other. Bond lengths can be thought of as simply the sum of the two atomic radii, and trends in atomic radii from the periodic table can also be used to predict bond lengths.

Example: Rank the following chemical bonds in order of increasing bond length.
(a) Br—Br, Cl—Cl, F—F; (b) C—Br, C—Cl, C—I; (c) C—H, N—H, O—H; (d) C—N, C=N, C≡N

Solution:
(a) The trend in atomic radii is that the radius of an atom increases as you look down a column in the periodic table. So, Cl is larger than F and Br is larger than both of them. Since the bond lengths are twice the atomic radii of the atoms, the order is F—F (shortest) < Cl—Cl < Br—Br (longest).
(b) Since each bond has a C atom, its effect on the bond length should be the same. Since the atomic radius of an atom increases as you look down a column in the periodic table, I is the largest atom and Cl is the smallest atom, which translates to the bond lengths as well: C—Cl (shortest) < C—Br < C—I (longest).
(c) The effect of the H atom should be the same. The atomic radius of an atom decreases as you look across a row in the periodic table. So, C is the largest atom and O is the smallest atom: O—H (shortest) < N—H < C—H (longest).
(d) Single bonds are longer than double bonds, which are longer than triple bonds. This is because the more electron pairs being shared between the two atoms, the stronger the electron-nucleus attractions are and the closer to the nuclei are pulled together. So, the trend is C≡N (shortest) < C=N < C—N (longest).

Relating Bond Enthalpies to Bond Lengths (8.6)

Bond enthalpies (also called **bond energies**) are the enthalpy changes that occur when two bonded atoms in the gas phase are separated—this is the energy needed to break the chemical bond. All bond enthalpies are positive, indicating that it takes energy (endothermic) to break any chemical bond. There is no simple relationship between bond enthalpies and bond lengths; however, when you are looking at the same types of atoms bonded together, bond enthalpies increase as the bond lengths get shorter—so, triple bonds are stronger than double bonds and double bonds are stronger than single bonds between the same two atoms.

Calculating Enthalpies of Reaction from Bond Enthalpies (8.6)

You can use bond enthalpies to estimate the enthalpy change of a reaction ($\Delta H°$). This is especially useful when working with chemicals that do not have tabulated $\Delta H°_f$ values. The formula for this calculation is written below, where $D(X\text{-}Y)$ is the symbol for bond energies.

$$\Delta H = \sum \{n_{broken}D(broken)\} - \sum \{n_{formed}D(formed)\}$$

Example: Use bond enthalpies to predict $\Delta H°$ for the following reaction. Is it endothermic or exothermic?

$$2 \text{ H—O——O—H} \quad \rightarrow \quad 2 \text{ H—O—H} \quad + \quad \text{O}=\text{O}$$

Solution: Although you can break all of the reactants' bonds and make all of the products' bonds, there are times when some bonds do not change and can be left alone. Here, the reactants have 4 H—O bonds and 2 O—O bonds and the products have 4 H—O bonds and 1 O=O bond. In this case, the 4 H—O bonds do not change and can be left out of the equation. Looking at the bonds that have changed, 2 O—O bonds must be broken and 1 O=O bond must be made. Using the bond enthalpies from p. 350 of the textbook, $\Delta H°$ is:

$$\Delta H = \left\{ 2 \text{ mol}_{O—O} \times \frac{146 \text{ kJ}}{\text{mol}_{O—O}} \right\} - \left\{ 1 \text{ mol}_{O=O} \times \frac{498 \text{ kJ}}{\text{mol}_{O=O}} \right\} = -206 \text{ kJ (exothermic)}$$

71

Predicting Bond Polarities from Electronegativities (8.7)

The difference between ionic bonds and covalent bonds is that the atoms (ions) in an ionic bond do not share electrons while atoms in a covalent bond do share electrons. However, these two atoms do not always share the electrons equally (i.e., one atom may have 'custody' of the electrons more often than the other atom). When there is equal sharing, the bond is referred to as a **nonpolar covalent bond**, but if the electron sharing is unequal it is called a **polar covalent bond**. Nonpolar covalent bonds result in '50-50' (equal) sharing while ionic bonds result in '0-100' (no) sharing; polar covalent bonds have electron sharing somewhere in between '0-100' and '50-50' sharing (unequal sharing).

Chemists use the concept of **electronegativity** (*EN*) to determine the type of bond formed between two atoms. Electronegativity is a measure of an atom's ability to attract electrons from a chemical bond to itself. Since *EN* values are based on chemical bonds and noble gases don't make chemical compounds, *EN* values are not usually reported for noble gases. The trends from the periodic table are that *EN* values decrease as you look down a column and increase as you look across a row in the periodic table (just like *IE* and *EA* values). *EN* values range from about 0.8 for alkali metals to 4.0 for fluorine.

In order to classify a chemical bond, chemists often calculate an *electronegativity difference* for the bond $(\Delta EN = EN_{larger} - EN_{smaller})$. When ΔEN is small (0.0-0.4), the atoms have similar *EN* values and they share the electrons equally (a nonpolar covalent bond). At the opposite extreme, when ΔEN is very large (2.1 or greater), the atom with a much higher *EN* value steals the electron (gaining a – charge) from the atom with a lower *EN* value (leaving it with a + charge)—an ionic bond. If the ΔEN value falls in between this range (0.5-1.4), the electrons are shared unequally (polar covalent bond)—the more *EN* atom has the electrons more often (resulting in a partial negative charge δ–) and the less *EN* atom has custody of the electrons less often (resulting in a partial positive charge δ+). When ΔEN falls between 1.5 and 2.0, the bond is polar covalent if both atoms are nonmetals and it is ionic if one atom is a metal and the other is a nonmetal.

Using Formal Charges to Compare Lewis Structures (8.8)

The ammonium ion (NH_4^+) has a positive charge, but where is the positive charge in this polyatomic ion? Is it on the N atom, one of the H atoms, or both? Chemists use formal charges to determine where the charges in polyatomic ions or molecules reside. There is a formula to calculate formal charges (described on p. 350 of the textbook), but it is easier to do these calculations in your head. The **formal charge** is the charge on each atom assuming that each bonding pair is shared <u>equally</u> (i.e., each atom has one of the two electrons in each bond). To determine the formal charge, figure out how many valence electrons are on the atom in the molecule or polyatomic ion and compare to the neutral atom. Just as with oxidation numbers, the sum of the formal charges must add to the total charge on the molecule (zero) or the polyatomic ion.

Example: The Lewis symbol for the ammonium ion is written to the left.
(a) Determine the formal charges on each atom in the ammonium ion.
(b) What are the oxidation numbers of each atom in this ion?
(c) How are formal charges and oxidation numbers different?

$$\left[\begin{array}{c} H \\ | \\ H - N - H \\ | \\ H \end{array} \right]^+$$

Solution:
(a) When the bonds are cut in half, the N atom has 4 e⁻ and each H atom has 1 e⁻. Since neutral N atoms have 5 e⁻, an N atom with 4 e⁻ would have a formal charge of +1 (1 e⁻ less than neutral). Since neutral H atoms have 1 e⁻ and each of these H atoms have 1 e⁻, the formal charge for each H atom is 0.
(b) From Rule 4, $O.N._N + 4 \times O.N._H = +1$. Using Rule 3c, $O.N._H = +1$, and solving for nitrogen gives you $O.N._N + 4 \times (+1) = +1$, and $O.N._N = +1 - 4 = -3$.
(c) Formal charges split the bonding electrons equally between the two atoms; oxidation numbers give the bonding electrons to the more electronegative atom. Formal charges and oxidation numbers are two different ways of counting electrons, and neither one is perfectly correct (they are the two extreme cases).

When there are two or more possible Lewis structures, formal charges can be used to determine which ones are more stable. In general, the Lewis structures with the lower or fewer formal charges are more stable, negative formal charges should reside on the atoms with greater *EN*, and the same charge should not appear on atoms that are bonded together.

Example: When the Lewis structure for carbonyl sulfide (OCS) is drawn, after Step 4 the C atom has only 4 e$^-$ and needs to make multiple bonds. There are 3 possible Lewis symbols that satisfy the octet for C. Which of these are the most stable structures?

Solution: In structure I the O atom shared two e$^-$ pairs, in structure II the O and S atoms each shared one pair, and in structure III, the S atom shared two e$^-$ pairs. Neutral C atoms have 4 e$^-$, and neutral O and S atoms have 6 e$^-$. In structure I, the O atom has 5 e$^-$ (+1 formal charge), the C atom has 4 e$^-$ (0 formal charge), and the S atom has 7 e$^-$ (–1 formal charge). In structure II, the O atom has 6 e$^-$ (0 formal charge), the C atom has 4 e$^-$ (0 formal charge), and the S atom has 6 e$^-$ (0 formal charge). In structure III, the O atom has 7 e$^-$ (–1 formal charge), the C atom has 4 e$^-$ (0 formal charge), and the S atom has 5 e$^-$ (+1 formal charge). Formal charges are usually written near the atom and encircled. Since structure II has the lowest formal charges (all 0), it is the most stable structure. Note that in all three structures, the formal charges all add to zero (the charge on the neutral OCS molecule).

Using Resonance Structures to Describe Bonding (8.9)

Sometimes there are molecules or polyatomic ions that have more than one stable Lewis structure. These alternative structures are referred to as **resonance structures**, and are drawn together using single double-headed arrows. The actual molecules does not look like either resonance structure; instead, it looks like an 'average' of these structures. This 'average' structure is called a **resonance hybrid**. Perhaps you can better understand a resonance hybrid using an analogy from Greek mythology. If you were to describe Pegasus to a child who has been to the zoo, how would you describe it using animals the child recognizes? You might say Pegasus was 'half horse, half eagle', but what does that mean? Does that mean half of the time it's a horse, then it magically changes into an eagle, and back and forth? Neither single description perfectly explains what Pegasus looks like (just like neither resonance structure is ever correct for the molecule), and it is best thought of as an average or composite of these individual descriptions.

Example: There are two equivalent resonance structures for the nitrite ion (NO_2^-). What is the N–O bond length in the actual ion? What is the N–O bond enthalpy in the actual ion? The N—O bond is 136 pm long and $D(N—O) = 201$ kJ/mol, and the N=O bond is 115 pm long and $D(N=O) = 598$ kJ/mol.

Solution: In each structure, there is an N–O single bond and an N–O double bond; however, in the actual ion, each of these bonds is an average of the bonds in each structure. So, the N–O bond on the left and the one on the right are an average of a single and double bond, and they are equivalent. The N–O bond length should be: 1/2(136 pm + 115 pm) = 126 pm. The bond enthalpy of each bond should also be an average of the single and double bond enthalpies: 1/2(201 kJ/mol + 598 kJ/mol) = 400 kJ/mol.

Exceptions to the Octet Rule (8.10)

There are two major exceptions to the octet rule—having less than eight valence electrons and having more than eight valence electrons. Although H never has an octet (it always has 2 e^- in its compounds), it still follows the 'noble gas rule' (get as many electrons as the nearest noble gas). Boron (B) is another atom that has difficulty getting 8 e^-. Because it has only 3 valence e^-, it can only make 3 bonds for a total of 6 e^-. To fix this problem, B prefers to bond to very electronegative atoms having lone pairs on them that can be shared, giving B an octet from a newly formed multiple bond.

Another common class of chemicals that have less than eight valence electrons are **free radicals**, chemicals that have an odd number of electrons in them. Free radicals are very chemically reactive (and destructive) because most atoms in molecules prefer to have an octet. So, these free radicals steal an electron from a stable molecule, turning it into a free radical (because it now has an odd number of electrons), which is now unstable and will steal an electron from another molecule, making it a free radical, and so on.

When drawing Lewis structures, you have seen that some atoms can have more than 8 valence electrons, and these atoms are said to have an *expanded octet*. Expanded octets are rare, and usually occur only for larger atoms (with outer electrons in $n = 3$ or larger orbits) bonded to very electronegative atoms (F or O).

Bonding and Constitutional Isomers in Aromatic Compounds (8.11)

Aromatic compounds are organic (C-H containing) molecules that have a C_6 ring with alternating single and double bonds. In aromatic rings, there are two resonance structures, and each C-C bond in these molecules is an average between a single and a double bond. Sometimes, the alternating single and double bonds are replaced with a circle. Each C atom in the ring has another atom bonded to it, but those have been omitted in these pictures for clarity.

Since there are six places around the ring for different atoms or groups of atoms to be bonded, this can lead to constitutional isomers. When four of the substituents around the ring are the same and two are different, the isomer is called *ortho* if the two different substituents are on neighboring C atoms, it is called *meta* if they are on C atoms that have one C atom between them, and it is called *para* if the substituents are on C atoms that are across from each other in the ring (2 C atoms between them).

Using Molecular Orbital Theory to Describe Bonding (8.12)

The theory used when drawing Lewis structures is called the *valence bond theory*, and it assumes that two atoms share their valence electrons to make covalent single and multiple bonds. This theory works pretty well, but there are times when it does not work (e.g., the need for resonance structures because no single Lewis structure can explain the bonding in certain molecules). Another example where the valence bond theory fails is in explaining why diatomic oxygen (O_2) is paramagnetic and has two unpaired electrons.

To address these problems, chemists use another bonding theory, called *molecular orbital (MO) theory*. MO theory says that atoms in a molecule use their atomic orbitals to bond to other atoms in the molecule. The resulting **molecular orbitals** are similar to atomic orbitals except they can have electron density over the entire molecule (**delocalized electrons**) instead of being limited to bonds that have electron density between only two atoms, as in Lewis structures.

MO theory is very complicated, so this book will only discuss simple examples. When the two atomic orbitals in H_2 (1s orbital from each H atom) interact, there will be two molecular orbitals. One of these orbitals has substantial electron density between the H atoms (**bonding molecular orbital**), and the other has almost no electron density between the atoms (**antibonding molecular orbital**). Bonding orbitals are lower in energy than antibonding orbitals due to the electron-nucleus attractions that are maximized in the bonding orbitals and the nuclear repulsions that are maximized in the antibonding orbitals. The orbitals in H_2 are called *sigma (σ) orbitals* (Greek letter 's') because they look like distorted s orbitals. An asterisk is added to the antibonding orbital (σ* or sigma-star).

σ orbital   σ* orbital

Double bonds in alkenes (like C_2H_4) are made of sigma orbitals (like those in H_2) and *pi orbitals*. Pi (π) orbitals (Greek letter 'p') are named because look like distorted p orbitals. The π orbital (bonding) has electron density in between the two C atoms but the π* orbital (antibonding) does not.

π orbital   π* orbital

Chapter Review — Key Terms

The key terms that were introduced in this chapter are listed below, along with the section in which they were introduced. You should understand these terms and be able to apply them in appropriate situations.

alkenes (8.5)	delocalized electrons (8.9)	octet rule (8.2)
alkynes (8.5)	double bond (8.4)	polar covalent bond (8.7)
aromatic compounds (8.11)	electronegativity (8.7)	resonance hybrid (8.9)
antibonding molecular orbital (8.12)	formal charge (8.8)	resonance structures (8.9)
bond enthalpy (bond energy) (8.6)	free radicals (8.10)	saturated hydrocarbons (8.3)
bond length (8.6)	functional group (8.3)	single covalent bond (8.2)
bonding electrons (8.2)	Lewis structure (8.2)	*trans* isomer (8.5)
bonding molecular orbital (8.12)	lone pair electrons (8.2)	triple bond (8.4)
cis isomer (8.5)	molecular orbital (8.12)	unsaturated hydrocarbons (8.5)
cis-trans isomerism (8.5)	multiple covalent bonds (8.4)	
covalent bond (8.1)	nonpolar covalent bond (8.7)	

Practice Test

After you have finished studying the chapter and the homework problems, the following questions can serve as a test to determine how well you have learned the chapter objectives.

1.   Draw the Lewis dot symbols for sulfur and bromine.
      (a)   Draw the Lewis dot symbols for S and Br. How many bonds will each atom make?
      (b)   Use these symbols to write two Lewis structures for compounds made from these atoms.
      (c)   Write the names and formulas for these compounds.

2. Draw the Lewis structures for the following molecules and polyatomic ions.
   (a) BBr_4^-
   (b) SCl_3^+
   (c)   HNS
   (d) $XeOF_4$

3. Which of the following alkenes can have *cis-trans* isomers? Which isomer is shown?

   (a) 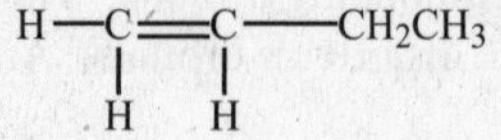   H—C=C—CH₂CH₃
              | |
              H   H

   (c)   H—C=C—CH₃
              | |
              H   CH₃

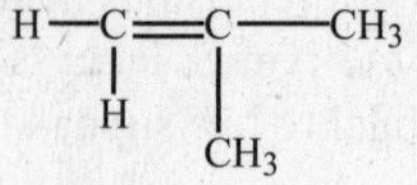

   (b) 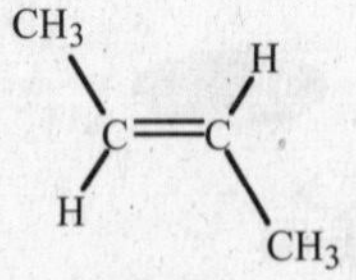   CH₃   H
            \\   /
             C=C
            /   \\
           H   CH₃

   (d) 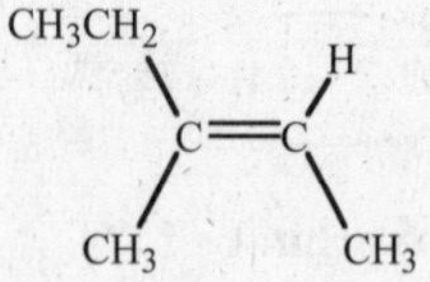   CH₃CH₂   H
               \\   /
                C=C
               /   \\
             CH₃   CH₃

4. Elemental sulfur exists as a ring of eight S atoms, but these rings can be converted to diatomic molecules by the following formula.

   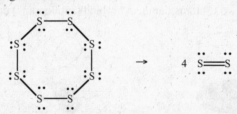

   (a)   Estimate $\Delta H°$ using bond enthalpies. $D(S—S)$ = 226 kJ/mol, $D(S=S)$ = 425 kJ/mol
   (b)   Is this reaction endothermic or exothermic? Which form of sulfur is more stable, S_8 or S_2?
   (c)   A similar reaction could be written for oxygen. Estimate $\Delta H°$ for this reaction for oxygen. $D(O—O)$ = 146 kJ/mol, $D(O=O)$ = 498 kJ/mol.
   (d)   Is the reaction for oxygen endothermic or exothermic? Which form of oxygen is more stable, O_8 or O_2? Why are these answers different for sulfur and oxygen?

5. Use the following bond enthalpies, $\Delta H°_f\{NOCl(g)\}$, and the following equation to estimate the bond enthalpy for an N-O double bond. Useful information: $D(Cl—Cl)$ = 242 kJ/mol, $D(N—Cl)$ = 193 kJ/mol, $D(O=O)$ = 498 kJ/mol, $D(N≡N)$ = 946 kJ/mol, $\Delta H°_f\{NOCl(g)\}$ = +53 kJ/mol.

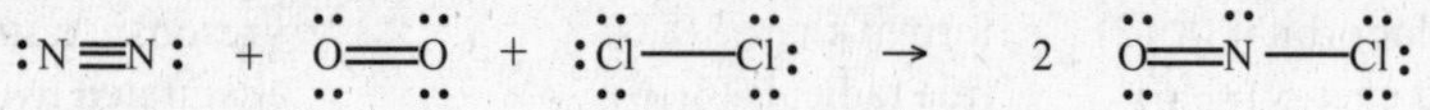

   :N≡N: + O=O + :Cl—Cl: → 2 O=N—Cl:

6. Rank the following elements in order of increasing electronegativity.
   (a)   Be, Ca, Mg
   (b)   Al, Cl, P

7. Classify the following bonds as ionic, nonpolar covalent, or polar covalent. What are the charges on the atoms in these bonds?
   (a)   B—O
   (b)   C—H
   (c)   F—Na
   (d)   N≡N
   (e)   O=S
   (f)   O—Zn

8. Determine the formal charges on each atom in the following Lewis structures.

   (a) 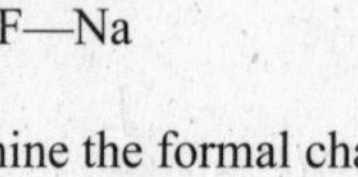   [:C≡N:]⁻

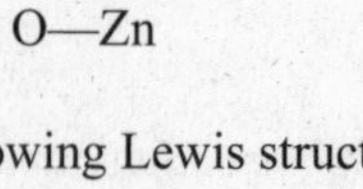

   (c)   [:N≡O:]⁺

   (b) 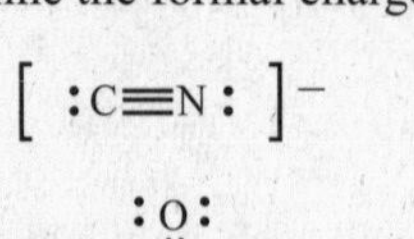   :O:
               ‖
        :O—N—O—H

   (d) 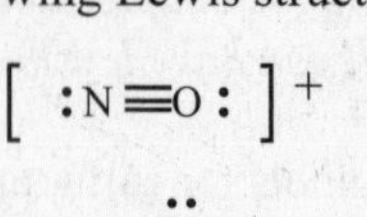   :O:
               |
        :Cl—S—Cl:

76

9. When the Lewis structure for cyanogen bromide (BrCN) is drawn, after Step 4 the C atom has 4 e⁻ and needs to make multiple bonds. There are 3 possible Lewis symbols that satisfy the octet for C:

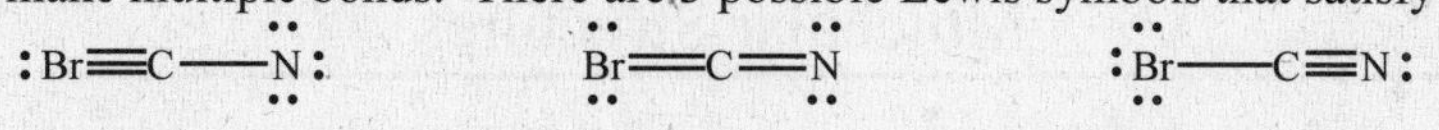

        I   II   III

(a) Determine the formal charges on each atom in these resonance structures.
(b) Which of these structures are most stable?
(c) Estimate the bond length and bond enthalpy of the C-N bond. The C—N bond is 147 pm long and $D(C—N) = 285$ kJ/mol, the C=N bond is 127 pm long and $D(C=N) = 616$ kJ/mol, and the C≡N bond is 115 pm long and $D(C≡N) = 866$ kJ/mol.

10. When the Lewis structure for the fluoroformate ion (CO_2F^-) is drawn, after Step 4 the C atom has only 6 e⁻ and needs to make a multiple bond. There are 3 possible Lewis symbols that satisfy the octet for C:

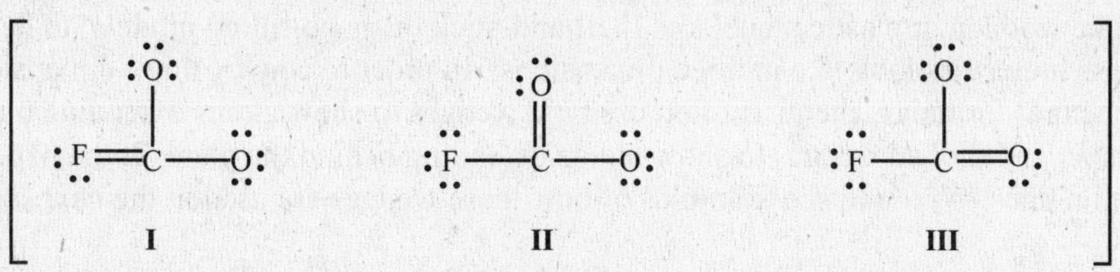

        I   II   III

(a) Determine the formal charges on each atom in these resonance structures.
(b) Which of these structures are the most stable?
(c) Estimate the bond length and bond enthalpy of the C-F bond. The C—F bond is 141 pm long and $D(C—F) = 486$ kJ/mol, and the C=F bond is 120 pm long and $D(C=F) = 845$ kJ/mol.
(d) Estimate the bond lengths and bond enthalpies of the C-O bonds. The C—O bond is 143 pm long and $D(C—O) = 336$ kJ/mol; the C=O bond is 122 pm long and $D(C=O) = 695$ kJ/mol.

11. Classify the following aromatic compounds as *ortho*, *meta*, *para*, or none of these.

(a)

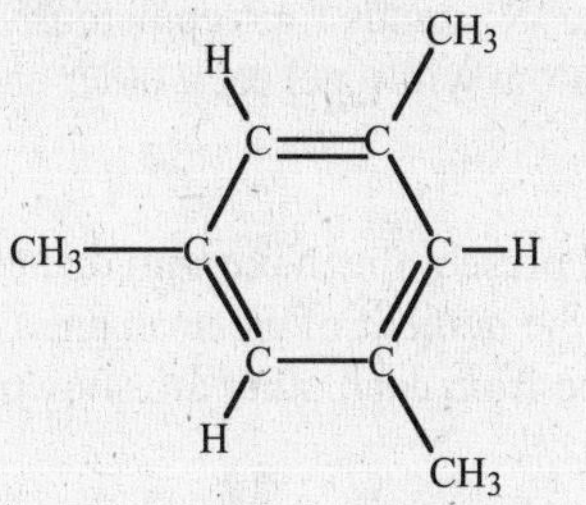

(c)

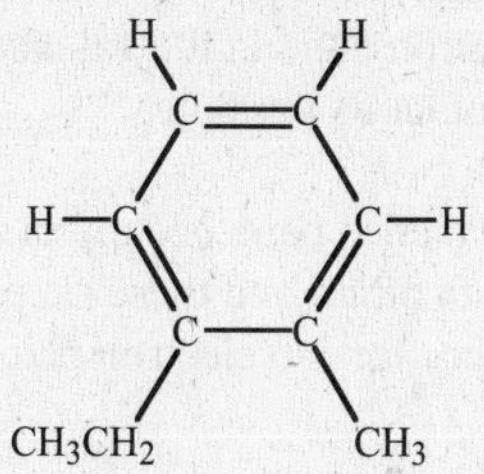

(b)

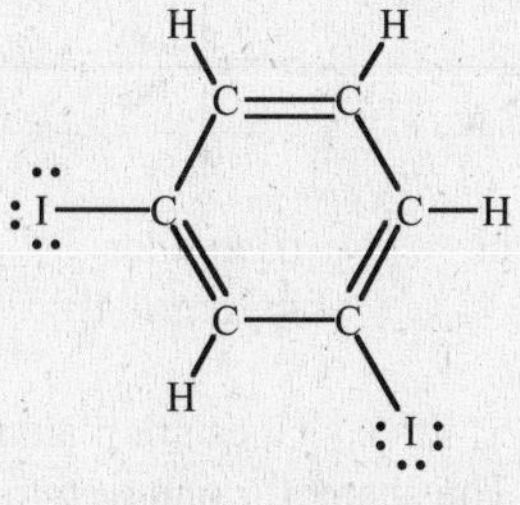

(d)

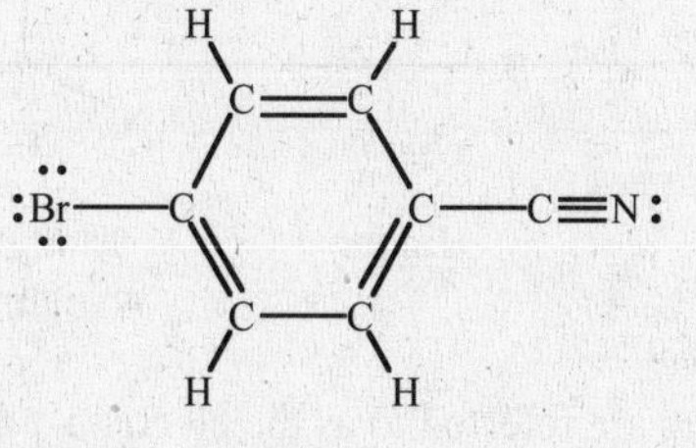

12. How does molecular orbital theory explain the fact that there are molecules whose bonding cannot be described by a single Lewis structure (i.e., the need for resonance structures and resonance hybrids)?

Chapter 9
Molecular Structures

The goal of this chapter is to introduce the 3-dimensional shapes of molecules and polyatomic ions using the valence-shell electron-pair repulsion model and Lewis dot structures. The chapter also describes how to determine the polarity of a molecule based on individual bond polarities within the molecule and the molecule's shape. In addition, this chapter describes the forces holding molecules together in a solid or liquid. Finally, this chapter describes the components present in DNA molecules.

Representing Molecular Shapes (9.1)

Structural formulas, ball-and-stick models, and space-filling models of molecules were introduced in Chapter 1 of the textbook. In addition to computer drawings in textbooks (which represent chemists' attempts to draw three-dimensional molecules on a two-dimensional sheet of paper), chemists often use physical models like wooden or plastic model kits (ball-and-stick or space-filling models) to help them visualize what these molecules look like in three dimensions. In order to convey three-dimensional information in structural formulas, chemists often use bold wedges to show atoms extending out in front of the plane of the paper and dashed wedges to show atoms extending behind the plane of the paper instead of using the simple thin lines. You will see examples of how these wedges are used in the next section.

Predicting Molecular Shapes using the VSEPR Model (9.2)

The shapes of molecules and polyatomic ions are usually predicted using the **valence-shell electron-pair repulsion (VSEPR) model**. The 'VS' stands for 'valence-shell'—this model assumes that only the outer (valence) electrons affect the shape of the molecule or ion—the inner core electrons tend to be spherically symmetric and are buried deep within the outer valence electrons, so their effect on shape should be minimal. The 'EP' stands for 'electron pair'—this model assumes that electrons will exist in pairs around the nucleus of the central atom. This assumption is also present in the **valence bond model** (also called the *valence bond theory*) that is the basis for drawing Lewis structures. Therefore, you can use Lewis structures and the VSEPR model to predict molecular shapes. The 'R' stands for 'repulsions'—this model assumes that the electron pairs in the valence shell on the central atom will repel each other and try to get as far away from each other as possible.

The 3-D spaces the valence pairs occupy in any molecule or ion represent the maximum distances these pairs can be from each other, and these shapes are determined by the number of electron pairs repelling each other. The shapes for 1-6 electron pairs at maximum distance from each other are shown below.

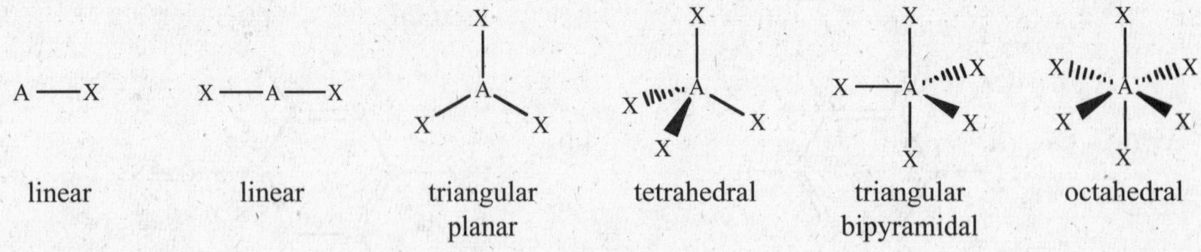

linear	linear	triangular planar	tetrahedral	triangular bipyramidal	octahedral

These shapes are referred to as the **electron-pair geometry**. The first three shapes are planar (2-D) shapes, and the last three are 3-D shapes that use bold and dashed wedges. The shape for one or two pairs around the central atom is *linear*; the angle between the two pairs on the second shape (**bond angle**) is 180°. The shape for three pairs is called *triangular planar* or *trigonal planar* and the pairs are 120° from each other ('triangular' means 'three angles' while 'trigonal' means 'three sides'—both are correct descriptions of a triangle). The shape for four pairs is called *tetrahedral* (this is the adjective form of the noun 'tetrahedron', meaning 'four faces') and the angles are 109.5°. The shape for five pairs is called *triangular bipyramidal* or *trigonal bipyramidal* (the adjective form of the noun 'bipyramid'). This term comes from the fact that if you ignore the bottom pair you have a triangular pyramid, and if you ignore the top pair you have another triangular pyramid; so, together they are a triangular bipyramid (two pyramids). The angles between the

78

top and bottom pairs (**axial positions**) is 180°, the angle between the three middle pairs (**equatorial positions**) is 120°, and the angle between an axial and equatorial pair is 90°. The shape for six pairs is called *octahedral* (the adjective form of the noun 'octahedron', meaning 'six faces') and the angles are 90°. In general, as the number of electron pairs around the central atom increases, the bond angles decrease.

In order to use these shapes, you must be able to determine the total number of electron pairs (both bonding pairs and lone pairs) around the central atom. Many chemists use the *AXE notation* for this step. The 'A' is the central atom, 'X' are the outer atoms bonded to the central atom (using a subscript to show how many outer atoms are present), and 'E' are lone pairs on the central atom (using a subscript to show the number of lone pairs present). Since both 'X' and 'E' represent an electron pair on the central atom, the total number of objects around the nucleus that need to get as far away from each other as possible is $X + E$.

Example: Write the AXE notation for the following molecules or ions. Which shape would the electron pairs on the central atom in these molecules or ions adopt?

(a) 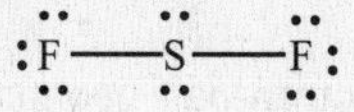 :F——S——F: (b) 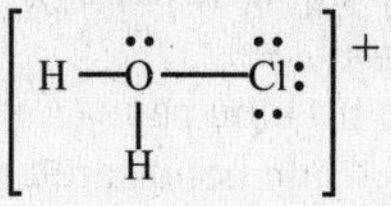 $\left[\text{H}—\text{O}—\text{Cl}: \right]^{+}$ (c) 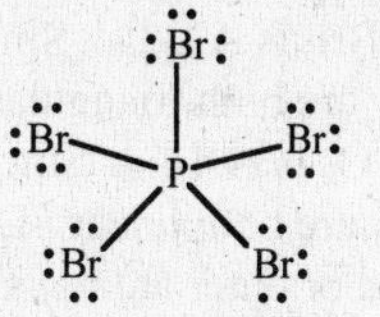 Br—P(Br)—Br   (d) 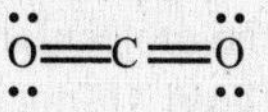 O=C=O

Solution:
(a) There are two outer atoms and two lone pairs on the central atom, so the notation is AX_2E_2. Do not count the electrons on outer atoms when writing AXE notations. Since $X + E = 2 + 2 = 4$, these electron pairs will use a tetrahedral shape.
(b) Since there are three outer atoms and one lone pair on the central atom, its notation is AX_3E_1 or AX_3E (the subscripted '1' can be left out). Note that the outer atoms do not have to be the same atom. In this case, $X + E = 3 + 1 = 4$, so these pairs will also use a tetrahedral shape.
(c) There are five outer atoms and no lone pairs on the central atom; the notation is AX_5E_0 or AX_5 (when E is zero, the 'E' can be omitted from the AXE notation). Since $X + E = 5 + 0 = 5$, these pairs will use a triangular bipyramidal shape.
(d) Since there are two outer atoms and no lone pairs on the central atom, its notation is AX_2E_0 or AX_2. When writing AXE notations (and when using the VSEPR model to predict molecular shapes), treat multiple bonds like single bonds; even though they have more pairs of electrons in them, multiple bonds still occupy only one spot around the central atom. Since $X + E = 2 + 0 = 2$, these pairs will be linear.

Although electron-pair geometries are often discussed, in actual molecules the electron pairs cannot be seen. Instead, chemists collect data regarding where the atoms (central and outer) appear; the shapes of the collection of atoms within the molecules are called **molecular geometries**. When the central atom has no lone pairs (i.e., when $E = 0$), the electron-pair geometry and the molecular geometry are the same; when the central atom has lone pairs, new shapes are encountered. The most common molecular geometries for central atoms with lone pairs (derived from the triangular planar and tetrahedral shapes) appear below. Additional molecular geometries can be derived from the triangular bipyramidal and octahedral shapes (Figure 9.6 on p. 384 in the textbook), but they are rare and will not be discussed in this chapter.

molecular shape		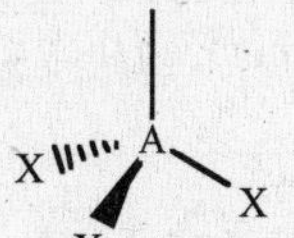	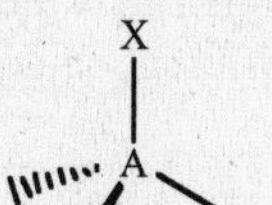
AXE notation	AX_2E_1	AX_3E_1	AX_2E_2
electron-pair geometry	triangular planar	tetrahedral	tetrahedral
molecular geometry	bent (V-shaped)	triangular (trigonal) pyramidal	bent (V-shaped)

Example: Using the Lewis structures and the AXE notation determined in the previous example, predict the shapes of the following molecules or ions.

(a) $:\ddot{F}-\ddot{S}-\ddot{F}:$ (b) $\left[H-\ddot{O}-\ddot{C}l:\atopH\right]^+$ (c) $\begin{array}{c}:\ddot{B}r:\\ :\ddot{B}r\quad P\quad \ddot{B}r:\\ :\ddot{B}r\quad \ddot{B}r:\end{array}$ (d) $\ddot{O}=C=\ddot{O}$

Solution:
(a) The AXE notation for SF_2 is AX_2E_2. The four pairs of electrons will use a tetrahedral shape, but since only the S and F atoms can be seen, the molecular geometry is bent. Note that it does not matter which of the four spots have F atoms or lone pairs (all possible pictures represent identical molecules).
(b) The AXE notation for H_2OCl^+ is AX_3E_1. The four pairs of electrons also use a tetrahedral shape, but since only the H, O, and Cl atoms can be seen, the molecular geometry is triangular pyramidal. Again, it does not matter which of the spots have atoms or lone pairs.
(c) The AXE notation for $PBrl_5$ is AX_5E_0. Since there are no lone pairs ($E = 0$), the electron-pair geometry and the molecular geometry are both triangular bipyramidal.
(d) The AXE notation for CO_2 is AX_2E_0. Since there are no lone pairs ($E = 0$), the electron-pair geometry and the molecular geometry are both linear. Double and triple bonds are usually drawn in these pictures, but lone pairs (on the central or outer atoms) are not.

(a)   (b)   (c)   (d)   O=C=O

Chiral Molecules and Enantiomers (12.4)

A compound that is **chiral** is not superimposable (identical) with its mirror image. An example of a chiral object is your left hand. Its mirror image (your right hand) is not identical to your left hand. An **achiral** compound, however, is superimposable with its mirror image. An example of an achiral object is your body as a whole because the left hand (leg, eye, ear, etc.) in your reflection would be superimposable with your right hand (leg, eye, ear, etc.). A chiral molecule and its mirror image are referred to as **enantiomers** (isomers that have the same number or types of bond connections, but differ only in their 3-D orientations).

Whenever an organic molecule has an atom (usually carbon) with four different groups attached to it, it is a chiral compound. Because enantiomers have different 3-D shapes, they can also have different chemical and physical properties. For example, one of the enantiomers of the chemical called carvone smells like spearmint and the other enantiomer smells like caraway seeds.

Example: Identify the chiral atoms in the ion to the right.

Solution: The N atom and the C atom to the right of it both have four different objects attached to them, so they are both chiral.

Orbital Hybridization and Molecular Geometry (9.3)

Triangular planar and triangular bipyramidal geometries have angles of 120° and tetrahedral geometries have angles of 109.5°. However, no sets of orbitals (s, p, d, etc.) are 109.5° or 120° apart. So, how can atoms bond at these odd angles? Chemists use the valence bond model to explain this discrepancy. The valence bond model assumes that the central atom uses atomic orbitals to make covalent bonds with their valence electrons. To explain the odd angles, the valence bond model assumes that the central atom will combine its existing atomic orbitals (*s*, *p*, etc.) into new orbitals that can be thought of as 'mixtures' or 'hybrids' of the atomic orbitals, called **hybrid orbitals**. This hybridization is much like resonance structures in that the actual hybrid orbitals have properties different from the atomic orbitals (shapes, energies) and are best thought of as averages or composites of atomic orbitals. When atomic orbitals are **hybridized**, the number of hybrid orbitals made is the same as the number of atomic orbitals used.

The number of atomic orbitals used by the atom when making hybridized orbitals equals the number of electron pairs around the central atom ($X + E$), and the lowest-energy orbitals are used first—for the main group elements, they use the *ns* orbital first and then the three *np* orbitals (if they exist). The relationship between the number of electron pairs on the central atom ($X + E$), the electron-pair geometry, and hybrid orbitals used by the central atom are listed below for atoms obeying the octet rule ($X + E \leq 4$).

$X + E$ value	1	2	3	4
atomic orbitals used	*s*	*s*, 1 *p*	*s*, 2 *p*	*s*, 3 *p*
hybrid orbitals made	—	*sp*	*sp*^{2}	*sp*^{3}
electron-pair geometry	linear	linear	triangular planar	tetrahedral

When an atom has two electron pairs around it will use ***sp* hybrid orbitals** (linear), when it has three electron pairs it will use ***sp*^{2} hybrid orbitals** (triangular planar), and when it has four electron pairs it will use ***sp*^{3} hybrid orbitals** (tetrahedral). In the past, chemists believed that when an atom has five electron pairs it will use *sp*^{3}*d hybrid orbitals* (triangular bipyramidal), and when it has six electron pairs it will use *sp*^{3}*d*^{2} *hybrid orbitals* (octahedral). While many instructors may still talk about *sp*^{3}*d* and *sp*^{3}*d*^{2} *hybrid orbitals*, the book explains that these molecules are better described using resonance structures with four bonds around the central atom, a positive charge on the central atom, and a negative charge on the outer atoms. Hydrogen will not form hybrid orbitals because its valence shell has only one orbital (1*s*).

Orbital Hybridization and Multiple Bonds (9.4)

Single, double, and triple bonds all have **sigma (σ) bonds** in them, which were introduced in the last chapter. Sigma bonds are bonds made by atoms using molecular orbitals that have electron density directly in between the atoms (and sometimes look like distorted *s* orbitals). For single bonds, there is only one bond and it is a sigma bond. Double and triple bonds have more than one bond; double bonds are made of one sigma bond and one **pi (π) bond** (which has electron density above and below the atoms and looks like a distorted *p* orbital) and triple bonds are made of one sigma bond and two perpendicular pi bonds.

$$:\ddot{O}=\ddot{O}: \qquad\qquad :N\equiv N:$$
$$\sigma + \pi \qquad\qquad\qquad\qquad \sigma + 2\pi$$

The double bond in diatomic oxygen is made of a sigma bond and a pi bond. The AXE notation of each O atom is AX_1E_2, so these atoms are *sp*^{2} hybridized ($X + E = 3$); two of these *sp*^{2} orbitals are filled with lone pairs and the other *sp*^{2} orbital is used by each O atom to make the sigma portion of the double bond. The remaining unhybridized 2*p* orbital is used to make the pi portion of the double bond.

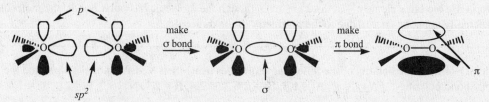

The triple bond in diatomic nitrogen is made of a sigma bond and two perpendicular pi bonds. The AXE notation of each N atom is AX_1E_1, and these atoms are *sp* hybridized ($X + E = 2$); one of these *sp* orbitals is filled with a lone pair and the other *sp* orbital is used by the N atoms to make the sigma portion of the triple bond. The remaining two unhybridized 2*p* orbitals are used to make the two pi portions of the triple bond.

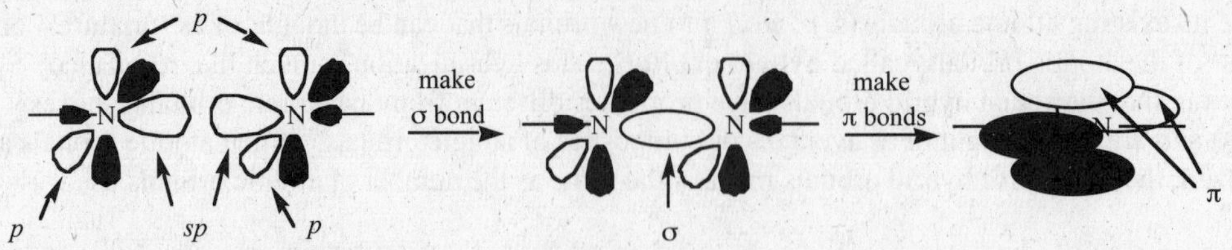

Describing Covalent Bonding in Terms of Sigma and Pi Bonds (9.3)

Covalent bonds in molecular compounds can be explained in terms of the hybridization of the central atoms and the formation of sigma and pi bonds. When a molecule has more than one central atom, the molecule has no simple shape, but its shape can be interpreted from the perspective of each central atom.

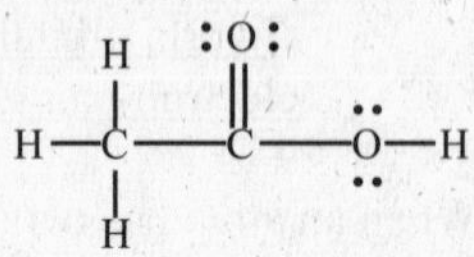

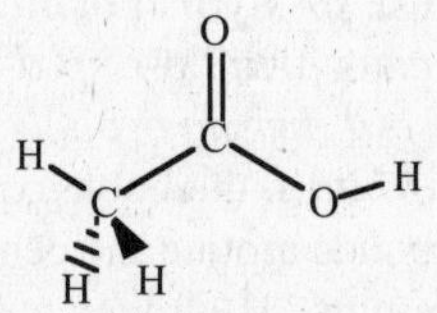

As an example, consider acetic acid (CH_3COOH). Its Lewis structure is shown on the right. The 'hybridization' of the H atoms is $1s$ since they have only one electron pair around them. The C atom in the CH_3 group has a notation of AX_4E_0, so its molecular geometry is tetrahedral and it is sp^3 hybridized (109.5° angles). The C and O atoms doubly-bonded to each other have an AXE notation of AX_3E_0, so their electron-pair geometry is triangular planar (120°) and they are sp^2 hybridized. The molecular geometry for the C atom is triangular planar and the O atom is linear. The O atom bonded to the H atom has a notation of AX_2E_2, so its electron-pair geometry is tetrahedral (109.5°), it is sp^3 hybridized, and its molecular geometry is bent (its angle is less than 109.5° because the lone-pair repulsions push the two bonded electron pairs closer together). All of the single bonds are sigma (σ) bonds. The C-O double bond is made of a σ component and a π component.

Predicting the Polarity of Molecules (9.5)

In the previous chapter, you were given guidelines for predicting whether a chemical bond was polar or nonpolar. Just as in chemical bonds, a **polar molecule** is a molecule with a net positive and a net negative end (but overall neutral or it would be an ion) and a **nonpolar molecule** has no net positive or negative ends. A **dipole moment** (μ), which is the charge separation (in Coulombs, C) times the distance of separation (in meters), can be calculated to determine whether a molecule is polar—polar molecules have positive μ values and nonpolar molecules have $\mu = 0$.

Instead of calculating dipole moments (which can be difficult and time-consuming), chemists use the polarities of the individual bonds and the VSEPR model to determine whether a molecule is polar. When a molecule has no polar bonds (all $\Delta EN < 0.5$), then the molecule is considered to be nonpolar (even though some of these molecules have μ values that are not zero, they behave as if they were completely nonpolar). If the molecule has at least one polar bond ($\Delta EN \geq 0.5$), then the molecule is generally considered to be polar. The one exception to this rule occurs when the inner atom has no lone pairs ($E = 0$) and all of the outer atoms are the same element—in this case, even though the molecule has polar bonds, the molecule itself is actually nonpolar ($\mu = 0$) due to the symmetry of the molecule.

Example: Determine the polarity of the following molecules using their VSEPR model shapes and the individual bond polarities. _EN_ values: Al = 1.4, Br = 2.6, C = 2.4, H = 2.1, N = 3.0, P = 2.0, Se = 2.4.

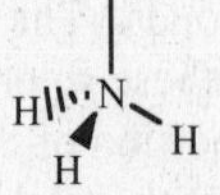

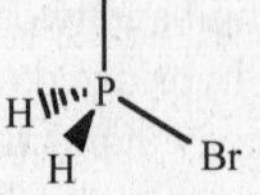

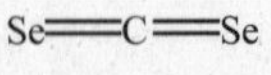

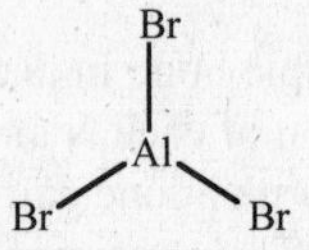

Solution:

(a) The ΔEN value is $3.0 - 2.1 = 0.9$ (polar covalent). Since the bonds are polar, the molecule will be polar unless it has no lone pairs and all outer atoms are the same. The N atom has a lone pair (the line on top with no atom attached), so the molecule is polar. The top region of the molecule is negative (the N atom) and the bottom region is positive (the H atoms). This is shown using a *dipole arrow*, where the tip points to the negative end of the molecule and the end with a + sign on it points to the positive end of the molecule.

(b) ΔEN is $2.1 - 2.0 = 0.1$ (nonpolar covalent) for the P-H bonds and $2.6 - 2.0 = 0.6$ (polar covalent) for the P-Br bond. Since there is as least one polar bond (P-Br), the molecule has a lone pair, and the outer atoms are different, the molecule is polar. The P atom is positive and the Br atom is negative.

(c) The ΔEN value is $2.4 - 2.4 = 0.0$ (nonpolar covalent). Since all of the bonds are nonpolar, you can determine that the molecule is nonpolar without looking at its shape.

(d) ΔEN is $2.6 - 1.4 = 1.2$ (polar covalent). Even though the bonds are polar (and the Al atom has a net positive charge and the Br atoms have a net negative charge), the molecule is nonpolar due to symmetry—all of the outer atoms are the same (Br) and the central atom has no lone pairs.

(a)   (b)   (c)   (d)

Noncovalent Interactions (9.6)

Compounds in the liquid and solid state are held together by **intermolecular forces** (also referred to as **noncovalent interactions**). These forces are called intermolecular forces since they represent the forces of attraction holding two or more different molecules together. Covalent bonds between atoms can be called *intramolecular forces* because they are the forces of attraction that hold the atoms in individual molecules together. Intermolecular forces represent the attractions of positive and negative objects to each other.

In polar molecules, the primary intermolecular force present is **dipole-dipole attractions** (*dipole-dipole forces*). Because these molecules have a permanent separation of + and − charges, the positive ends of these molecules will be attracted to the negative ends of other molecules, and vice versa. Dipole-dipole attractions are typically about 5-25 kJ/mol; dipole-dipole attractions are larger for molecules that are more polar, and weaker for those that are less polar.

Example: Consider the following liquid diatomic molecules: Br_2, BrCl, BrF, BrH, and BrI. *EN* values: Br = 2.6, Cl = 2.7, F = 4.0, H = 2.1, I = 2.1.

(a) Which of the following diatomic molecules is expected to have dipole-dipole attractions?
(b) Which molecule is expected to have the strongest dipole-dipole attractions?
(c) Draw a picture of the attractions present in each of the polar liquids.

Solution:

(a) Since these are diatomic molecules, the molecular polarities are the same as the bond polarities. The ΔEN values are 0.0 for Br_2, 0.1 for BrCl, 1.4 for BrF, 0.5 for BrH, and 0.5 for BrI. So, Br_2 will not have dipole-dipole forces, BrCl is expected to have no (or very small) dipole-dipole forces since this bond is considered to be nonpolar ($\Delta EN < 0.5$), and BrH, BrI, and BrF will have the stronger dipole-dipole attractions since their bonds are polar ($\Delta EN \geq 0.5$).

(b) Since BrF has the largest ΔEN value, it is the most polar molecule and should have the strongest forces.

(c) In BrH and BrI, the Br has the partial negative charge and H or I has the partial positive charge ($Br^{\delta-}H^{\delta+}$ and $Br^{\delta-}I^{\delta+}$), so the negative Br atoms will be attracted to the positive H or I atoms (the dipole-dipole attractions are shown by dashed lines). In BrF, the Br is partially positive and the F is partially negative ($Br^{\delta+}F^{\delta-}$), so the positive Br atoms will be attracted to the negative F atoms.

83

Hydrogen bonds are classified as another type of intermolecular forces, but they are really just a special case of dipole-dipole attractions. Hydrogen bonds occur in molecules that have H atoms bonded to small electronegative atoms (F, O, or N). These electronegative atoms form very polar bonds with hydrogen, gaining a net negative charge and giving the H atom a net positive charge. Because these atoms (H and the nonmetal) are very small, these atoms can get very close to each other, resulting in a particularly strong dipole-dipole attraction. Hydrogen bonds are typically about 10-40 kJ/mol.

Although nonpolar molecules do not have net positive or negative charges, they are still held together by intermolecular forces called **London forces**, (also known as *dispersion forces* or *induced dipole-induced dipole attractions*). Although every molecule experiences London forces, they are most important for nonpolar molecules because nonpolar molecules have only London forces. In any molecule, electrons from the lone pairs and covalent bonds are randomly moving throughout the entire molecule. At any instant, the distribution of the electrons could be temporarily unbalanced (more e⁻ on one side and fewer e⁻ on the other). This temporary imbalance of electrons causes a temporary dipole (separation of + and − charges) in the molecule, making it temporarily polar. When a molecule with a temporary dipole approaches another molecule, electrons in the negative region of the first molecule repel electrons in the new molecule to the opposite side of the molecule, causing a temporary dipole (**induced dipole**) in the new molecule.

London forces (0.05-40 kJ/mol) are usually weaker than dipole-dipole attractions or hydrogen bonds since they result from temporary dipoles that are created and dissipate once the electrons are re-balanced several million times a second. However, the bigger the molecule is and the more electrons it has, the more likely the electrons are to be temporarily balanced and the larger these London forces will be. So, even though water (H_2O) has very strong hydrogen bonds, the London forces in iodine (I_2, a nonpolar molecule) are stronger because water has only 18 e⁻ while iodine has 106 e⁻ and its overall London forces are stronger. Therefore, direct comparisons of dipole-dipole attractions and hydrogen bonds within different molecules should be done with molecules of comparable size (molar mass, total number of electrons).

Melting and boiling points measure the energy needed to break the intermolecular forces holding a solid or a liquid together, respectively. So, molecules with higher melting or boiling points must have stronger intermolecular forces than molecules with lower melting or boiling points, and these values can be used to estimate the relative strengths of the intermolecular forces in liquids or solids.

The Major Components in DNA (9.7)

Deoxyribonucleic acid (DNA), the molecule that stores the genetic codes of all living creatures, is a **polymer**—a molecule containing several small repeating units (**nucleotides**) bonded together. The nucleotides in DNA are made of three basic components. The backbone of DNA is made of alternating *sugar groups* and *phosphate groups* linked together. The sugar in DNA is deoxyribose, which is where DNA gets its name (*ribonucleic acid, RNA*, looks just like DNA except the deoxyribose sugar in DNA is replaced with a ribose sugar in RNA). Attached to the sugar groups are **complementary base pairs** consisting of *adenine* (A), *cytosine* (C), *guanine* (G) or *thymine* (T).

The structure of DNA is a *double helix*, where two strands (sugar-phosphate-base pair polymer) are joined together by the base pairs to form a twisted strand. Although the sugar-phosphate backbone is held together by covalent bonds, the two DNA strands are held together by hydrogen bonds occurring between the base pairs on the two DNA strands. The structures of adenine and thymine are complementary (as are the structure of cytosine and guanine), so C bases on one DNA strand will only hydrogen bond to G bases in another DNA strand and T bases will only hydrogen bond to C bases.

Using Spectroscopy to Determine Molecular Structures (*Tools of Chemistry*)

Spectroscopy is a technique for probing the interaction of matter (molecules) with electromagnetic radiation (light). In *infrared (IR) spectroscopy*, molecules absorb energy that causes the atoms within the molecule to vibrate in specific patterns; in *ultraviolet (UV) spectroscopy*, molecules absorb energy that causes electrons in molecules with double and triple bonds to move to higher orbitals (π to π^*). In both methods, atoms in different chemical environments absorb different frequencies of light, and the frequencies of light that they absorb tell chemists something about their molecular structure.

Chapter Review — Key Terms

The key terms that were introduced in this chapter are listed below, along with the section in which they were introduced. You should understand these terms and be able to apply them in appropriate situations.

achiral (9.2)
axial positions (9.2)
bond angle (9.2)
chiral (9.2)
complementary base pairs (9.7)
deoxyribonucleic acid (DNA) (9.7)
dipole-dipole attraction (9.6)
dipole moment (9.5)
electron-pair geometry (9.2)
enantiomers (9.2)
equatorial positions (9.2)
gene (9.7)
hybrid orbital (9.3)

hybridized (9.3)
hydrogen bond (9.6)
hydrophilic (9.6)
hydrophobic (9.6)
induced dipole (9.6)
intermolecular forces (9.6)
lipid bilayer (9.6)
London forces (9.6)
molecular geometry (9.2)
noncovalent interactions (9.6)
nonpolar molecule (9.5)
nucleotide (9.7)
pi bond, π bond (9.4)

polar molecule (9.5)
polarization (9.6)
polymer (9.7)
replication (9.7)
sigma bond, σ bond (9.3)
sp hybrid orbital (9.3)
sp^2 hybrid orbital (9.3)
sp^3 hybrid orbital (9.3)
spectroscopy (pp. 386, 401)
valence bond model (9.3)
valence-shell electron-pair
  repulsion (VSEPR) model (9.2)

Practice Test

After you have finished studying the chapter and the homework problems, the following questions can serve as a test to determine how well you have learned the chapter objectives.

1.   What is the general theory behind the VSEPR model? Why is the tetrahedral shape more stable than the *square planar* shape where all four electron pairs lie in a plane at 90° angles?

2.   Use the Lewis structures to determine AXE notations, electron-pair and molecular geometries, bond angles, hybridizations of the central atom, and drawings of the following molecules or ions.

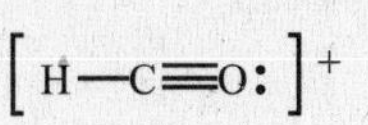

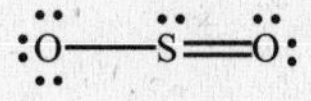

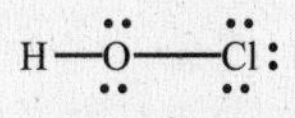

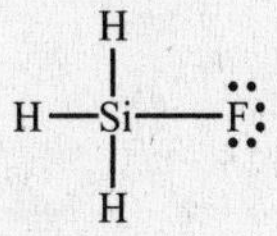

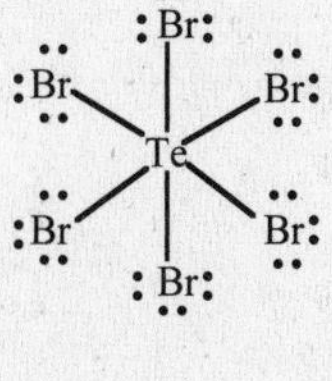

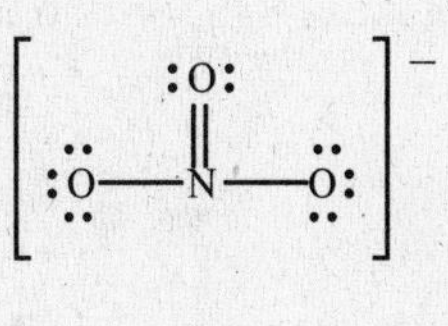

3.   Identify the chiral atoms in the following chemicals.

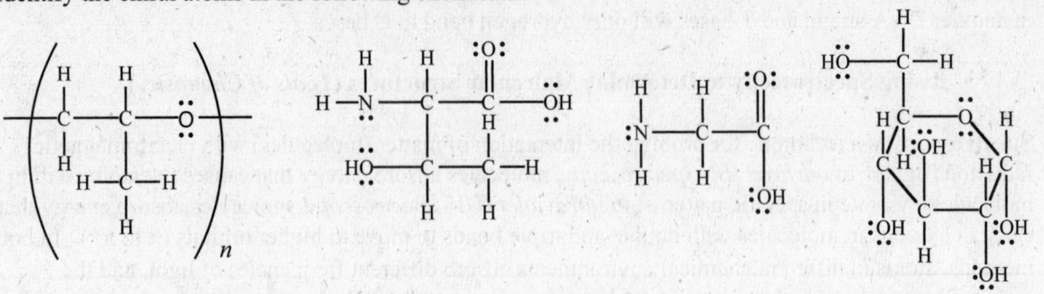

4.   Determine the polarity of the molecules or ions listed in question 2, and list the intermolecular forces that would be present in a liquid or solid sample of each of these compounds. _EN_ values: Br = 2.6, C = 2.4, Cl = 2.7, F = 4.0, H = 2.1, N = 3.0, O = 3.4, S = 2.3, Si = 1.6, Te = 1.9.

5.   The dipole moment for carbon monoxide (:C≡O:) is very small (0.112 D), much smaller than would be predicted for a C-O bond.
       (a)   Determine the molecular polarity of CO. Which atom is negative and which is positive?
       (b)   Use the Lewis structure above to determine the formal charges on each atom.
       (c)   Explain why CO has such a small dipole moment.

6.   The melting points of F_2, and HCl are –220°C and –115°C, respectively. Which of these molecules has the stronger intermolecular forces? Explain this trend.

7.   The boiling points of NH_3, H_2O, and HF are –33°C, 100°C, and 20°C, respectively. Which of these molecules has the strongest intermolecular forces and which has the weakest? Explain this trend.

8.   Polar molecules like hydrogen fluoride will dissolve in water.
       (a)   What intermolecular forces are present between two HF molecules? between two H_2O molecules? between an HF molecule and an H_2O molecule?
       (b)   Explain why hydrogen fluoride molecules dissolve in water.

9.   Nonpolar molecules like hydrogen do not readily dissolve in water.
       (a)   What intermolecular forces are present between two H_2 molecules? between two H_2O molecules? between an H_2 molecule and an H_2O molecule?
       (b)   Explain why hydrogen molecules do not dissolve readily in water.

Chapter 10
Gases and the Atmosphere

The goal of this chapter is to introduce the macroscale, symbolic, and nanoscale properties of gases. This chapter also discusses qualitative and quantitative relationships between pressures, volumes, numbers of particles, and temperatures of gases. This chapter also uses these relationships to perform density, molar mass, and stoichiometry calculations. Finally, the chapter describes several important chemical processes occurring in earth's gaseous atmosphere including ozone depletion, air pollution, and global warming.

Components of the Atmosphere (10.1)

The earth's *atmosphere*, the thin layer of gaseous molecules that surround its surface, consists of about 78% nitrogen, 21% oxygen, 1% argon, and trace amounts of other gases like carbon dioxide, neon, hydrogen, and helium in dry (nonhumid) air. Gaseous water vapor is responsible for the *humidity* in the air; humid or wet air is about 1-5% water vapor and 95-99% dry air. The earth's atmosphere is divided into several layers. The layer that is closest to the earth's surface (0-12 km above the surface), and where life on the planet exists, is called the **troposphere**. The layer above the troposphere (12-50 km above the surface) is called the **stratosphere**. Due to earth's gravity, most of the molecules in the atmosphere exist in the troposphere (about 75%) and the stratosphere (about 25%); less than 0.1% exists above the stratosphere.

Macroscopic Properties of Gases (10.2)

Early chemists studies gases, particularly their qualitative macroscopic properties. In general, all gases have similar properties regardless of their chemical identity. These properties include the fact that gases can be compressed into smaller volumes, that gases can expand to fill larger volumes, that gases will completely mix with each other to form homogeneous mixtures, that gases are invisible to the human eye, and that gases exert pressures on their surroundings. These properties are in contrast to liquids and solids which do not change volumes when placed in larger or smaller containers, do not always mix together to make homogeneous mixtures, are not invisible to human eyes, and do not exert pressures on surroundings.

The mathematical definition of **pressure** is the force exerted on an object per unit area ($P = F/A$). An example of force is an object's weight, which equals its mass times the earth's gravitational constant ($g = 9.8$ m/s^2). Since mass has units of kg and g has units of m/s^2, force has units of kg m/s^2, which is called a **newton** (N) after Isaac Newton. Pressure (force per unit area) has units of N/m^2, which is called a **pascal** (Pa) after mathematician Blaise Pascal. Other units of pressure are the **bar**, defined to be 100,000 Pa, the **standard atmosphere** (atm), defined as the pressure exerted by earth's atmosphere at sea level, and **millimeters of mercury** (mm Hg, also called the **torr**), defined as the pressure exerted by a column of mercury liquid that is 1 mm high. The common English unit for pressure is *pounds per square inch* (lb/in^2 or psi). The relationships between these pressure units appear in Table 10.2 of the textbook (p. 428).

Example: Use the concept of pressure to explain why stepping on the pointed end of a thumbtack causes more damage to your foot than stepping on the flat end.

Solution: The formula for pressure is $P = F/A$. The force exerted on the thumbtack is your weight, which does not change when the tack's orientation changes. However, the area does change considerably. The flat end has a much larger area over which your weight is exerted, while the pointed end has a very small area over which your weight is exerted. So, the pressure you exert on the flat end is less than that exerted on the pointed end (dividing by a large area gives a small pressure, dividing by a small area gives a large pressure); this difference in pressure explains why the tack punctures your skin only with its pointed end.

From a molecular point of view, the pressure exerted by a gas is a measure of the number and strength of collisions of the gas molecules with its surroundings (container walls, your skin, etc.).

Using the Kinetic-Molecular Theory to Explain Gas Behavior (10.3)

In order to explain the nanoscale (molecular) behaviors of gas particles, you must revisit the kinetic-molecular theory (KMT), introduced briefly in Chapter 1. This chapter describes five ideas from the KMT that explain particle behavior at the nanoscale level: (1) Gas particles are much smaller than the distances between them—this means that the gas particles are isolated in space from each other (i.e., they are not close to each other); (2) Gas particles move randomly in all directions and at various speeds—gas particles do not all have the same speed and at any instant some are moving faster and some are moving slower; (3) The forces of interaction (attraction or repulsion) between gas particles are small enough to ignore—this is primarily due to the fact that gas particles are so far apart from each other; (4) Collisions between gas particles are elastic—this is a representation of the law of conservation of energy. Energy can be gained or lost by one gas particle in the collision only if the other gas particle loses or gains that energy; and (5) The average speed of the gas particles and their average kinetic energy (KE) are directly proportional to the gas sample's temperature—as temperature increases, the average speed and KE increase, and vice versa.

Example: Use the kinetic-molecular theory to explain the following macroscopic gas behaviors.
(a) Gases can be compressed into smaller volumes, but solids and liquids cannot.
(b) Gases are invisible to the human eye, but solids and liquids are visible.
(c) Gases exert pressure on every wall in a container, but solids and liquids exert pressure on the bottom.

Solution:
(a) Because gas particles are much smaller than the distance between them (Rule 1), they have lots of empty space between them. When a gas sample is compressed, some of this empty space is removed and the gas particles move slightly closer to each other, but still far enough away from each other so that they are not touching. Solids and liquids are not compressible because there is not a lot of empty space between these particles, and these particles are already close enough to be touching and can't get closer.
(b) Because gas particles are not attracted to each other, they remain isolated as single molecules. Solids and liquids, on the other hand, have substantial intermolecular forces holding them together as clusters consisting of huge numbers of particles. Human eyes are not sensitive enough to see the individual gas molecules, but they can see the huge macroscale clusters of solid or liquid particles.
(c) Because gas particles move randomly in all directions, they will eventually hit all sides of the entire container. These collisions represent pressure being exerted on every part of the container's walls. Solid and liquid particles do not move throughout the entire container and are forced to stay in the volume occupied by the solid or liquid. Their only contact with the container is the bottom surface, so they exert their weight (force) on this surface alone.

Rule 2 of the kinetic-molecular theory states that the gas particles within a sample do not travel at the same speed. A plot of the number of gas particles traveling at any speed versus that speed is called a *Boltzmann distribution*. As the temperature of a particular gas sample is increased, the average kinetic energy and speed of the gas particles increase, and the distribution of particle speeds in the Boltzmann distribution shifts to the right toward higher speeds and flattens out. When the temperature is decreased, the distribution shifts to the left toward lower speeds and becomes more pointed. When you look at the distribution of particle speeds for two or more different gases, the plots are shifted to the right toward higher speeds when the particles have small masses (molar masses) and are shifted to the left toward lower speeds when the particles have higher masses (molar masses).

Effusion is a measure of how fast a gas sample escapes from a tiny hole into a *vacuum* (an area with no gas particles), and **diffusion** is a measure of how fast a gas sample moves throughout a container that is already filled with a gas. In general, both properties are related to the inverse square root of the molar mass.

Example: Why is effusion related to the inverse square root of a gas's molar mass ($rate \propto 1/\sqrt{M}$)?

Solution: The rate of a gas sample's escape from a container into empty space is related to the speed of the gas particles. If the gas particles are moving faster they will escape faster, and if they are moving slower they will escape slower. Solving for the gas particles' speed, you get:

$$KE = \frac{1}{2} mv^2, \quad v = \sqrt{\frac{2\,KE}{m}}$$

At any given temperature, the average KE for each gas sample is the same (Rule 5 of the KMT). So, the mass of one gas molecule is simply its molar mass (in units of amu), and the rate of gas escape, which is directly related to the gas's KE value, is inversely proportional to the square root of its molar mass.

Solving Problems Using Gas Laws (10.4)

The properties of an ideal gas sample can be defined by looking at four parameters—its pressure (P), its volume (V), the number of particles present in the sample (n), and its temperature (T). At the nanoscale level, P measures the number of particle collisions with the container's walls, V measures the volume of the gas particles, n measures the number of moles of gas particles present, and T measures the average kinetic energy and speed of the gas particles.

Avogadro's law relates the volume and amount of a gas sample (assuming P and T are constant). This law states that V and n are proportional to each other (as n increases V increases, and vice versa): $V \propto n$ or V/n = constant. When you have two sets of conditions for a gas sample, the equation can be written $V_1/n_1 = V_2/n_2$ (the subscripts can also be 'B' and 'A' for '*before*' and '*after*' or 'i' and 'f' for '*initial*' and '*final*').

This relationship can be explained at the nanoscale using the KMT. When the number of particles is increased, there are more particle collisions on the side of the container. If the volume is fixed and cannot change (like a steel tank), this would lead to an increase in the gas pressure (P/n = constant, *Dalton's law*, which will be discussed in section 10.7). If the container walls are flexible (like a rubber balloon), then the increased number of collisions will push the walls outward (increasing V but leaving P constant) until the number of collisions inside the balloon pushing out are balanced by the number of collisions on the outside of the balloon pushing in. This can also be argued in reverse (as n decreases, V decreases).

Related to Avogadro's law is *Avogadro's hypothesis*—all gases containing the same number of particles will have the same volume, if P and T are held constant. This leads to the **law of combining volumes** which states that volumes of gases will react in the same whole number ratio as the balanced equation, if P and T are held constant. This hypothesis helped chemists decide that many elemental gases are diatomic.

Example: When reacting, 1.0 L of nitrogen reacts with 3.0 L of hydrogen to make 2.0 L of ammonia (NH_3). How did this tell chemists that nitrogen and hydrogen are diatomic?

Solution: The law of combining volumes says that the coefficients in the balanced equation must match the volumes of gases, so we get:

$$1\ \text{"N"}(g) + 3\ \text{"H"}(g) \rightarrow 2\ NH_3(g)$$

In order for this equation to be balanced, there must be 2 N on both sides. The only way for the coefficient to be 1 (required by the law of combining volumes) is for the formula of nitrogen to be N_2. Similarly, the only way to have 6 H on both sides of the equation and a coefficient of 3 for hydrogen is for it to be H_2.

Charles's law relates the volume and temperature of a gas sample (assuming P and n are constant). This law states that V and T are proportional to each other (V increases as T increases, and vice versa): $V \propto T$ or V/T = constant, and $V_1/T_1 = V_2/T_2$. However, this relationship is true only when the temperature is listed in the **Kelvin temperature scale** (also called the **absolute temperature scale**). The relationship between the

Celsius and Kelvin temperature scales is: $T(K) = T(°C) + 273.15$. The Kelvin scale is called the absolute scale because zero Kelvin ($0\ K = -273.15°C$) is the lowest temperature possible (**absolute zero**).

This law can be explained at the nanoscale because KMT states that when the gas temperature is increased, the gas particles will move faster and have more energy. This results in more particle collisions with the walls of the container (because the particles are moving faster) and harder collisions with the container's walls (because the particles have more kinetic energy). If the volume is fixed, this leads to an increase in the gas pressure (P/T = constant, *Gay Lussac's law*, which is not discussed in the textbook). If the walls of the container are flexible, then the increased number of collisions push the walls outward (increasing V but leaving P constant) until the number of collisions inside the balloon pushing out and the number outside the balloon pushing in are balanced. This can also be argued in reverse (as T decreases, V decreases).

Although the individual gas laws are useful, they require that two of the four gas parameters be held constant. A more general (and more useful) equation would allow all four of these parameters to change. When these individual gas laws are combined, you get the **ideal gas law**, $PV = nRT$, which does not require any of the parameters to be held constant. R is the **ideal gas constant**, a proportionality factor that relates the four gas parameters to each other (0.082057 L atm/mol K). When you have two sets of conditions, the ideal gas law can be written as the combined gas law (some textbooks will leave the n_1 and n_2 terms out of the combined gas law, but then it is a requirement for this new formula that n be held constant).

$$\frac{P_1V_1}{n_1T_1} = \frac{P_2V_2}{n_2T_2}, \text{ or } \frac{P_1V_1}{P_2V_2} = \frac{n_1T_1}{n_2T_2}$$

Example: How many moles of HBr gas are present in 16.5 L at 3.24 atm and –25°C?

Solution: First you must decide whether to use the ideal gas law or the combined gas law. In general, if you are given two values for the same parameter (like two P values or two T values), you should use the combined gas law. If you are given only one of each parameter, use the ideal gas law.

$$PV = nRT, \ n = \frac{PV}{RT} = \frac{(3.24 \text{ atm})(16.5 \text{ L})}{(0.082057 \text{ L atm/mol K})(273.15 + -25 \text{ K})} = 2.63 \text{ mol}$$

Remember that T must be in Kelvin units; if you did not change –25°C to 248.15 K, you would have calculated a negative n value, which is impossible (P, V, n, and $T(K)$ can only be positive numbers).

Example: When a balloon is filled with H_2 gas at 745 mm Hg, its volume is 5.00 L. What will its volume be if the pressure is decreased to 280 mm Hg?

Solution: You are given two pressures, so you should use the combined gas law. Since nothing was said about the number gas particles present or the temperature of the gas, you must assume they have not changed (held constant). When a parameter is constant, their values are canceled to give a ratio of 1.

$$\frac{P_1V_1}{P_2V_2} = \frac{nT}{nT} = 1, \text{ or } P_1V_1 = P_2V_2. \ V_2 = \frac{P_1}{P_2} \times V_1 = \frac{745 \text{ mm Hg}}{280 \text{ mm Hg}} \times 5.00 \text{ L} = 13.3 \text{ L}$$

In the combined gas law, you may use different units for P (mm Hg, psi, bar, etc.), V (mL, gal, m^3, etc.), and n (molecules) as long as they cancel out. However, you must still use Kelvin units for temperatures.

Stoichiometry Calculations and Gas Laws (10.5)

These calculations represent the combination of two different calculations you have already encountered—stoichiometry calculations (mass-mole conversions, limiting reactants, etc.) and ideal gas law calculations. The factor common to both calculations is the amount (n) of a gas sample. If you are given three of the four gas parameters or are asked to find the mass of a chemical, perform the gas law calculation first; if you are not given three of the four gas parameters or are asked to find one of the gas law parameters, perform a stoichiometry calculation to determine n for a gas sample, then perform the gas law calculation.

Example: High-purity nickel is made using the carbonyl process. In this process, nickel(II) oxide reacts with carbon monoxide to make gaseous nickel tetracarbonyl, which is later decomposed into nickel metal. How much solid nickel(II) oxide (in g) will react if the CO gas pressure is 2.56 atm, the temperature is 50.3°C, and the tank volume is 128.0 L?

$$NiO(s) + 5\,CO(g) \rightarrow Ni(CO)_4(g) + CO_2(g)$$

Solution: You are given the volume, pressure, and temperature (3 of the 4 gas parameters), so perform the gas law calculation first (solving for n), then use stoichiometry to solve for g NiO. Since you are told that the volume is the volume of CO gas, the n value that you solved for is n for CO gas (not $Ni(CO)_4$ or CO_2).

$$PV = nRT, \quad n = \frac{PV}{RT} = \frac{(2.56 \text{ atm})(128.0 \text{ L})}{(0.082057 \text{ L atm/mol K})(273.15 + 50.3 \text{ K})} = 12.346 \text{ mol CO (unrounded)}$$

$$12.346 \text{ mol CO} \times \frac{1 \text{ mol NiO}}{5 \text{ mol CO}} \times \frac{74.71 \text{ g NiO}}{\text{mol NiO}} = 184 \text{ g NiO}$$

Determining Densities and Molar Masses of Gases (10.6)

Since density is defined as $d = m/V$, you can use the ideal gas law to get the density formula $d = mP/nRT$. This equation can also be written as $d = MP/RT$, since the molar mass of a gas is $M = m/n$. The molar mass can be also calculated from these formulas—$M = mRT/PV = dRT/P$.

Example: What is the density of the CO gas in the previous example?

Solution: Since you are given P and T, use the equation $d = MP/RT$ (M for CO is 28.01 g/mol).

$$d = \frac{MP}{RT} = \frac{(28.01 \text{ g/mol})(2.56 \text{ atm})}{(0.082057 \text{ L atm/mol K})(273.15 + 50.3 \text{ K})} = 2.70 \frac{\text{g}}{\text{L}}$$

Note that you did not need to use the V value, and that all of the units except g and L cancel.

Example: A sample of chlorine gas in a 2.50 L flask at 25.1°C and 735 mm Hg weighs 7.00 g. What is the molecular weight (and molecular formula) of this gas?

Solution: Since you are given V, T, P, and m, use the equation $M = mRT/PV$.

$$M = \frac{mRT}{PV} = \frac{(7.00 \text{ g})(0.082057 \text{ L atm/mol K})(273.15 + 25.1 \text{ K})}{\left(735 \text{ mm Hg} \times \dfrac{1 \text{ atm}}{760 \text{ mm Hg}}\right)(2.50 \text{ L})} = 70.9 \frac{\text{g}}{\text{mol}}$$

Since $M = 70.9$ g/mol and each Cl atom weighs 35.45 g/mol, the molecular formula for chlorine must be Cl_2. This is another way chemists determined that many elemental gases are diatomic.

Performing Calculations Using Partial Pressures (10.7)

Dalton's law of partial pressures relates the pressure and amount of a gas sample (assuming V and T are constant). This law states that P and n are proportional to each other (as n increases P increases, and vice versa): $P \propto n$ or P/n = constant. When you have two sets of conditions for a gas sample, the equation can be written $P_1/n_1 = P_2/n_2 = P_{total}/n_{total}$. This last term is used because pressure gauges measure the total pressure exerted in the container. The individual pressures (P_1, P_2, etc.) are called **partial pressures**, and the total pressure is equal to the sum of the partial pressures. The nanoscale explanation for this law appears in the Avogadro's law section of this book (p. 88).

Example: You have 4.62 mol HCl, 3.85 mol Ne, and 0.57 mol Br_2 present in a 15.0 L tank at 3.4°C. What is the total pressure and what are the partial pressures for each gas?

Solution: You can solve this problem in two different ways. (1) Add all of the mole values together to get n_{total}, solve for P_{total} using the ideal gas law, and then use Dalton's law of partial pressures to determine each partial pressure; or (2) Calculate the partial pressure for each of the three gases using the ideal gas law, then add these values together to get the total pressure. In this example, the first method will be used.

$$P_{total} = \frac{n_{total}RT}{V} = \frac{(9.04 \text{ mol})(0.082057 \text{ L atm/mol K})(273.15 + 3.4 \text{ K})}{15.0 \text{ L}} = 13.68 \text{ atm}$$

$$\frac{P_{HCl}}{n_{HCl}} = \frac{P_{total}}{n_{total}}, \quad P_{HCl} = \frac{n_{HCl}}{n_{total}} \times P_{total} = \frac{4.62 \text{ mol HCl}}{9.04 \text{ mol total}} \times 13.68 \text{ atm} = 6.99 \text{ atm}$$

$$\frac{P_{Ne}}{n_{Ne}} = \frac{P_{total}}{n_{total}}, \quad P_{Ne} = \frac{n_{Ne}}{n_{total}} \times P_{total} = \frac{3.85 \text{ mol Ne}}{9.04 \text{ mol total}} \times 13.68 \text{ atm} = 5.82 \text{ atm}$$

$$\frac{P_{Br_2}}{n_{Br_2}} = \frac{P_{total}}{n_{total}}, \quad P_{Br_2} = \frac{n_{Br_2}}{n_{total}} \times P_{total} = \frac{0.57 \text{ mol Br}_2}{9.04 \text{ mol total}} \times 13.68 \text{ atm} = 0.86 \text{ atm}$$

Note that the partial pressures add up to the total pressure (6.99 atm + 5.82 atm + 0.86 atm = 13.67 atm). The ratios of n_X/n_{total} are referred to as **mole fractions** and represent the fraction of all particles that are X.

Real (Nonideal) Gases (10.8)

Although no gas is a perfectly **ideal gas** (i.e., no gas perfectly follows the ideal gas law), most gases follow the ideal gas law reasonably well. However, at higher pressures and lower temperatures most gases deviate from the ideal gas law. There are two ways that *real gases* behave differently than ideal gases. First, the particles in real gases have a finite volume. So, when real gases are compressed to very small volumes chemists cannot ignore the volume occupied by the gas particles. Second, the particles in real gases are slightly attracted to each other. This results in a somewhat lower pressure than expected in an ideal gas because when gas particles collide with the container's walls, they are actually attracted away from the wall by the other gas particles and hit the walls with a smaller force than expected.

These nonideal behaviors can be corrected for by using the **van der Waals equation**—$(P + n^2a/V^2)(V - nb) = nRT$, where a and b are measured (tabulated) corrective factors. The n^2a/V^2 term is added to P to correct for the decreased P value due to the attractions of gas particles and the nb term is subtracted from V to correct for the increased V value due to the nonzero volume of the gas particles. Values for a and b values for several gases appear in Table 10.5 in the textbook (p. 453). Since the a values correct for particle attractions, it should not be surprising that large polarizable molecules like Cl_2 and very polar molecules with strong hydrogen bonding forces like H_2O and NH_3 have the highest a values. Since the b values correct for nonzero particle volumes, it should not be surprising that the bigger molecules like Cl_2, CO_2, and CH_4 have the highest b values.

Ozone Depletion and Chlorofluorocarbons (10.9)

The biological relevance of the ozone layer is that it absorbs high-energy ultraviolet (UV) light, converting this light into lower energy light (visible, infrared, etc.) and preventing this high-energy light from reaching the earth's surface where it could cause damage to living cells. When ozone (O_3) absorbs ultraviolet light with λ < 242 nm, it undergoes **photodissociation**, in which one of the O-O bonds is broken, producing an O_2 molecules and an O atom with 2 unpaired electrons.

Chlorofluorocarbons (CFC) are molecules made of Cl, F, and C atoms. These chemicals are normally very unreactive, which is an advantage for using them as *refrigerants* (gases that can be compressed and expanded to cool other objects). However, because these gases are unreactive, they last a long time in the earth's atmosphere, and eventually reach the earth's stratosphere, where they encounter the **stratospheric ozone layer** that is protecting the earth. CFC molecules can also absorb UV light, breaking one or more of the C-Cl bonds and producing Cl atoms with unpaired electrons. These Cl atoms catalytically react to convert ozone molecules into O_2 (a *catalyst* allows a reaction to occur without being used up itself; since Cl atoms are not used up, they can catalyze the reaction of many ozone molecules without being changed).

The buildup of Cl atoms in the stratosphere, and the destruction of ozone molecules, has led to an **ozone hole** above Antarctica. As a way to reduce the ozone hole, international treaties have been created to ban the use and emission of CFC molecules. It appears that these bans are having a positive effect on shrinking the ozone hole, but it may take several decades for the hole to completely disappear.

Industrial Pollution and Urban Pollution (10.10)

Industrial pollution is primarily caused by sulfur dioxide (SO_2). Sulfur dioxide is produced by burning chemicals containing sulfur, including gasoline or coal. **Air pollution** containing SO_2 is referred to as *chemically reducing smog* because SO_2 can reduce other chemical (being oxidized to SO_3 or H_2SO_4). **Photochemical smog**, polluted air containing nitrogen monoxide NO and nitrogen dioxide NO_2 (known collectively as **NO_x**), is referred to as *chemically oxidizing smog* because the NO_x can oxidize other chemicals (being reduced from NO_2 to NO, or from NO_2 or NO to N_2). Nitrogen oxides come primarily from automobile emissions since N_2 and O_2 are converted to NO in car engines. *Catalytic converters* in automobiles are intended to convert most of the NO molecules produced back into N_2 and O_2 molecules, but automobiles still generate large amounts of NO_x. Photochemical smog contains NO_2 and O_3, which are both known to affect living creatures' abilities to breathe.

The Greenhouse Effect and Enhanced Global Warming (10.11)

Much of the energy from the sun that reaches earth is absorbed by the surface (land and oceans), which warms the planet. The surface also reradiates this energy back to space as infrared light. Gases in the atmosphere (like water vapor and carbon dioxide) absorb some of this energy, preventing it from escaping into space and keeping earth warmer than expected (called the **greenhouse effect**). This is a good thing— if earth's atmosphere did not absorb some of this energy, the average temperature of the planet would be −18°C instead of +15°C. If this were the earth's temperature the oceans would be frozen solid. This might have prevented the formation of life as it now exists, which requires liquid water as a solvent.

However, the addition of large amounts of carbon dioxide in the earth's atmosphere as a result of the world's industrial revolution (which required the burning of huge amounts of fuel like wood, coal, and petroleum) may be leading to an even higher increase in the earth's average temperature (**global warming**). Clearly, the earth's average temperature is rising (leading to more severe weather patterns and increased melting of the polar ice caps), and the amount of carbon dioxide in the earth's atmosphere is increasing (based on careful measurements over the past few decades). What is not clear is whether the increased CO_2 levels are causing the increased temperature, or if this is a natural fluctuation in the earth's temperature.

Chapter Review — Key Terms

The key terms that were introduced in this chapter are listed below, along with the section in which they were introduced. You should understand these terms and be able to apply them in appropriate situations.

absolute temperature scale (10.4)	air pollutant (10.10)	barometer (10.2)
absolute zero (10.4)	Avogadro's law (10.4)	Boyle's law (10.4)
aerosols (10.10)	bar (10.2)	Charles's law (10.4)

93

chlorofluorocarbons (CFCs) (10.9)	law of combining volumes (10.4)	primary pollutant (10.10)
combined gas law (10.4)	millimeters of mercury (10.2)	secondary pollutant (10.10)
Dalton's law of partial pressures (10.7)	mole fraction (10.7)	smog (10.10)
	newton (10.2)	standard atmosphere (10.2)
diffusion (10.3)	NO_x (10.10)	standard molar volume (10.4)
effusion (10.3)	ozone hole (10.9)	standard temperature and pressure (STP) (10.4)
ideal gas (10.4)	partial pressure (10.7)	
ideal gas constant (10.4)	particulates (10.10)	stratosphere (10.1)
ideal gas law (10.4)	pascal (Pa) (10.2)	stratospheric ozone layer (10.9)
global warming (10.11)	photochemical smog (10.10)	torr (10.2)
greenhouse effect (10.11)	photodissociation (10.9)	troposphere (10.1)
Kelvin temperature scale (10.4)	pressure (10.2)	van der Waals equation (10.8)

Practice Test

After you have finished studying the chapter and the homework problems, the following questions can serve as a test to determine how well you have learned the chapter objectives.

1. Use the concept of pressure to explain why a sharp knife cuts better than a dull one.

2. In the United States, barometric pressure is usually measured in inches of mercury (in Hg). On a rainy day, the barometric pressure is 30.00 in Hg. What is this pressure in mm Hg and atm?

3. Use the kinetic-molecular theory to explain the following macroscopic gas behaviors.
   (a) Gases can expand to occupy larger volumes, but solids and liquids cannot.
   (b) Gases always form homogeneous mixtures with each other.

4. A rubber tire on an automobile with a volume of 24.0 L was filled with air during the summer with 1.09 mol of gas at 34.7°C.
   (a) In the winter, the temperature is –3.2°C, and the tire seems flat. What is the tire volume?
   (b) Use the kinetic-molecular theory to explain why the tire's volume has changed.
   (c) To counteract the flat tire, you fill the tire with additional air to reach a volume of 24.0 L. How many moles of air are now in the tire?
   (d) Use the kinetic-molecular theory to explain why adding air changed the tire's volume.

5. What is the pressure (in atm) when a 312.8 g sample of solid iodine (I_2) placed in a 10.00 L steel container at 253.4°C is allowed to sublime (change from solid to gas)?

6. What is the density (in g/L) of the iodine gas in Problem 5?

7. Hydrogen gas can be made from the reaction of solid diboron hexahydride (diborane) and water. How many liters of hydrogen gas can be made from 47.47 g of diboron hexahydride if the hydrogen gas is collected at 18.8°C and 789.2 mm Hg?

$$B_2H_6(g) + 3 H_2O(\ell) \rightarrow B_2O_3(s) + 6 H_2(g)$$

8. A sample of an unknown alkaline earth metal (M) is heated in the presence of oxygen gas, and it reacts to make the alkaline earth metal oxide (MO). When 500.0 g of the metal and 500.0 g of oxygen are heated at 174.3°C in a 30.0 L flask, the final pressure (after the reaction) is 12.14 atm.

$$2 M(s) + O_2(g) \rightarrow 2 MO(s)$$

   (a) What is the pressure of oxygen gas under these conditions before the reaction occurs?
   (b) How many moles of oxygen gas have taken part in the chemical reaction?
   (c) What is the molar mass and identity of the alkaline earth metal?

94

9. When white phosphorus (an allotrope of elemental phosphorus that is a solid at room temperature) is heated, it will turn into a gas. When 100.0 g of white phosphorus is heated to 374°C in a 25.0 L container, its pressure is 1.71 atm. What is the molar mass and formula of white phosphorus?

10. Use the ideal gas law and the van der Waals equation to determine the pressure of water vapor present when 15.00 mol of water in a 89.5 L tank is heated to 180°C (the van der Waals constants for water vapor are $a = 5.46$ L^2 atm/mol^2 and $b = 0.0305$ L/mol). Which factor (finite volume of water particles or attractions of the water particles) is more important in this sample?

11. When chlorofluorocarbons (CFCs) in the stratosphere absorb UV light, a C-Cl bond is broken instead of a C-F bond. Why is this? (*Hint:* Refer to Table 8.2 on p. 350 of the textbook).

$$CF_2Cl_2(g) \xrightarrow{hv} CF_2Cl\cdot(g) + Cl\cdot(g), \text{ not } CFCl\cdot(g) + F\cdot(g)$$

12. Suppose a scientist has found a way to remove all of the carbon dioxide (a greenhouse gas) from the earth's atmosphere. Should the scientist do this?

Chapter 11
Liquids, Solids, and Materials

The goal of this chapter is to introduce the physical properties of liquids and solids. Liquids are different from gases since they have intermolecular forces holding them together (Chapter 9). This chapter describes properties that can be used as relative measures of the strengths of intermolecular forces present in liquids. Solids are unique because they have long-range repeating patterns, and this chapter describes these patterns using unit cells. The chapter also describes the four types of solids and their physical properties. Finally, the chapter also introduces the concept of materials science (the process of designing solids to meet specific needs) and several modern man-made solid materials like semiconductors, cement, ceramics, and glass.

Physical Properties of Liquids and Intermolecular Forces (11.1 & 11.2)

Even though steel has a density that is roughly eight times the density of water, a steel paper clip can float on the surface of water. This is due to **surface tension**. The molecules in liquid water are attracted to each other by strong hydrogen bonding forces, and these forces pull the water molecules close together. In order for the paper clip to fall through the surface of the water and sink to the bottom of the container, it must break the strong hydrogen bonds holding the water molecules together. In general, the stronger the liquid's intermolecular forces, the higher the surface tension present in the liquid. **Viscosity** (the resistance of the flow of a liquid) also measures a liquid's intermolecular forces—the stronger the intermolecular forces present in the liquid, the harder it is for the molecules within the liquid to move past one another (due to their strong attractions for each other) and the higher the viscosity of the liquid will be.

When the molecules in a liquid absorb energy, they undergo **evaporation** (the conversion from the liquid state to the gaseous state). This process is endothermic, and the energy absorbed is used by the molecules to break the intermolecular forces holding them together in the liquid state, allowing them to escape into the gas phase. The tendency to evaporate is called the **volatility** of the liquid. The volatility of the liquid is also a measure of its intermolecular forces—if the liquid has strong intermolecular forces, a lot of energy is needed to evaporate the liquid and it evaporates slowly; if it has weak intermolecular forces, little energy is needed to evaporate the liquid and it evaporates quickly. However, the process of evaporation is also affected by other parameters like the liquid's surface area and temperature.

Example: How does a liquid's surface area and temperature affect evaporation?

Solution: As an example of the effect of surface area, you probably know that water evaporates more quickly off of the sidewalk than it does from a puddle. The difference here is that the water on the sidewalk has a greater surface area than the puddle does. Because liquid molecules must escape into the gas phase at the liquid's surface, a liquid having a greater surface area provides the molecules with a greater opportunity to be at the surface so they can escape the liquid.

Liquids also evaporate more quickly at higher temperatures. When the liquid is heated, the molecules in the liquid move faster and have more energy. So, more of these liquid molecules will have enough energy to escape from the intermolecular forces holding them in the liquid (the *minimum escape energy*), more molecules will escape the liquid, and the rate of evaporation increases.

When a liquid is placed inside a sealed container, some of the molecules in the liquid will evaporate into the gas phase. As this reaction proceeds, a point is reached where the number of liquid molecules that are evaporating is matched by the number of gas particles colliding with the liquid's surface and rejoining the liquid. The pressure exerted by the gas phase molecules is called the **vapor pressure** (or the **equilibrium vapor pressure**). A liquid's vapor pressure is also a measure of the intermolecular forces present in the liquid—liquids with strong intermolecular forces have few molecules with enough energy to break the strong attractions in the liquid and their vapor pressures are low; liquids with weak intermolecular forces have many molecules with enough energy to break the weak attractions in the liquids, so many molecules escape the liquid and become gas particles (high vapor pressure). The surface area of a liquid does not affect its vapor pressure—a larger surface means more liquid molecules can escape the liquid, but it also means that more gas molecules can collide with the surface and become trapped in the liquid, and these two factors offset each other. However, the liquid's temperature does effect its vapor pressure: As the liquid is heated, the molecules move faster and more liquid molecules have enough energy to escape into the gas phase, causing the number of gas particles and the liquid's vapor pressure to increase. Vapor pressures can be used to compare the intermolecular forces in two liquids as long as their temperatures are the same.

As a liquid is heated, its vapor pressure increases. Once the vapor pressure reaches atmospheric pressure, molecules do not have to be at the surface to be converted into gases and gas bubbles can form throughout the liquid. This process is called **boiling**, and the temperature where the vapor pressure equals atmospheric pressure is called the liquid's **boiling point**. The temperature where the vapor pressure of the liquid equals the normal atmospheric pressure (1 atm = 760 mm Hg) is called the **normal boiling point** of the liquid. A liquid's boiling point is not affected by surface area or temperature (water boils at 100°C whether it is a hot summer day or a cold winter night), but it is affected by the intermolecular forces present in the liquid.

Example: Predict whether benzene (C_6H_6) or toluene (C_7H_8) has the higher boiling point.

Solution: Because both molecules have only C and H atoms, these compounds are hydrocarbons, which are known to be nonpolar (no net positive or negative ends). So, these molecules will have only London forces, which increase as the molecules get bigger (heavier). The molar masses are 78.11 g/mol for C_6H_6 and 92.13 g/mol for C_7H_8, so toluene is expected to have stronger London forces than benzene. Since toluene has stronger intermolecular forces, it should evaporate less easily than benzene and benzene should have a higher vapor pressure at any given temperature than toluene does ($P_{benzene}$ = 75 mm Hg and $P_{toluene}$ = 22 mm Hg at 25°C). The boiling point of a liquid is the temperature where its vapor pressure equals 760 mm Hg (atmospheric pressure). Since benzene has a higher vapor pressure at 25°C, it will take less heat (a lower temperature) to get its vapor pressure to 760 mm Hg, and since toluene has a lower vapor pressure, it will take more heat (a higher temperature) to get its vapor pressure to 760 mm Hg. So, you should expect toluene to have a higher boiling point than benzene ($bp_{benzene}$ = 80.1°C and $bp_{toluene}$ = 110.6°C).

Describing Phase Changes (11.3)

Figure 11.8 in the textbook (p. 485) describes the six phase changes that can occur between the three states of matter. When a solid is converted into a liquid, the solid absorbs energy (endothermic) as it *melts*. In the reverse process, a liquid releases energy (exothermic) when it *freezes* (*solidifies*, *crystallizes*, undergoes **crystallization**) into a solid. When a liquid is converted into a gas, it absorbs energy (endothermic) as it *boils* (*evaporates*, undergoes *evaporation*, *vaporizes*, undergoes **vaporization**). In the reverse process, a gas releases energy (exothermic) when it *condenses* (undergoes **condensation**) into a liquid. When a solid is converted directly into a gas, it absorbs energy (endothermic) as it *sublimes* (undergoes **sublimation**) to the gas phase. In the reverse process, a gas releases energy (exothermic) when it *deposits* (undergoes **deposition**) into a solid.

You might expect a plot of the temperature of a solid versus energy (heat) added to increase linearly. This curve does show the expected temperature increases, but there are two flat regions in this **heating curve** where the added heat does not change the substance's temperature. In the angled regions of this curve, only one state of matter is present (solid in the first one, liquid in the middle one, and gas in the last one). When the curve is flat, two states of matter are present (solid and liquid for the first one, and liquid and gas for the second one). When heat is added to the pure states of matter, their particles

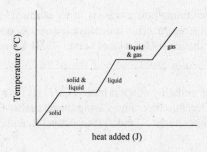

move faster, resulting in an increased temperature. When the solid's temperature reaches the *melting point*, it starts to melt. Then, the energy added to the substance is not going to heat it (as shown by the constant temperature); instead, this energy is used to break the intermolecular forces holding the substance in its solid state. Once all of the substance has melted, the energy added to the substance can then go into warming the liquid, and the temperature begins to rise again. When the liquid's temperature reaches the boiling point, it starts to boil and the added energy is used to break the liquid's intermolecular forces (and the temperature levels off) until the liquid has completely boiled.

Calculating Energy Transfers Associated with Phase Changes (11.3)

Energy (heat) changes associated with states of matter changes can be calculated using tabulated enthalpy values and the equation $q = n\Delta H°$ from chapter 6; energy changes when a pure state of matter is heated can be calculated using the equation $q = mc\Delta T$. The enthalpy required to melt a solid is called the *enthalpy of fusion* ($\Delta H°_{fus}$) and the enthalpy required to boil a liquid is called the *enthalpy of vaporization* ($\Delta H°_{vap}$).

Example: When 20.0 g of water vapor (gaseous water) at 100°C is mixed with 120.0 g of ice (solid water) at 0°C, these samples are both converted to liquid water. What is the final temperature of the water? $\Delta H°_{fus} = 6.01$ kJ/mol, $\Delta H°_{vap} = 40.7$ kJ/mol, and $c(liquid\ H_2O) = 4.184$ J/g °C.

Solution: This is a calorimetry problem, so the heat change for ice and steam must add to zero. The steam loses heat when it condenses to liquid water at 100°C, then this liquid water is cooled even further (to some unknown temperature). The ice gains heat when it melts to liquid water at 0°C, then this liquid water is also warmed by the steam to the same unknown temperature.

$$q_{steam} + q_{ice} = 0$$

$$q_{steam} = q_{cond.} + q_{cool}$$

$$q_{cond.} = n\Delta H = \left(20.0\ \text{g H}_2\text{O} \times \frac{\text{mol H}_2\text{O}}{18.02\ \text{g H}_2\text{O}}\right)\left(-40.7\ \frac{\text{kJ}}{\text{mol H}_2\text{O}}\right) \times \frac{10^3\ \text{J}}{1\ \text{kJ}} = -45{,}172\ \text{J}$$

$$q_{cool} = mc\Delta T = (20.0\ \text{g})\left(4.184\ \frac{\text{J}}{\text{g C}}\right)(T_f - 100\ \text{C}) = 83.68\ \frac{\text{J}}{\text{C}} \times T_f - 8{,}368\ \text{J}$$

$$q_{ice} = q_{melt} + q_{heat}$$

$$q_{melt} = n\Delta H = \left(120.0\ \text{g H}_2\text{O} \times \frac{\text{mol H}_2\text{O}}{18.02\ \text{g H}_2\text{O}}\right)\left(6.01\ \frac{\text{kJ}}{\text{mol H}_2\text{O}}\right) \times \frac{10^3\ \text{J}}{1\ \text{kJ}} = 40{,}022\ \text{J}$$

$$q_{heat} = mc\Delta T = (120.0\ \text{g})\left(4.184\ \frac{\text{J}}{\text{g C}}\right)(T_f - 0\ \text{C}) = 502.08\ \frac{\text{J}}{\text{C}} \times T_f$$

$$(-45{,}172\ \text{J} + 83.68\ \frac{\text{J}}{\text{C}} \times T_f - 8{,}368\ \text{J}) + (40{,}022\ \text{J} + 502.08\ \frac{\text{J}}{\text{C}} \times T_f) = 0$$

$$(83.68 + 502.08)\ \frac{\text{J}}{\text{C}} \times T_f = 45{,}172\ \text{J} + 8{,}368\ \text{J} - 40{,}022\ \text{J} = 13{,}518\ \text{J}$$

$$T_f = \frac{13{,}518\ \text{J}}{585.76\ \text{J/ C}} = 23\ \text{C}$$

Even though there was six times as much ice present as steam, the final temperature is right around room temperature. This is because the energy released by the steam is so much greater than the energy absorbed by the ice (–40.7 kJ/mol versus +6.01 kJ/mol).

The enthalpy of vaporization ($\Delta H°_{vap}$) for a liquid can be calculated from its vapor pressures at two different temperatures using the **Clausius-Clapeyron equation**: P_1 and P_2 are the vapor pressures at temperatures T_1 and T_2 (respectively), and R is the ideal gas constant with a value of 8.314 J/mol K.

$$\ln\left(\frac{P_2}{P_1}\right) = -\frac{\Delta H_{vap}}{R}\left(\frac{1}{T_2} - \frac{1}{T_1}\right)$$

Using Phase Diagrams to Predict Changes in States of Matter (11.3)

A **phase diagram** is a temperature-pressure plot that shows the states of matter present at any given temperature and pressure for a chemical. The chemical is a solid at low T and high P values (the top left part of the curve) and a gas at high T and low P values (the bottom right part of the curve). The liquid state of matter appears between them at intermediate T and P values. The lines between two states of matter represent the conditions where both states of matter are stable (the solid-liquid line is the melting point, the liquid-gas line is the boiling point, and the solid-gas line is the sublimation point for the substance). The single point in a phase diagram where all three states of matter are stable is called the **triple point**.

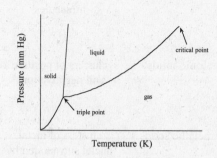

Critical Temperature and Critical Pressure (11.3)

Above a specific temperature (the **critical temperature**) and pressure (the **critical pressure**), collectively called the **critical point**, the line between the liquid and gaseous states disappears. In this region the substance, called a **supercritical fluid**, has properties that match both liquids (density, particles very close to each other, etc.) and gases (free flowing, exerts pressures, expands to fill its container, etc.).

The Unusual Properties of Water (11.4)

Although water is the most common liquid on the earth, it has several very unique properties that are not shared by most liquids. Most of these unique properties are due to water's incredibly strong intermolecular forces, which are the result of water's very small and very polar nature. Water has a very high melting and boiling point for a molecule of its size. Molecules of similar size have much lower melting and boiling points—for CH_4 (16.0 g/mol) $mp = -182°C$ and $bp = -164°C$, for H_2O (18.0 g/mol) $mp = 0°C$ and $bp = 100°C$, and for Ne (20.2 g/mol) $mp = -249°C$ and $bp = -246°C$. This makes water a liquid at much higher temperatures than would be expected (which was crucial for the formation of life on earth). Liquid water also has a very high specific heat, enthalpy of fusion, enthalpy of vaporization, and surface tension. These large values reflect the fact that it takes large amounts of energy to separate water molecules in the liquid due to the incredibly strong hydrogen-bonding intermolecular forces present in liquid water. In addition, water is unique in that its solid form (ice) is less dense than its liquid form (solids are normally more dense than their liquid states). This is due to the fact that in solid water, the molecules are aligned for maximum formation of hydrogen bonds, and this results in empty spaces within the ice crystal. When the ice crystal melts, the empty spaces are eliminated and the liquid particles move closer together.

All of these unique properties affect life on earth in many ways. The large amounts of energy absorbed and released by water when changing states of matter leads to the weather patterns on earth and the evaporative cooling of animals (called sweating). The fact that ice floats in water prevents lakes from freezing solid.

The Major Types of Solids (11.5)

Most solids can be classified into one of four major categories—molecular solids, ionic solids, metallic solids, and covalent network solids (*amorphous* or *glassy solids* are simply a non-crystalline subset of the covalent network solids). The properties of these solids appear in the table on the next page.

Molecular solids are made from neutral molecules. Even though each molecule has strong covalent bonds holding it together, the molecules are attracted to each other by weak intermolecular forces (London forces, dipole-dipole attractions, and hydrogen bonds). Because these forces are rather weak, it takes very little energy to disrupt these attractions, and these solids have low melting points (in fact, many are liquids or gases at room temperature). Since these molecules do not have charges, they will never conduct electricity.

99

solid type	description of solid	forces holding solid together	melting point	electrical conductivity
molecular solids	clusters of neutral molecules	intermolecular forces between neutral molecules (very weak)	very low (< 200°C)	never conducts electricity
ionic solids	clusters of positive and negative ions	attractions of the cations and anions (strong)	high (>500°C)	does not conduct electricity as a solid but does as a molten salt
metallic solids	clusters of metal nuclei surrounded by a 'sea of electrons'	attractions of the metal nuclei and electrons (strong)	variable (from −40°C to 3400°C)	always conducts electricity
covalent network solids	clusters of nonmetal atoms covalently bonded together	covalent bonds (very strong)	very high (>1000°C)	never conducts electricity

Ionic solids are made from cations and anions. The attractions of the cations and anions holding these solids together are strong, so they have high melting points (generally above 500°C). Because the ions are not free to move in the solid, solid ionic solids do not conduct electricity; however, when the ions are free to move (in its melted or molten state, or when dissolved in water) these solids will conduct electricity.

Metallic solids (which will be described in detail later) are held together by the attractions of the metal nuclei (which are positive) and their surrounding electrons (which are negative). These forces are generally strong, but the melting points of metals vary tremendously. Since the electrons within a metal solid are always free to move, metallic solids always conduct electricity.

Covalent network solids are made from individual nonmetal atoms covalently bonded together into a three-dimensional structure. Because covalent network solids are held together by covalent bonds (which are very strong), they have very high melting points. Because covalent network solids have no charges (only neutral atoms), they will never conduct electricity.

Example: Determine the solid type for the following elements in row 4 of the periodic table.
(a) Solid calcium has a melting point of 839°C and always conducts electricity.
(b) Solid germanium has a melting point of 937°C and never conducts electricity.
(c) Solid selenium has a melting point of 217°C and never conducts electricity.
(d) Solid vanadium has a melting point of 1890°C and always conducts electricity.

Solution:
(a) Since Ca always conducts electricity, it is a metallic solid (Ca is an alkaline earth metal).
(b) Since Ge never conducts electricity, it is either a molecular solid or a covalent network solid. Since germanium has a melting point close to 1000°C, it should be a covalent network solid.
(d) Since Se never conducts electricity, it is either a molecular solid or a covalent network solid. Selenium has a relatively low melting point (close to 200°C), so it should be a molecular solid.
(a) Since V always conducts electricity, it is a metallic solid (V is a transition metal).

Calculations Involving Simple Unit Cells (11.6)

Most solids have regular repeating patterns of particles (atoms, molecules, or ions); these solids are called **crystalline solids**. Solids that do not have these regular patterns are called **amorphous solids**. The smallest three-dimensional repeat pattern in a crystalline solid is referred to as the solid's **unit cell**. Just as electromagnetic waves can be reconstructed using the smallest one-dimensional repeating unit (*wavelength*) and wallpaper or bathroom tile patterns can be reconstructed by their smallest two-dimensional repeating patterns, a single solid crystal can be reconstructed from its three-dimensional unit cell.

The simplest of these unit cells are the **cubic unit cells**, in which the unit cell has all three sides of equal length and all angles at 90°. There are three types of cubic unit cells: (1) *primitive* (or *simple*) *cubic* unit cells, where there is one atom at the eight corners of the unit cell; (2) *body-centered cubic (bcc)* unit cells, where there is one atom at the eight corners of the unit cell and one atom in the center (body) of the unit cell; and (3) *face-centered cubic (fcc)* unit cells, where there is one atom at the eight corner of the unit cell and one atom at the center of each of the six faces on the unit cell. The table below shows the number of atoms in each unit cell and the relationship between the atomic radius (r) and the length of the sides in the unit cell (ℓ).

unit cell type	atom positions	# atoms per unit cell	relationship of r and ℓ
simple cubic	corners of unit cell	$8 \times \dfrac{1}{8} = 1$	$2r = \ell$
body-centered cubic	corners of the unit cell center of the unit cell	$8 \times \dfrac{1}{8} + 1 = 2$	$4r = \sqrt{3}\ell$
face-centered cubic	corners of the unit cell faces of the unit cell	$8 \times \dfrac{1}{8} + 6 \times \dfrac{1}{2} = 4$	$4r = \sqrt{2}\ell$

Example: Show that the face-center cubic unit cell is 74% occupied by atoms (and 26% empty space).

Solution: Since the percent of space filled in an *fcc* unit cell is independent of atom (sphere) size, you may use any sphere size you like—in this example, use a sphere with a radius of 1.00 cm. The total volume of the cube will be $V_{cube} = \ell^3$; the value of ℓ will be determined by the formula $4r = \sqrt{2}\ell$.

$$4r = \sqrt{2}\ell, \; \ell = \frac{4r}{\sqrt{2}} = \frac{4(1.00 \text{ cm})}{\sqrt{2}} = 2.83 \text{ cm}$$

$$V_{cube} = \ell^3 = (2.83 \text{ cm})^3 = 22.6 \text{ cm}^3$$

The volume occupied by the atoms will simply be the volume of each sphere ($4/3 \, \pi r^3$) times the number of spheres in the unit cell (4).

$$V_{atoms} = \frac{4}{3}\pi r^3 \times 4 = \frac{4}{3}(3.14159)(1.00 \text{ cm})^3 \times 4 = 16.8 \text{ cm}^3$$

$$\% \; occupied = \frac{V_{atoms}}{V_{cube}} \times 100\% = \frac{16.8 \text{ cm}^3}{22.6 \text{ cm}^3} \times 100\% = 74\%$$

This is the most efficient packing possible with spherical objects (like atoms). Because spherical objects cannot occupy more than 74% of an object's total volume, any packing that results in the atoms occupying 74% of the total volume is referred to as **closest packing**.

Bonding in Covalent Network Solids (11.7)

Covalent network solids (or **network solids**) consist of nonmetal atoms that are covalently bonded to each other. In *three-dimensional network solids*, the atoms are bonded to each other in a three-dimensional structure. Three-dimensional network solids can be thought of as a single molecule made from individual atoms that have incredible strength in all directions. Examples of 3-D network solids are diamond (C), silicon carbide (SiC), and quartz (SiO_2), which are all known for their incredible strengths and are used as industrial abrasives because they are strong enough to scratch other objects.

Two-dimensional network solids have atoms covalently bonded together in two of the three dimensions. These solids can be thought of as sheets of very strong materials that are loosely attached to each other by weaker intermolecular forces. Examples of 2-D network solids are graphite (C), talc ($Mg_3Si_2O_{12}H_2$), and mica ($KAl_3Si_3O_{10}F$), which are all known for the ease with which these 2-D sheets slide past one another, and are often used as lubricants to keep metal parts from grinding as they touch.

One-dimensional network solids have atoms covalently bonded together in one of the three dimensions. These solids can be thought of as chains or fibers of very strong materials that are loosely attached to each other by weaker intermolecular forces. Examples of 1-D network solids are **nanotubes** (C), asbestos ($Mg_3Si_2O_9H_4$) and simple polymers like polyethylene (C_2H_4) or silicone (C_2H_6SiO).

The Basis of Materials Science (11.8)

As toolmakers, humans have become very good at adapting naturally-occurring materials for their own uses. The manipulation of chemical substances (and their resulting properties) is often referred to as **materials science**. Materials science focuses on the relationship between a chemical's structure and its physical and chemical properties. In particular, chemists can often design materials to have specific properties simply by changing their chemical structures or their chemical identities. Examples of modern materials made from materials science principles include metals (alloys), polymers, ceramics, glasses, semiconductors, and composites.

Metallic Bonding, Metals, and Semiconductors (11.9)

Metallic solids (containing atoms with a few more electrons than a noble gas configuration) have many characteristic properties—high electrical and thermal conductivity, high ductility and malleability (the ability to be deformed without breaking), and a shiny luster. You know from previous discussions that metal atoms can readily lose electrons. In fact, metallic solids are often thought of as metal cations (that have lost one or more electrons) surrounded by a 'sea of electrons' (or like several 'beehives' surrounded by a 'swarm of electrons').

From a molecular orbital perspective, each atom provides one or more electrons and one or more atomic orbitals to the 'swarm'. When the atoms are very close to each other, their atomic orbitals overlap and create 'molecular' orbitals that encompass the entire solid. This results in a huge number of molecular orbitals with very small energy differences between them (called an **energy band**). An energy band is made of two types of bands—the **valence band** is a series of orbitals that are filled with the atoms' valence electrons, and the **conduction band** is a series of empty orbitals in the solid.

When the valence band and the conduction band overlap, electrons can move freely from one band to the other with very little added energy and the solid will conduct electricity (an electrical **conductor**). When the energy difference between the valence and conduction bands is very large, very few electrons will have enough energy to jump from one orbital to another and the solid will not conduct electricity (an electrical **insulator**). If the energy gap between the valence and conduction bands is somewhat small, some electrons will have enough energy to jump from the lower valence band to the higher conduction band; these solids are called **semiconductors**. In general, metal atoms tend to be conductors, nonmetal atoms tend to be insulators, and metalloids (semimetals) tend to be semiconductors.

Superconductivity (11.9)

In general, the electrical conductivity of any metal increases as the metal is cooled because the *electrical resistance* (the resistance of electrons to flow freely) of the metal decreases with decreasing temperature. For most solids, this resistance decreases slowly as the solid is cooled; however, for some solids the resistance suddenly drops to zero. The temperature where this happens is called the *critical temperature* (which is different than the critical temperature used in phase diagrams). Any object that has no electrical resistance is called a **superconductor**. Superconductors can be used for transmitting electricity through power lines, high-efficiency motors, computers, and other devices including levitated railways.

102

Semiconductors (11.10)

Because semiconductors have a small *band gap* (energy difference between the valence and conduction bands), semiconductors will conduct electricity when the electrons are given additional energy, like an applied electrical voltage. The conductivity of some semiconductors can be improved by **doping** the semiconductors with a tiny amount of another substance. When a semiconductor is doped with atoms containing more electrons than the semiconductor, an ***n*-type semiconductor** is made; when it is doped with atoms containing fewer electrons than the semiconductor, a ***p*-type semiconductor** is made. When *p*-type and *n*-type semiconductors are connected together, a ***p-n* junction** is formed. This type of junction acts as a *rectifier*, an object that allows current to flow one way, but not the other.

The Lack of Regular Structure and Amorphous Solids (11.11)

Amorphous (*noncrystalline*) solids are solids that do not have regular repeating patterns. Many modern materials (cement, concrete, ceramics, and glasses) are amorphous solids. **Cement** is a powdered mixture of several ionic and covalent oxides (CaO, Al_2O_3, Fe_2O_3, SiO_2, etc.); when mixed with sand and crushed rock (*aggregate*), cement forms **concrete**. Cement is not really a mixture of these oxides as much as it is a complex network of ions within the solid. Although concrete can resist large forces pushing on it without cracking (high *compressive strength*), it cannot resist cracking when it is pulled apart (low *tensile strength*).

Ceramics are made from clays mixed with water, which are subsequently hardened by a baking (firing) process. **Glasses** are made from amorphous silica (SiO_2) with element oxide impurities added to change the color, strength, and heat expansion properties. In general, ceramics and glasses are very brittle, and can easily break due to the lack of a regular crystalline pattern.

Chapter Review — Key Terms

The key terms that were introduced in this chapter are listed below, along with the section in which they were introduced. You should understand these terms and be able to apply them in appropriate situations.

amorphous solids (11.5)	deposition (11.3)	*p*-type semiconductor (11.10)
boiling (11.2)	dew point (11.2)	phase diagram (11.3)
boiling point (11.2)	doping (11.10)	relative humidity (11.2)
capillary action (11.1)	energy band (11.9)	semiconductor (11.9)
cement (11.11)	equilibrium vapor pressure (11.2)	solar cell (11.10)
ceramics (11.11)	evaporation (11.3)	sublimation (11.3)
Clausius-Clapeyron equation (11.2)	fullerenes (11.7)	superconductor (11.9)
closest packing (11.6)	glass (11.11)	supercritical fluid (11.3)
concrete (11.11)	heating curve (11.3)	surface tension (11.1)
condensation (11.3)	hexagonal close packing (11.6)	triple point (11.3)
conduction band (11.9)	insulator (11.9)	unit cell (11.6)
conductor (11.9)	materials science (11.8)	valence band (11.9)
critical point (11.3)	meniscus (11.1)	vapor pressure (11.2)
critical pressure (11.3)	metallic bonding (11.9)	vaporization (11.3)
critical temperature (11.3)	*n*-type semiconductor (11.10)	viscosity (11.1)
crystal lattice (11.6)	nanotube (11.7)	volatility (11.2)
crystalline solids (11.5)	network solids (11.7)	X-ray crystallography (p. 510)
crystallization (11.3)	normal boiling point (11.2)	zone refining (11.10)
cubic close packing (11.6)	optical fiber (11.11)	
cubic unit cell (11.6)	*p-n* junction (11.10)	

Practice Test

After you have finished studying the chapter and the homework problems, the following questions can serve as a test to determine how well you have learned the chapter objectives.

1. When two students were asked what gases are present in the bubbles formed when liquid water boils, one student said "water vapor—$H_2O(g)$" while the other student said "hydrogen and oxygen —$H_2(g)$ and $O_2(g)$".
    (a) Which student is right? Can you provide any evidence for your answer?
    (b) Use the strengths of intermolecular forces and covalent bonds to justify your answer.

2. The vapor pressures (in mm Hg) for eight liquids containing C, H, and O atoms are listed below.

name	condensed formula	P	name	condensed formula	P
pentane	$CH_3CH_2CH_2CH_2CH_3$	651	methanol	CH_3OH	347
hexane	$CH_3CH_2CH_2CH_2CH_2CH_3$	182	ethanol	CH_3CH_2OH	132
heptane	$CH_3CH_2CH_2CH_2CH_2CH_2CH_3$	89	1-propanol	$CH_3CH_2CH_2OH$	73
water	HOH	18	1-butanol	$CH_3CH_2CH_2CH_2OH$	18

   (a) Look at the three alkanes (containing only C and H atoms, and ending in '-*ane*'). Explain the trend in vapor pressures. What intermolecular forces are responsible for this trend?
   (b) Look at the four alcohols (containing OH groups, and ending in '-*anol*'). Explain the trend in vapor pressures. What intermolecular forces are responsible for this trend?
   (c) Which molecules should be compared to see the effect of hydrogen bonds on the vapor pressure? What is this effect?
   (d) Water has OH groups just like the alcohols. Why does it have a much lower vapor pressure than the alcohols even though it has a much lower molar mass?

3. Water (H_2O) has a boiling point of 100°C while elemental phosphorus (P_4) boils at 280°C.
    (a) What types of intermolecular forces are present in the two liquids?
    (b) Which of these liquids has the stronger intermolecular forces?
    (c) Which one has a higher vapor pressure at 60°C?

4. Use the heating curve to the right to answer the following questions.
    (a) In which regions are phase changes occurring?
    (b) In which regions are solids present?
    (c) In which regions are liquids present?
    (d) In which regions are gases present?
    (e) In which regions is the added heat going to raise the temperature of the substance?

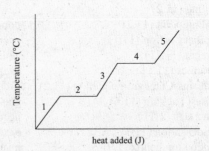

5. Many cooks will tell you that steam burns are more dangerous than being burned by boiling water. *c(liquid H_2O)* = 4.184 J/g °C, $\Delta H°_{vap}$ = 40.7 kJ/mol.
    (a) Explain this observation.
    (b) How much energy (in kJ) is transferred to the cook's hand when 10.0 g of boiling water cools from 100°C to 37°C (skin temperature)?
    (c) How much energy (in kJ) is transferred to the cook's hand when 10.0 g of steam at 100°C cools to 37°C (skin temperature)?

6. Use the Clausius-Clapeyron and the following information to determine the vapor pressure of liquid carbon disulfide (CS_2) at 25.0°C. $\Delta H°_f\{CS_2(g)\}$ = 117.36 kJ/mol; $\Delta H°_f\{CS_2(\ell)\}$ = 89.70 kJ/mol, *bp(CS_2)* = 46.2°C.

7.   The triple point for oxygen occurs at 1 mm Hg and 54 K, the melting point of oxygen at 760 mm Hg is 55 K, the normal boiling point of oxygen is 90 K, and the critical temperature and pressure of oxygen are 37,823 mm Hg and 155 K.
      (a)   Draw a phase diagram for oxygen (you may leave T in Kelvin units).
      (b)   What state of matter is present at room temperature and pressure?
      (c)   What state of matter is present at 63 K and 1,000 mm Hg?
      (d)   What state of matter is present at 7 K and 350 mm Hg?
      (e)   What state of matter is present at 170 K and 40,000 mm Hg?
      (f)   Under what conditions will solid oxygen sublime?

8.   Determine the type of solid the following compounds represent.
      (a)   Carbon dioxide (CO_2), $mp = -78.5°C$ (actually a sublimation point) and never conduct electricity.
      (b)   Silicon dioxide (SiO_2), $mp = 1723°C$ and never conduct electricity.
      (c)   Tin dioxide (SnO_2), $mp = 1630°C$ and conducts electricity only when melted.

9.   You have a sample of a shiny gold solid and want to know whether it is real gold (Au) or fool's gold (iron pyrite, FeS_2).
      (a)   What kind of solid would you expect gold to be? Explain your answer.
      (b)   What kind of solid would you expect iron pyrite to be? Explain your answer.
      (c)   How could you tell whether your solid is gold or fool's gold?

10.   Above 1000°C, solid iron (Fe) changes from a body-centered cubic unit cell (with an edge length of $\ell = 0.287$ nm) to a face-centered cubic unit cell (with an edge length of $\ell = 0.363$ nm).
      (a)   What is the radius of the iron atoms in each unit cell?
      (b)   Determine the density (in g/cm^3) for the two forms of iron.
      (c)   Explain why these densities are slightly different.

11.   The compound zinc selenide (ZnSe) is a useful semiconductor.
      (a)   With which element is ZnSe isoelectronic? Would you expect this element to be a semiconductor?
      (b)   If you were interested in making a p-type semiconductor from ZnSe, which atom would you use in excess?
      (c)   If you were interested in making an n-type semiconductor from ZnSe, which atom would you use in excess?

Chapter 12
Fuels, Organic Chemicals, and Polymers

The goal of this chapter is to introduce specific concepts related to the field of organic chemistry. Organic chemistry was originally thought of as the chemistry of compounds derived from living creatures, but it is now thought of more as the chemistry of compounds containing carbon, hydrogen, and other nonmetals (like oxygen, nitrogen, the halogens, etc.). This chapter also describes several organic (hydrocarbon) fuels like natural gas, gasoline, and coal. Finally, the chapter introduces polymers and discusses the properties and uses of several synthetic and natural polymers.

Petroleum Refining and Reformulating (12.1)

Crude oil (petroleum pumped from the ground) contains many different hydrocarbons that have very different properties. Chemists use *petroleum refining* to separate crude oil into different **petroleum fractions** based on differences in boiling points (a physical property). This process, called *fractional distillation*, results in fractions that are gases at room temperature (1-4 C atoms), gasoline (5-12 C atoms), kerosene (12-16 C atoms), fuel oil (15-18 C atoms), lubricating oil (16-20 C atoms), and a solid residue (asphalt, more than 20 C atoms).

Petroleum chemists introduced the concept of an **octane number** to describe how efficiently a sample of gasoline will burn in an engine. They arbitrarily set the octane numbers of heptane (a chemical that does not burn very efficiently) to 0 and the octane number of isooctane (a chemical that does burn efficiently) to 100. A sample of gasoline with an octane number (*octane rating*) of 89 burns with the same efficiency as a mixture that is 89% isooctane and 11% heptane.

In general, unbranched (straight-chain) alkanes burn less efficiently than branched alkanes, alkenes, or aromatics. Chemists can increase the octane rating of a gasoline sample by converting unbranched alkanes into these other compounds. The first method (called **catalytic cracking**) converts the long unbranched alkanes into two smaller components—a shorter unbranched alkane and an alkene. The second method (called **catalytic reforming**) converts unbranched alkanes into its branched alkane isomers or into aromatic compounds. The octane rating of a gasoline sample can also be increased by adding compounds with higher octane number (*octane enhancers*). These compounds can include compounds containing lead (tetraethyl lead, $C_8H_{20}Pb$, used in *Ethyl* gasoline) or oxygen (alcohols), and aromatic compounds like toluene. Oxygen-containing compounds like ethanol (C_2H_5OH) or methyl tertiarybutyl ether (MTBE, $C_5H_{12}O$) are also added to gasoline to reduce the amount of carbon monoxide produced by automobiles.

Major U. S. Energy Sources (12.2)

The United States gets most of its energy from fossil fuels, including coal, crude oil, and natural gas. Other sources include nuclear power, biomass, hydroelectric, geothermal, solar, and wind power.

Organic Chemicals Obtained from Coal (12.2)

Coal represents the solid form of hydrocarbons stored in the earth. Coal contains very large molecules containing C, H, N, and S atoms. Coal can be converted into smaller molecules (that are liquids or gases at room temperature) by *pyrolysis*, the heating of a chemical in the absence of oxygen (if oxygen were present, the heated coal would burn, producing CO_2, H_2O, N_2, and SO_2). When coal is pyrolyzed, several products are produced including coke (mostly solid C in its graphite form) and aromatic compounds like benzene (C_6H_6), toluene (C_7H_8), and xylenes (C_8H_{10}).

Major Organic Chemicals of Industrial and Economic Importance (12.3)

Many organic chemicals that are used in important industrial processes are derived from petroleum fuels or are produced in the catalytic cracking and reforming processes of liquid petroleum or the pyrolysis of solid coals. These compounds include acetylene (C_2H_2), ethylene (C_2H_4), propylene (C_3H_6), butylene (C_4H_8), and aromatic compounds like benzene (C_6H_6), toluene (C_7H_8), and xylenes (C_8H_{10}).

Oxidation of Alcohols (12.4)

When alcohols (organic compounds containing -OH groups) are oxidized, the products that are formed depend on the chemical structure of the alcohol. When the C atom attached to the -OH group has two H atoms attached to it (a *primary alcohol*), the alcohol is converted to an **aldehyde** (an organic compound containing a -C(=O)H group), which then reacts with the oxidizing agent to produce a **carboxylic acid** (an organic compound containing a -C(=O)OH group). In the first step, the oxidizing agent (which is usually thought of as providing an O atom to the chemical being oxidized) removes two H atoms from the alcohol (one from the O atom and one from the C atom attached to the O atom) and makes a C-O double bond. In the second step, the oxidizing agent converts the H atom attached to the C=O group into an -OH group.

primary alcohol   aldehyde   carboxylic acid

Chemists use the symbol 'R' to represent alkyl (hydrocarbon, C_xH_y-) groups. When the C atom attached to the -OH group has one H atom attached to it (a *secondary alcohol*), the alcohol is converted to a **ketone** (an organic compound with the general formula RC(=O)R'). This reaction is just like the first step of oxidizing primary alcohols—the oxidizing agent removes two H atoms from the alcohol (one from the O atom and one from the C atom attached to the O atom) and makes a C-O double bond. When the alcohol has no H atom on the C atom attached to the -OH group (a *tertiary alcohol*), an oxidation reaction does not occur.

Example: Explain why carbon dioxide (CO_2) is the product when methanol (CH_3OH) is oxidized.

Solution: Since methanol is a primary alcohol, the first product should be an aldehyde (formaldehyde, CH_2O). This compound can then be oxidized, converting both H atoms attached to the -C(=O)- group into -OH groups (carbonic acid, HOC(=O)OH, or H_2CO_3), which decomposes into water and carbon dioxide.

Uses of Some Important Alcohols (12.4)

Table 12.3 in the textbook (p. 546) summarizes the names, formulas, and common uses of many alcohols. Many of these alcohols have only one -OH group in them like methanol (CH_3OH, a fuel and gasoline additive), ethanol (C_2H_5OH, the alcohol that humans drink, and a gasoline additive and chemical solvent), and 2-propanol (C_3H_7OH, rubbing alcohol). Others have two -OH groups (diols) like ethylene glycol ($C_2H_4(OH)_2$, automobile antifreeze) or three -OH groups (triols) like glycerol ($C_3H_5(OH)_3$, a moisturizer). Many biologically relevant molecules (like cholesterol and the male and female sex hormones) have -OH group in them, making them alcohols.

Nuclear Magnetic Resonance (*Tools of Chemistry*)

Nuclear magnetic resonance (NMR) used in chemistry laboratories and *magnetic resonance imaging* (MRI) used in hospitals utilize the same technique. It should be noted that the word *nuclear* simply means 'of or pertaining to the nucleus of an atom' and this is not a radioactive process at all. These techniques involve the use of radio waves (low-energy electromagnetic radiation) to distinguish between atoms in different chemical environments (bonded to different types of atoms or appearing in different compounds) or biological environments (bone, muscle, cancer cells). One advantage to these techniques is that they are non-destructive (i.e., the object is not destroyed in the process of making these measurements).

Carboxylic Acids and the Formation of Esters (12.5)

Carboxylic acids have the general formula RC(=O)OH. As their names imply, carboxylic acids can react with bases (like NaOH) to produce water and a salt containing a *carboxylate ion* (RCO_2^-). Because these acids have H atoms bonded to O atoms, they have very strong hydrogen bonding forces, which leads to low vapor pressures and high boiling points compared to other molecules of the same approximate size. Organic compounds containing two –C(=O)OH groups are called *dicarboxylic acids* or *diacids*, and are commonly used to make polymers.

When carboxylic acids react with alcohols, water and **esters** (organic compounds with the general formula RC(=O)OR') are formed. The R' group from the alcohol ends up attached to the O atom of the ester, and the R group from the carboxylic acid ends up attached to the C atom of the ester.

Triglyceride Formation (12.5)

An example of esters from a biological context is a class of compounds called **triglycerides**. These *triesters* are made from a single molecule of glycerol (a triol containing three -OH groups) and three molecules containing a carboxylic acid group (called *fatty acids*).

When esters react with water (a **hydrolysis** reaction), the esters are converted back into alcohols and carboxylic acids. This is just the reverse of the reaction used to create esters (the reaction listed above).

Saturated, Monounsaturated, and Polyunsaturated Fats (12.5)

When the carboxylic acids (fatty acids) used to make triglycerides have no C-C double bonds in their R groups, these acids are referred to as **saturated fats**, and if they have at least one C-C double bond, they are called **unsaturated fats**. Unsaturated fats can be split into two categories—acids with only one C-C double bond are called *monounsaturated fats*, and acids with more than one C-C double bond are called *polyunsaturated fats*. In general, saturated fats have structures that enable the triglycerides to form solids (*fats*) while unsaturated fats do not form solids as easily and can remain as liquids (*oils*). These unsaturated oils are healthier for humans to ingest because they are less likely to form solids that can cause blockages in the arteries or veins.

Addition and Condensation Polymerization (12.6)

Polymers are large chain-like molecules made from the reaction of smaller molecules, called **monomers**. There are two major types of polymers—**addition polymers**, where the monomers react and all of the atoms in the monomers end up in the polymer, and **condensation polymers**, where the monomers react but release small molecules (like water) that do not end up in the polymer.

Most addition polymers are made from monomers containing double bonds (usually C=C bonds), in which the doubly-bonded atoms break their double bond in favor of making two single bonds. The physical and chemical properties of addition polymers can be changed by changing the atoms or groups bonded to the atoms in the double bond (i.e., using different monomers). These properties can be further modified by making **copolymers** (polymers made from two different monomers containing double bonds).

Condensation polymers are usually made from reactions of molecules containing two different **functional groups** (groups of atoms that allow molecules to react with each other) like carboxylic acids, alcohols, or **amines** (organic compounds containing N atoms attached to C atoms). Condensation polymers can be made by the reaction carboxylic acids and alcohols to make esters (**polyester**). An analogous reaction is the reaction of carboxylic acids and amines to make **amides** (organic compounds with the general formula RC(=O)NR'R", where the R groups can be an H atom or an alkyl group) and **polyamide** chains.

Condensation polymers can be made from a single monomer (if it contains a carboxylic acid and either an alcohol or an amine), or from two different monomers (one with two carboxylic acids and one with either two alcohols, two amines, or an alcohol and an amine).

Examples of Synthetic Polymers (12.6)

Using the descriptions of addition and condensation polymers described above, you can draw the structure of a polymer if you are given its monomer, or if you are given the polymer you can deduce what the structure of the monomer must be.

Example: Can only one type of addition polymer be made from vinyl chloride ($ClCH=CH_2$)?

Solution: Since the C atoms have different atoms on them, different forms of poly(vinyl chloride) can be made. If the CH_2 part of one monomer reacts with the $CHCl$ group in another, you get a structure where the Cl atoms appear on every other C atom. If the CH_2 parts in the monomers react together (and the $CHCl$ parts react), you get a structure where the Cl atoms appear on two adjacent atoms, and then there are two C atoms without Cl atoms. Finally, they could react randomly, so that there is no pattern of Cl atoms.

—$(CH_2—CHCl—CH_2—CHCl)_n$—   or   —$(CH_2—CH_2—CHCl—CHCl)_n$—   or a random pattern.

Example: Can only one type of condensation polymer be made from methanediol ($HOCH_2OH$) and oxalic acid ($HOC(=O)C(=O)OH$)?

Solution: Methanediol has two -OH groups (a diol). Similarly, oxalic acid has two carboxylic acid groups (a diacid). Since the alcohols will react only with the carboxylic acids (and vice versa), you should expect only one kind of condensation polymer, in which the diol and the diacid alternate back and forth.

Identifying Functional Groups (12.4-12.6)

The reason organic chemists focus on the functional groups in a molecule is that they represent the most reactive portions of these molecules. Therefore it is important to be able to identify the functional groups (alcohols, aldehydes, aromatic rings, amines, carboxylic acids, esters, and ketones) in organic molecules.

Example: Identify the functional groups in tyrosine.

Solution: Tyrosine contains (from left to right), an alcohol (the -OH group), an aromatic ring (the six-membered ring of alternating C-C single and double bonds), an amine (the N atom which is bonded to 2 H atoms and a C atom), and a carboxylic acid (the –C(=O)OH group).

Types of Plastics and Recycling Efforts (12.6)

Although there are six major categories of plastics (polyethylene terephthalate, high-density polyethylene, poly(vinyl chloride), low-density polyethylene, polypropylene, and polystyrene), only the first two (PET and HDPE) are currently being recycled to a large extent. PET is widely used to make clear plastic drink bottles, and recycled PET is used in sleeping bags, carpet fibers, and tennis balls. HDPE is commonly used to make opaque drink bottles like milk, juice, and water containers. Recycled HDPE is used to make sportswear, insulating wrap for building construction, and shipping envelopes.

Polysaccharides (12.7)

Polysaccharides are polymers made from monomers of simple sugars (**monosaccharides**). Simple sugar molecules have the general formula $C_x(H_2O)_y$ (which explains why they are called *carbohydrates* since they can be view as being made of carbon and water). When monosaccharides are joined together, the -OH groups on the two sugars are replaced by a C-O-C bond (an *ether linkage* or **glycosidic linkage**).

Starches and cellulose are the most common natural polysaccharides. The major difference between starches (which humans can metabolize) and cellulose (which humans cannot) is the orientation of the glycosidic linkages—the human body can break the *cis* linkages in starches into its monomer units (metabolizing the monomer sugars) but cannot break the *trans* glycosidic linkages in cellulose.

Protein Structures and Peptide Linkages (12.7)

Proteins are one example of biologically relevant polymers (*biopolymers*). Proteins are made from monomer units called **amino acids**. Amino acids contain a carboxylic acid group (-C(=O)OH) and an amine group (-NH$_2$ group) attached to the same carbon (the **alpha carbon**). There are 20 amino acids used by living creatures that differ only by the R group attached to the alpha carbon atom, and all but one of these amino acids are chiral. When amino acid monomers react together, they make **amide linkages** (also called **peptide linkages**) by a condensation reaction that releases a water molecule.

Primary, Secondary, and Tertiary Protein Structures (12.7)

The 3-D shape of any protein is determined by the interactions of the amino acid groups within the protein chain. The **primary protein structure** is the sequence of amino acids that are linked together by amide

110

(peptide) linkages. All proteins have the same primary structure because they are all made of amino acid monomers linked together by amide bonds. The **secondary protein structure** is determined by hydrogen bonds formed between H and O atoms of the amide backbone (-NH-C(=O)-CR-) from amino acids that are not directly bonded together. Since all protein chains have the same amide backbone, the secondary structures formed are generally applicable to all proteins. Two of these secondary structures are the *alpha helix* (that looks like a corkscrew) and the *beta-pleated sheet* (that looks like a piece of paper that has been folded to have a zig-zag shape).

The **tertiary protein structure** is based on the interactions of the R groups attached to the alpha carbons of the amino acids. Some of these R groups are nonpolar (containing mostly C, H, and sometimes S atoms), while other groups are polar (containing N or O atoms along with the C and H atoms) or ionic (containing positive or negative charges). In general, the polar and ionic R groups tend to be strongly attracted to each other, and the protein will distort its shape to place these groups near each other. The nonpolar groups also tend to cluster together, forming hydrophobic regions within the protein. Since every protein has a different order of amino acids (and different 3-D arrangements of these R groups based on their attractions to each other), the tertiary structures of individual proteins tend to be very different.

Chapter Review — Key Terms

The key terms that were introduced in this chapter are listed below, along with the section in which they were introduced. You should understand these terms and be able to apply them in appropriate situations.

addition polymer (12.6)	disaccharide (12.7)	polyester (12.6)
aldehyde (12.4)	ester (12.5)	polymer (12.6)
alpha carbon (12.7)	functional group (12.3)	polypeptide (12.7)
amide (12.6)	glycosidic linkage (12.7)	polysaccharide (12.7)
amide linkage (12.6)	hydrolysis (12.5)	primary protein structure (12.7)
amine (12.6)	ketone (12.4)	saponification (12.5)
amino acid (12.7)	macromolecule (12.6)	saturated fats (12.5)
carboxylic acid (12.5)	monomer (12.6)	secondary protein structure (12.7)
catalyst (12.1)	monosaccharide (12.7)	tertiary protein structure (12.7)
catalytic cracking (12.1)	nuclear magnetic resonance	thermoplastics (12.6)
catalytic reforming (12.1)	(NMR) (p. 553)	thermosetting plastics (12.6)
condensation polymer (12.6)	octane number (12.1)	triglycerides (12.5)
condensation reaction (12.5)	peptide linkage (12.7)	unsaturated fats (12.5)
copolymer (12.6)	petroleum fractions (12.1)	
degree of polymerization, *n* (12.6)	polyamides (12.6)	

Practice Test

After you have finished studying the chapter and the homework problems, the following questions can serve as a test to determine how well you have learned the chapter objectives.

1. Explain why smaller molecules (those with lower molar masses and fewer number of C atoms) have lower boiling points than larger molecules (this property that is used in the fractional distillation of petroleum).

2. Why is it possible to have chemicals with octane numbers lower than 0 or higher than 100?

3. Explain the difference between coal, crude oil, and natural gas.

4. How would you make the following compounds using only alcohols?

(a)

$$:\ddot{C}l-\underset{\underset{H}{|}}{\overset{\overset{H}{|}}{C}}-\underset{\underset{H}{|}}{\overset{\overset{H}{|}}{C}}-\underset{\underset{H}{|}}{\overset{\overset{H}{|}}{C}}-\overset{\overset{:\ddot{O}:}{\|}}{C}-H$$

(c)

$$:\ddot{F}-\underset{\underset{:\ddot{F}:}{|}}{\overset{\overset{:\ddot{F}:}{|}}{C}}-\overset{\overset{:\ddot{O}:}{\|}}{C}-\ddot{O}H$$

(b)

$$H-\underset{\underset{H}{|}}{\overset{\overset{H}{|}}{C}}-\overset{\overset{:\ddot{O}:}{\|}}{C}-\underset{\underset{H}{|}}{\overset{\overset{H}{|}}{C}}-H$$

(d)

$$H-\underset{\underset{H}{|}}{\overset{\overset{H}{|}}{C}}-\overset{\overset{:\ddot{O}:}{\|}}{C}-\ddot{O}-\underset{\underset{H}{|}}{\overset{\overset{H}{|}}{C}}-H$$

5. How would you make the following triglyceride molecule? Classify each of the reactants you use as a saturated fat, a monounsaturated fat, a polyunsaturated fat, or none of these.

$$H_2C-\ddot{O}-\overset{\overset{:\ddot{O}:}{\|}}{C}-CH_2CH_2CH_2CH_2CH_2CH_2CH_2CH=CHCH_2CH_2CH_2CH_2CH_2CH_3$$

$$HC-\ddot{O}-\overset{\overset{:\ddot{O}:}{\|}}{C}-CH_2CH_2CH_2CH_2CH_2CH_2CH_2CH_2CH_2CH_2CH_2CH_2CH_2CH_2CH_3$$

$$H_2C-\ddot{O}-\overset{\overset{:\ddot{O}:}{\|}}{C}-CH_2CH_2CH_2CH_2CH_2CH_2CH_2CH=CHCH_2CH=CHCH_2CH=CHCH_2CH_3$$

6. Acetylene ($HC\equiv CH$) can also be used to make addition polymers. Draw the repeating units for two possible repeating units that can exist for polyacetylene.

7. Determine the monomer(s) that can be used to make the following polymers.

(a)

$$-\left(\underset{\underset{H}{|}}{\overset{\overset{CH_3}{|}}{C}}-\underset{\underset{H}{|}}{\overset{\overset{H}{|}}{C}}\right)_n-$$

(c)

$$-\left(\underset{\underset{H}{|}}{\overset{\overset{H}{|}}{N}}-\underset{\underset{H}{|}}{\overset{\overset{H}{|}}{C}}-\underset{\underset{H}{|}}{\overset{\overset{H}{|}}{C}}-\underset{\underset{H}{|}}{\overset{\overset{H}{|}}{N}}-\overset{\overset{:\ddot{O}:}{\|}}{C}-\underset{\underset{H}{|}}{\overset{\overset{H}{|}}{C}}-\overset{\overset{:\ddot{O}:}{\|}}{C}\right)_n-$$

(b)

$$-\left(\underset{\underset{:\ddot{F}:}{|}}{\overset{\overset{:\ddot{F}:}{|}}{C}}-\underset{\underset{:\ddot{F}:}{|}}{\overset{\overset{:\ddot{F}:}{|}}{C}}-\underset{\underset{H}{|}}{\overset{\overset{H}{|}}{C}}-\underset{\underset{H}{|}}{\overset{\overset{H}{|}}{C}}\right)_n-$$

(d)

$$-\left(\overset{\overset{:\ddot{O}:}{\|}}{C}-\underset{\underset{H}{|}}{\overset{\overset{H}{|}}{C}}-\underset{\underset{H}{|}}{\overset{\overset{H}{|}}{C}}-\ddot{O}\right)_n-$$

8. Identify the functional groups in the following molecules.

(a) (benzene ring structure with a carboxyl group $-C(=\ddot{O}):\ddot{O}H$ and $:\ddot{O}H$ substituent)

(b) (benzene ring structure with a $-CH=CH-CH=\ddot{O}:$ group)

9. How many trimers (three-chain polymers) can be made from the amino acids alanine, methionine, and serine? Draw two of them. The structures of these amino acids are in the textbook (p. 579).

10. Both tyrosine and phenylalanine have benzene groups in their R group. Explain why tyrosine's R group is considered to be polar but phenylalanine's R group is considered to be nonpolar.

112

Chapter 13
Chemical Kinetics: Rates of Reactions

The goal of this chapter is to introduce the concept of **chemical kinetics**, the study of the time-dependence of chemical reactions (how long it takes for reactions to occur). This chapter introduces reaction rates and how these rates are affected by reactant concentrations, time, molecular orientations, and temperature. This chapter also introduces reaction mechanisms, which describe the nanoscale interactions of the reactants as they are converted into products. Finally, the chapter introduces the concept of catalysts and some examples of industrial and biological catalysts (enzymes).

Calculating Average Rates (13.1)

The **rate** of a chemical reaction (also called the **reaction rate**) is simply the speed at which a chemical reaction occurs. The rate of travel in a car (its *speed* or *velocity*) is determined by dividing the change in distance by the change in time ($v = \Delta d/\Delta t$); for a chemical reaction, the reaction rate is calculated as the change in concentration over the change in time.

$$\text{Rate}_X = \frac{\Delta[X]}{\Delta t} = \frac{[X]_f - [X]_i}{t_f - t_i}$$

Example: Hydrogen and chlorine react according the following equation: $H_2(g) + Cl_2(g) \rightarrow 2\,HCl(g)$. The boxes below show how H_2 and Cl_2 react after 1 second. Calculate the reaction rate for each chemical (assume that each molecule represents a concentration of 1 M).

◆ = H
□ = Cl

Solution: The concentrations at $t = 0$ s are 4 M for H_2, 3 M for Cl_2 and 0 M for HCl. At $t = 1$ s, the concentrations are 1 M for H_2, 0 M for Cl_2 and 6 M for HCl.

$$\text{Rate}_{H_2} = \frac{\Delta[H_2]}{\Delta t} = \frac{[H_2]_f - [H_2]_i}{t_f - t_i} = \frac{1\,M - 4\,M}{1\,s - 0\,s} = -3\,\frac{M}{s}$$

$$\text{Rate}_{Cl_2} = \frac{\Delta[Cl_2]}{\Delta t} = \frac{[Cl_2]_f - [Cl_2]_i}{t_f - t_i} = \frac{0\,M - 3\,M}{1\,s - 0\,s} = -3\,\frac{M}{s}$$

$$\text{Rate}_{HCl} = \frac{\Delta[HCl]}{\Delta t} = \frac{[HCl]_f - [HCl]_i}{t_f - t_i} = \frac{6\,M - 0\,M}{1\,s - 0\,s} = +6\,\frac{M}{s}$$

A negative sign for the reaction rate tells you that the amount of chemical is decreasing over time because it is reacting; a positive sign tells you that the amount of chemical is increasing over time because it is being made. Also, note that the amounts of H_2 and Cl_2 reacting (their rates) are the same, and the amount of HCl being made is twice the amount of H_2 and Cl_2 reacting. These values are consistent with the coefficients in the balanced equation (when 3 H_2 and 3 Cl_2 molecules react, 6 HCl molecules will be produced).

Once you have calculated the rate for any chemical in a reaction, you can use this value and the balanced equation to determine the reaction rates for any other chemical in the reaction. The textbook introduces the following mathematical equation (for the balanced equation $a\,A + b\,B \rightarrow c\,C + d\,D$), but these calculations are easier to understand using simple mole ratios from the balanced equation.

$$-\frac{1}{a}\frac{\Delta[A]}{\Delta t} = -\frac{1}{b}\frac{\Delta[B]}{\Delta t} = +\frac{1}{c}\frac{\Delta[C]}{\Delta t} = +\frac{1}{d}\frac{\Delta[D]}{\Delta t}$$

Example: If NH₃ is made at a rate of 12 M/s from this reaction, what are the rates of N_2H_4 and N_2?

$$3\,N_2H_4(g) \rightarrow 4\,NH_3(g) + N_2(g)$$

Solution: Use mole ratios from the balanced equation to convert the rate of NH_3 to the rates for N_2H_4 and H_2 (it helps to write the units of concentration as mol_X/L instead of M). In these calculations, the mole ratios are written with negative signs when the chemical is reacting (and being used up). Also, note that the rates of the products (NH_3 and N_2) are positive, and the rate of the reactant (N_2H_4) is negative.

$$Rate_{N_2} = +12\,\frac{mol\,NH_3/L}{s} \times \frac{-3\,mol\,N_2H_4}{4\,mol\,NH_3} = -9\,\frac{mol\,N_2H_4/L}{s} = -9\,\frac{M}{s}$$

$$Rate_{H_2} = +12\,\frac{mol\,NH_3/L}{s} \times \frac{1\,mol\,N_2}{4\,mol\,NH_3} = +3\,\frac{mol\,N_2/L}{s} = +3\,\frac{M}{s}$$

The Effect of Concentration on Reaction Rates and Rate Laws (13.2)

For most chemical reactions, the **initial rate** (the rate when the reaction begins) is the largest rate, and the magnitude of the rate for any chemical in the reaction slowly decreases (and may reach 0 M/s). This is because most reaction rates depend on the concentration of the reactants but not the products. A **rate law** is a mathematical equation showing the concentration dependence of the reaction rate. The general formula of the rate law (for the balanced equation $a\,A + b\,B \rightarrow c\,C + d\,D$) is: $Rate = k[A]^m[B]^n$. The k term is the **rate constant**. Note that only the reactant concentrations appear in the rate law and also notice that the exponents on the reactant concentrations are <u>not</u> always the same as the reaction coefficients (they *can* be the same but they are not *required* to be the same).

Example: Use the kinetics data listed below to answer the following questions.
(a) What is the general formula for the rate law? What is the specific rate law for this reaction?
(b) What is the rate constant for this reaction?

$$2\,NO(g) + 2\,CO(g) \rightarrow N_2(g) + 2\,CO_2(g)$$

run #	[NO] (M)	[CO] (M)	initial rate (M/s)
1	2.40	3.90	0.0076
2	2.40	1.30	0.0076
3	1.20	1.30	0.0019

Solution:
(a) The general formula for the rate law is: $Rate = k[NO]^m[CO]^n$. To determine the value of m, look at two runs where [NO] is the only concentration that changes (runs 2 and 3). When the concentration is doubled from run 3 to run 2, the rate quadruples. Since $[NO]^m$ is proportional to the rate, these data give you the relationship that $[2]^m = 4$ and $m = 2$.
To determine the value of n, look at two runs where [CO] is the only concentration that changes (runs 1 and 2). When the concentration is tripled from run 2 to run 1, the rate doesn't change (multiplied by 1). Since $[CO]^n$ is proportional to the rate, these data give you the relationship that $[3]^n = 1$ and $n = 0$. The rate law is $Rate = k[NO]^2[CO]^0 = k[NO]$ (since anything raised to the zero power is 1). Note that the exponents of the rate law do not match the reactant coefficients (CO has a coefficient of 2, but an exponent of 0).
(b) To solve for the rate constant, rearrange the rate law. Since k is a constant, its value will be the same no matter what run you use (and if you calculate two values, they should be the same). Although rates always have units of M/s and concentrations are always M, rate constants do not always have the same units.

$$Rate = k[NO]^2, \quad k = \frac{Rate}{[NO]^2} = \frac{0.0076\,M/s}{[2.40\,M]^2} = 1.3 \times 10^{-3}\,M^{-1}\,s^{-1}$$

Determining Reaction Orders and the Integrated Rate Law (13.3)

The exponents for the concentrations of each reactant in the rate law are called the **order of the reaction** for that reactant. The rate law in the previous example is second-order in NO ($m = 2$) and zero-order in CO ($n = 0$). The **overall reaction order** is simply the sum of all of the individual orders (in this case, the reaction is second-order overall because $m + n = 2 + 0 = 2$). The importance of the overall reaction order is that reactions with the same overall reaction order follow the same mathematical relationships.

Chemists use a form of the rate law (the *integrated rate law*) to show the time-dependence of the reactant concentration. This table shows the rate laws, integrated rate laws, and linear plot information for *zero-order*, *first-order*, and *second-order reactions* (reactions with overall orders of 0, 1, and 2). The rates are listed with a negative sign because $\Delta[A]/\Delta t$ is negative (the reactant is disappearing), but concentrations must be positive. The symbols $[A]_0$ and $[A]_t$ represent the reactant concentrations at time 0 and t.

order	rate law	integrated rate law	linear plot	slope	y-intercept
0	$-\dfrac{\Delta[A]}{\Delta t} = k[A]^0 = k$	$[A]_t = [A]_0 - kt$	$[A]$ vs. t	$-k$	$[A]_0$
1	$-\dfrac{\Delta[A]}{\Delta t} = k[A]^1 = k[A]$	$\ln[A]_t = \ln[A]_0 - kt$	$\ln[A]$ vs. t	$-k$	$\ln[A]_0$
2	$-\dfrac{\Delta[A]}{\Delta t} = k[A]^2$	$\dfrac{1}{[A]_t} = \dfrac{1}{[A]_0} + kt$	$\dfrac{1}{[A]}$ vs. t	$+k$	$\dfrac{1}{[A]_0}$

One way to determine the reaction order is to plot the concentration data versus time. If the $[A]$ vs. time plot is linear, the reaction is overall zero-order; if the $\ln[A]$ vs. time plot is linear, the reaction is overall first-order; and if the $1/[A]$ vs. time plot is linear, the reaction is overall second-order.

Calculations Using Integrated Rate Laws (13.3)

Once you know the overall order of a chemical reaction, you can use the appropriate integrated rate law to calculate the rate constant, concentrations, or times.

Example: For the data listed below for the reaction $3O_2(g) \rightarrow 2\,O_3(g)$, a plot of $1/[O_2]$ vs. time is linear.

(a) What is the overall order and rate law for this reaction?
(b) What is the rate constant for this reaction?
(c) What will the concentration be after 90 s?
(d) At what time will the concentration be 0.100 M?

t (s)	$[O_2]$ (M)
0	0.200
15	0.192
30	0.185

Solution:
(a) Since $1/[O_2]$ versus t is linear, this reaction is second-order and the rate law is Rate $= -\Delta[O_2]/\Delta t = k[O_2]^2$.
(b) To solve for the rate constant, plug two sets of data into the integrated rate law.

$$\frac{1}{[O_2]_t} = \frac{1}{[O_2]_0} + kt, \quad kt = \frac{1}{0.192\ M} - \frac{1}{0.200\ M} = 0.21\ M^{-1}, \quad k = \frac{0.21\ M^{-1}}{15\ s} = 0.014\ M^{-1}\ s^{-1}$$

(c) To solve for concentration at 90 s, use the $t = 0$ data and the rate constant.

$$\frac{1}{[O_2]_{90}} = \frac{1}{[O_2]_0} + kt = \frac{1}{0.200\ M} + (0.014\ M^{-1}\ s^{-1})(90\ s) = 6.25\ M^{-1}, \quad [O_2]_{90} = \frac{1}{6.25\ M^{-1}} = 0.160\ M$$

(d) To find the time needed to make the concentration 0.100 M, use the $t = 0$ data and the rate constant.

$$kt = \frac{1}{[O_2]_t} - \frac{1}{[O_2]_0} = \frac{1}{0.100\ M} - \frac{1}{0.200\ M} = 5.00\ M^{-1}, \quad t = \frac{5.00\ M^{-1}}{0.014\ M^{-1}\ s^{-1}} = 360\ s$$

So, it would take 360 s (6.0 minutes) for the concentration to reach 0.100 M.

For some chemical reactions (especially first-order reactions), the **half-life** ($t_{1/2}$) for the reaction is reported. The half-life is the time needed for half of the reactant to disappear; it is also the time needed for half of the reactant to be remaining.

Example: What is the relationship between $t_{1/2}$ and k for first-order reactions?

Solution: To determine this relationship, use the definitions of $[A]_0$ and $t_{1/2}$, and the integrated rate law for the first-order reaction. If the concentration at $t = 0$ is $[A]_0$, then the concentration at $t = t_{1/2}$ will be $1/2[A]_0$.

$$\ln[A]_t = \ln[A]_0 - kt, \text{ or } \ln\frac{[A]_t}{[A]_0} = -kt$$

$$\ln\frac{1/2[A]_0}{[A]_0} = \ln(1/2) = -\ln 2 = -kt_{1/2}, \ kt_{1/2} = \ln 2 = 0.693, \ t_{1/2} = \frac{\ln 2}{k}$$

Unimolecular and Bimolecular Reactions (13.4)

Unimolecular reactions are reactions that occur when a single chemical species (atom, molecule, or ion) reacts on its own to cause the chemical reaction. Examples of chemical reactions that are unimolecular include $CH_3NC(g) \rightarrow CH_3CN(g)$ and *cis*-2-butene $\rightarrow$ *trans*-2-butene. **Bimolecular reactions**, on the other hand, occur when two different chemical species collide with each other and react. Examples of bimolecular reactions include $H_2(g) + Cl_2(g) \rightarrow 2 HCl(g)$ and $2 NO(g) \rightarrow N_2(g) + O_2(g)$.

Energy Diagrams of Chemical Reactions (13.4)

When reactants undergo a chemical reaction, some of the bonds within the reactants must be broken before the bonds in the products can be formed, and this requires energy. An *energy diagram* (which is also called an *energy profile*) plots the energy of the chemicals in a reaction as the reactants change into products.

The energy difference between the products and the reactants is $\Delta H°$ for the reaction. If the products are lower in energy than the reactants, the reaction is exothermic and vice versa. The energy needed to get to the top of the 'energy hill' is called the **activation energy** (E_a), and is the energy needed for the reactants to be converted into products. The activation energy can be determined for the reaction in either direction as E_a(forward) or E_a(reverse). The chemical at the highest energy in the diagram is the **transition state** (or **activated complex**) and is the chemical species that is intermediate in structure and energy between the reactant and product molecules.

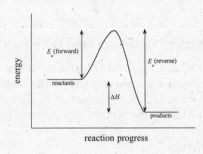

reaction progress

Example: When the molecules involved in the reaction $H_2(g) + Cl_2(g) \rightarrow 2 HCl(g)$ are frozen in time, the following species are seen. Which of these represent possible transition states for this reaction?

♦ = H
□ = Cl
   I ♦□□♦   **II** □♦   **III** ♦□   **IV** □□♦♦

Solution: For a species to be a successful transition state, you must be able to break it into the reactant molecules and the product molecules. Chemical I can break apart into H_2 and Cl_2 but cannot form 2 HCl without rearranging the atoms. Chemical II can break apart into 2 HCl but cannot form H_2 and Cl_2. Since chemicals III and IV can break apart into either H_2 and Cl_2 or into 2 HCl, they could be transition states for this reaction, but chemicals I and II could not be transition states for this reaction.

116

The Effect of Temperature on Reaction Rates and Rate Constants (13.5)

When the temperature of a system is increased, the rates of all chemical reactions tend to increase. This is because when the temperature is increased, the particles have more kinetic energy and more of the reactants will have enough energy to reach the top of the 'energy hill' (E_a) where they can be converted to products. The temperature-dependence of a reaction rate (Rate = $k[A]^m[B]^n$) appears in the rate constant—rate constants are only constant when the temperature remains constant; if the temperature changes, so does the k value. The mathematical formula relating the rate constant and temperature is called the **Arrhenius equation**— $k = Ae^{-E_a/RT}$. The A term is the **frequency factor**, which depends on the number of molecular collisions and the fact that not every collision between reactants leads to a successful chemical reaction.

The activation energy and the frequency factor for most chemical reactions are determined by measuring the rate constant at several different temperatures and using the Arrhenius equation to solve for A and E_a. If you have two data points, you can use the two-point Arrhenius equation listed below to solve for E_a. If you have more than two points, you can plot $\ln k$ versus $1/T$ (in Kelvin units). The slope of the line will be $-E_a/R$ (where R is the gas constant) and the y-intercept will be $\ln A$.

$$\ln \frac{k_1}{k_2} = \frac{E_a}{R}\left(\frac{1}{T_2} - \frac{1}{T_1}\right)$$

Example: When the rate constants for the reaction $3\,O_2(g) \rightarrow 2\,O_3(g)$ are plotted versus temperature, the slope is -4.23×10^4 K. The rate constant at 517°C is $k = 0.0042$ $M^{-1}\,s^{-1}$.
(a) What are the values for the activation energy and the frequency factor for this reaction?
(b) What is the value for the rate constant of this reaction at 565°C?

Solution:
(a) E_a can be calculated from the slope, and A can be calculated using the Arrhenius equation.

$$\text{slope} = -\frac{E_a}{R} = -4.23 \times 10^4 \text{ K}, \quad E_a = (-4.23 \times 10^4 \text{ K})\left(-8.314 \frac{J}{mol\ K}\right) = 3.52 \times 10^5 \frac{J}{mol}$$

$$k = Ae^{-E_a/RT}, \quad A = ke^{+E_a/RT} = (0.0042 \text{ M}^{-1}\text{ s}^{-1})e^{\frac{3.52 \times 10^5 \text{ J/mol}}{(8.314 \text{ J/mol K})(790.15 \text{ K})}} = 7.8 \times 10^{20} \text{ M}^{-1}\text{ s}^{-1}$$

(b) To solve for the rate constant at $T = 838.15$ K (565°C), use the two-point Arrhenius equation.

$$\ln \frac{k_{565}}{k_{517}} = \frac{E_a}{R}\left(\frac{1}{T_{517}} - \frac{1}{T_{565}}\right) = \frac{3.52 \times 10^5 \text{ J/mol}}{8.314 \text{ J/mol K}}\left(\frac{1}{790.15 \text{ K}} - \frac{1}{838.15 \text{ K}}\right) = 3.069$$

$$\ln \frac{k_{565}}{k_{517}} = 3.069, \quad \frac{k_{565}}{k_{517}} = e^{3.069} = 21.5, \quad k_{565} = 21.5 \times k_{517} = 21.5(0.0042 \text{ M}^{-1}\text{ s}^{-1}) = 0.090 \text{ M}^{-1}\text{ s}^{-1}$$

Deriving Rate Laws for Elementary Reactions (13.6)

An **elementary reaction** is a reaction that occurs in one step. Chemists use an extension of the kinetic-molecular theory called *collision theory* to provide a nanoscale explanation of the reaction rate law derived from empirical data. Collision theory states that the rate law of an elementary reaction is simply the rate constant times the concentrations of the reactants in the reaction. For unimolecular reactions, the reaction rate is first-order (Rate = $k[A]$) and requires the single reactant to convert into the products on its own; for bimolecular reactions, the reaction rate is second-order (Rate = $k[A][B]$, or $k[A]^2$ if two A molecules react). If the exponents in the rate law don't match the reactant coefficients, the reaction cannot be elementary.

Example: Write the rate law for the following elementary (one-step) reactions.
(a) $N_2O_4(g) \rightarrow 2\,NO_2(g)$, (b) $C_2H_2(g) + H_2(g) \rightarrow C_2H_4(g)$, (c) $2\,NO(g) \rightarrow N_2(g) + O_2(g)$

Solution: For elementary reactions, the rate law is the rate constant times the reactant concentrations.
(a) Rate = $k[N_2O_4]$, (b) Rate = $k[C_2H_2][H_2]$, (c) Rate = $k[NO][NO] = k[NO]^2$.

Multiple-Step Reaction Mechanisms (13.7)

Many chemical reactions do not occur in a single elementary step. This is especially true for reactions involving three or more reactant molecules. Although there are millions of molecular collisions within a gas or an aqueous solution, the probability that three molecules will collide at the same instant with the exact orientation in space needed for the chemical reaction to take place is incredibly small. Instead, these reactions tend to occur by a series of unimolecular or bimolecular reactions that occur one after the other.

For multi-step reactions, the rate laws for these series of reactions (called the **reaction mechanism**) are simply the rate constant times the concentrations of the reactants in the slowest, or **rate-limiting step**. Rate laws for multi-step mechanisms should contain concentrations of reactants (and occasionally products), but not **intermediates** (also called **reaction intermediates**, which are chemicals that are made in one step of the mechanism and then used up in another step). If the rate law contains an intermediate, you must perform a calculation to replace its concentrations with the concentrations of reactants and/or products.

Example: The rate law for the reaction $2 NO(g) + O_2(g) \rightarrow 2 NO_2(g)$ was experimentally measured to be Rate = $k[NO]^2[O_2]$. When performing the experiment, the compound $N_2O_2(g)$ was detected.
(a) Why is it unlikely that this is an elementary (one-step) reaction?
(b) Propose a mechanism for this reaction and identify the intermediates.
(c) What is the rate-limiting step of this reaction?

Solution:
(a) Since this reaction involves three reactant molecules, it is very unlikely that the two NO molecules and the O_2 molecule will collide at the same time and hit with enough energy and the proper orientation to make two NO_2 molecules. It is much more likely that two of the molecules will hit and react, and these products can then react with the last molecule (i.e., two bimolecular steps instead of one *termolecular* step).
(b) Since N_2O_2 was seen, it makes sense to propose it as an intermediate from the reaction of 2 NO molecules. The second step will simply be the reaction of N_2O_2 (the only intermediate) with the O_2 molecule.

Step 1: $2 NO(g) \rightarrow N_2O_2(g)$

Step 2: $N_2O_2(g) + O_2(g) \rightarrow 2 NO_2(g)$

c) If step 1 was rate-limiting, the rate law would be Rate = $k[NO]^2$, which does not match the actual rate law. So, step 1 cannot be the rate-limiting step. If step 2 was rate-limiting, the rate law would be Rate = $k[N_2O_2][O_2]$, which also doesn't match the actual rate law.
However, intermediates should not appear in the rate law. If step 1 is faster, N_2O_2 would be produced quickly and its concentration would build up. Step 2 must have a larger E_a value than step 1, so it is more energetically favorable for N_2O_2 to form 2 NO (the reverse of step 1) than to react with O_2 (step 2). This leads to an *equilibrium reaction* (discussed in detail in the next chapter) where the forward and reverse rates are equal (Rate$_f$ = Rate$_r$).

$$\text{Step 1': } 2 NO(g) \underset{k_{-1}}{\overset{k_1}{\rightleftharpoons}} N_2O_2(g)$$

$$\text{Rate}_f = k_1[NO]^2 = \text{Rate}_r = k_{-1}[N_2O_2]. \text{ So, } [N_2O_2] = \frac{k_1}{k_{-1}}[NO]^2$$

$$\text{Rate} = k_2[N_2O_2][O_2] = k_2\frac{k_1}{k_{-1}}[NO]^2[O_2] = k'[NO]^2[O_2]$$

Since this formula matches the actual rate law, step 2 is the rate-limiting step.

Identifying Consistent Mechanisms from the Rate Law (13.7)

If you are given different mechanisms for the same reaction, you can use the experimentally-determined rate law to determine which of these mechanisms could be possible. If the rate laws for each step in a mechanism do not match the actual rate law, this mechanism cannot be correct; if the rate law for one step in a mechanism matches the actual rate law, the mechanism <u>could</u> be correct but you cannot prove that it <u>is</u> the correct mechanism.

<u>Example</u>: The rate law for the reaction $CH_3OH(aq) + H^+(aq) + Cl^-(aq) \rightarrow CH_3Cl(aq) + H_2O(\ell)$ was experimentally measured to be Rate = $k[CH_3OH][Cl^-]$.
(a) Identify the intermediates in the following two proposed mechanisms.
(b) Which of these mechanisms is consistent with the actual rate law? What is the rate-limiting step?

Mechanism I	Mechanism II
$CH_3OH(aq) + Cl^-(aq) \rightarrow CH_3Cl(aq) + OH^-(aq)$	$CH_3OH(aq) + H^+(aq) \rightarrow CH_3OH_2^+(aq)$
$H^+(aq) + OH^-(aq) \rightarrow H_2O(\ell)$	$CH_3OH_2^+(aq) + Cl^-(aq) \rightarrow CH_3Cl(aq) + H_2O(\ell)$

<u>Solution</u>:
(a) In mechanism I, $OH^-(aq)$ is the only intermediate; in mechanism II, $CH_3OH_2^+(aq)$ is the intermediate.
(b) For mechanism I, if the first step was rate-limiting, the rate law would be Rate = $k[CH_3OH][Cl^-]$, which matches the actual rate law. If the second step was rate-limiting, the rate law would be Rate = $k[H^+][OH^-]$, but $[OH^-]$ should not appear in the rate law since it is an intermediate. Using the equilibrium conditions for the first step (Rate$_f$ = Rate$_r$), the concentration of $[OH^-]$ and the rate law are:

$$\text{Rate}_f = k_1[CH_3OH][Cl^-] = \text{Rate}_r = k_{-1}[CH_3Cl][OH^-]. \text{ So, } [OH^-] = \frac{k_1}{k_{-1}}\frac{[CH_3OH][Cl^-]}{[CH_3Cl]}$$

$$\text{Rate} = k_2[H^+][OH^-] = k_2\frac{k_1}{k_{-1}}[H^+]\frac{[CH_3OH][Cl^-]}{[CH_3Cl]} = k'\frac{[CH_3OH][H^+][Cl^-]}{[CH_3Cl]}$$

For mechanism II, if the first step was rate-limiting, the rate law would be Rate = $k[CH_3OH][H^+]$, which doesn't match the actual rate law. If the second step was rate-limiting, the rate law would be Rate = $k[CH_3OH_2^+][Cl^-]$, but $[CH_3OH_2^+]$ should not appear in the rate law since it is an intermediate. Using the equilibrium conditions for the first step, the concentration of $[CH_3OH_2^+]$ and the rate law are:

$$\text{Rate}_f = k_1[CH_3OH][H^+] = \text{Rate}_r = k_{-1}[CH_3OH_2^+]. \text{ So, } [CH_3OH_2^+] = \frac{k_1}{k_{-1}}[CH_3OH][H^+]$$

$$\text{Rate} = k_2[CH_3OH_2^+][Cl^-] = k_2\frac{k_1}{k_{-1}}[CH_3OH][H^+][Cl^-] = k'[CH_3OH][H^+][Cl^-]$$

Mechanism I is consistent with the actual rate law, but only if the first step is the rate-limiting step; mechanism II is not consistent with the actual rate law, so it is not a possible reaction mechanism.

The Effect of Catalysts on Reaction Rates (13.8)

Catalysts are chemicals that cause a reaction to happen more quickly (increases the reaction rate). If the catalyst exists in the same state of matter as the reactants and products (usually gas or aqueous), it is called a **homogeneous catalyst**; if the catalyst exists in a different state of matter than the reactants and products (usually solid), it is called a **heterogeneous catalyst**. In reaction mechanisms, catalysts appear as reactants that are used up but are then regenerated in a subsequent step of the mechanism.

The way that a catalyst increases the rate of a chemical reaction is by providing an alternative mechanism (pathway) that has a lower-energy requirement (lower E_a value). Because it takes less energy for the catalyzed reaction to occur, more reactant molecules have enough energy to reach the top of the 'energy hill' in order to react. Although the addition of a catalyst lowers the activation energy for the forward and reverse reactions, it does not change the $\Delta H°$ value for the reaction since the energy of the reactants and the products are still the same. The energy diagrams for an uncatalyzed and a catalyzed reaction appear below.

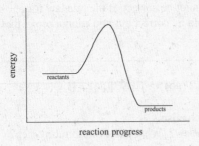 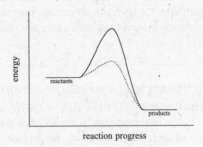

Enzyme-Catalyzed Reactions (13.9)

The rates of many chemical reactions within biological systems are controlled by the presence of biological catalysts, called **enzymes**. The reactant molecule in a biological system is called the **substrate**. When the substrate interacts with the enzyme, it binds to the enzyme at the **active site** (where the reaction occurs) and an **enzyme-substrate complex** is formed. Often the molecular shape of the enzyme and/or the substrate changes when the enzyme-substrate complex is formed; this is called **induced fit** because the shapes of the molecules must be forced (induced) to change in order to make the complex.

Enzymes tend to be highly selective, catalyzing only one type of chemical reaction. Although the rates of enzyme-catalyzed reactions generally increase with increasing temperature, enzyme reactions ultimately fail completely (and revert to the slower rates of the uncatalyzed reaction) if they are heated too much. This is because the increased temperature causes the enzyme to become **denatured**, in which the 3-D structure of the enzyme changes, and this results in an enzyme that has lost all of its catalytic behavior.

Chemicals that slow the rate of a reaction are called **inhibitors**. In biological systems, inhibitors are usually chemicals that interfere with the catalytic behavior of an enzyme, deactivating the enzyme and forcing the reaction to occur at the slower rate of the uncatalyzed reaction.

Examples of Industrial Catalyzed Reactions (13.10)

Most industrial processes use catalysts to speed up the rates of these reactions, and so that they require less energy to proceed. Examples of catalysts described in the textbook are rhodium(III) iodide (RhI_3) to produce acetic acid from methanol and carbon monoxide, a platinum-nickel(II) oxide mixture (Pt-NiO) to produce nitrogen and oxygen from nitrogen monoxide, and platinum-rhodium mixture (Pt-Rh) to produce methanol from carbon monoxide and hydrogen.

Chapter Review — Key Terms

The key terms that were introduced in this chapter are listed below, along with the section in which they were introduced. You should understand these terms and be able to apply them in appropriate situations.

activated complex (13.4)
activation energy (13.4)
active site (13.9)
Arrhenius equation (13.5)
average reaction rate (13.1)
bimolecular reaction (13.4)
chemical kinetics (*Introduction*)
cofactor (13.9)
denaturation (13.9)

elementary reaction (13.4)
enzyme (13.9)
enzyme-substrate complex (13.9)
frequency factor (13.5)
half-life (13.3)
heterogeneous catalyst (13.10)
heterogeneous reaction (13.1)
homogeneous catalyst (13.8)
homogeneous reaction (13.1)

induced fit (13.9)
inhibitor (13.9)
initial rate (13.2)
instantaneous reaction rate (13.1)
intermediate (13.7)
order of reaction (13.3)
overall reaction order (13.3)
rate (13.1)
rate constant (13.2)

rate law (13.2)
rate-limiting step (13.7)
reaction intermediate (13.7)

reaction mechanism (13.7)
reaction rate (13.1)
steric factor (13.4)

substrate (13.9)
transition state (13.4)
unimolecular reaction (13.4)

Practice Test

After you have finished studying the chapter and the homework problems, the following questions can serve as a test to determine how well you have learned the chapter objectives.

1. Dinitrogen pentoxide decomposes according to the following reaction:
   $2 N_2O_5(g) \rightarrow 4 NO_2(g) + O_2(g)$.
   (a) Calculate the average rate of N_2O_5 decomposition from 0 to 15 s, and from 45 to 60 s.
   (b) Explain why the values for these two rates are different.
   (c) What is the rate of NO_2 formation from 0 to 15 s? What is the rate of O_2 formation from 45 to 60 s?

t (s)	$[N_2O_5]$ (M)
0	0.500
15	0.430
30	0.370
45	0.318
60	0.274

2. Use the kinetic data to the right to answer the following questions regarding the reaction
   $CHCl_3(g) + Cl_2(g) \rightarrow CCl_4(g) + HCl(g)$.
   (a) What is the rate law for this reaction?
   (b) What is the overall order for this chemical reaction?
   (c) What is the rate constant for this chemical reaction?
   (d) What is the initial reaction rate for run 4?

run #	$[CHCl_3]$ (M)	$[Cl_2]$ (M)	initial rate (M/s)
1	0.110	0.250	0.00132
2	0.330	0.250	0.00396
3	0.110	1.000	0.00264
4	0.180	0.360	_____

3. For the data listed below for the reaction $NaCl(s) \rightarrow Na^+(aq) + Cl^-(aq)$, a plot of *mass* vs. t is linear.

   (a) What is the overall order and rate law for this reaction?
   (b) What is the rate constant for this reaction?
   (c) What is the mass of NaCl after 50.0 s?
   (d) At what time will the mass of NaCl be 0.00 g?
   (e) Calculate the mass lost every 12.0 s. How are these values related?

t (s)	m(NaCl) (g)
0.0	2.25
12.0	2.00
24.0	1.75
36.0	1.50

4. For the data listed below for the reaction $2 O_3(g) \rightarrow 3 O_2(g)$, a plot of $\ln[O_3]$ vs. t is linear.

   (a) What is the overall order and rate law for this reaction?
   (b) What is the rate constant for this reaction?
   (c) What is the concentration of O_3 after 10.0 s?
   (d) At what time will the concentration of O_3 be 1.00 M?
   (e) Calculate the proportion of O_3 lost after every 8.0 s. How are these values related?

t (s)	$[O_3]$ (M)
0.0	3.00
8.0	2.40
16.0	1.92
24.0	1.54

5. Use the integrated rate law formulas to determine the relationship between k and $t_{1/2}$ for zero-order reactions and for second-order reactions. Why aren't these value tabulated in reference books like the half-lives for first-order reactions?

6. For the reaction in question 2 above, $CHCl_3(g) + Cl_2(g) \rightarrow CCl_4(g) + HCl(g)$, the rate constant k is $0.000201 M^{-1/2} s^{-1}$ at 333°C and $0.112 M^{-1/2} s^{-1}$ at 425°C.
   (a) Determine values for the activation energy and frequency factor for this reaction.
   (b) Explain why the reaction rate is larger at the higher temperature.
   (c) What is the value for the rate constant at 400°C?

7. One possible mechanism for the reaction that appears in the questions 2 and 6, $CHCl_3(g) + Cl_2(g) \rightarrow CCl_4(g) + HCl(g)$, appears below. The rate law was determined to be Rate $= k[CHCl_3][Cl_2]^{1/2}$.
   (a) What is the final step in this mechanism?
   (b) What are the intermediates formed in this mechanism?
   (c) Which of these steps is the rate-limiting step?

   Step 1: $Cl_2(g) \rightarrow 2\ Cl^{\cdot}(g)$
   Step 2: $CHCl_3(g) + Cl^{\cdot}(g) \rightarrow CCl_3^{\cdot}(g) + HCl(g)$
   Step 3: $CCl_3^{\cdot}(g) + Cl_2(g) \rightarrow CCl_4(g) + Cl^{\cdot}(g)$
   Step 4:   _____

8. The experimentally-determined rate law for the reaction $2\ NO(g) + 2\ CO(g) \rightarrow N_2(g) + 2\ CO_2(g)$ was found to be Rate $= k[NO]^2[CO]$.
   (a) What are the individual and overall orders for this reaction?
   (b) What are the units for the rate constant k?
   (c) What are the intermediates present in each mechanism?
   (d) Which of the mechanisms listed below are consistent with the actual rate law? What is the rate-limiting step for these mechanisms (all chemicals are in the gas phase)?

Mechanism I	Mechanism II	Mechanism III
$2\ NO \rightarrow N_2O_2$	$2\ CO \rightarrow C_2O_2$	$NO + CO \rightarrow CNO_2$
$N_2O_2 + CO \rightarrow N_2O + CO_2$	$C_2O_2 + NO \rightarrow NCO + CO_2$	$CNO_2 + NO \rightarrow N_2 + CO_3$
$N_2O + CO \rightarrow N_2 + CO_2$	$NCO + NO \rightarrow N_2 + CO_2$	$CO_3 + CO \rightarrow 2\ CO_2$

9. Suppose you have a reaction that involves a two-step mechanism. Draw the energy diagram for this reaction assuming the first step is rate-limiting and then assuming the second step is rate-limiting.

10. What is the difference between a homogeneous and a heterogeneous catalyst? Why do industrial chemists prefer to use heterogeneous catalysts in their reactions?

122

Chapter 14
Chemical Equilibrium

The goal of this chapter is to introduce the concept of equilibrium. Most chemical reactions you have studied so far have been described as happening in only one direction—reactants become products, written left to right. However, these reactions actually occur in both directions (reacting from left to right and from right to left). In order to determine whether the forward reaction or the reverse reaction is more likely to occur, chemists use equilibrium constants, which are described in detail in this chapter. The chapter also introduces Le Chatelier's principle, a qualitative way to predict how changing properties like volume, concentrations, or temperature will change a reaction at equilibrium. Finally, the chapter introduces entropy and describes how enthalpy and entropy determine the effectiveness of a chemical reaction.

Properties of Equilibrium Systems (14.1)

A chemical reaction where both the forward and reverse reactions are occurring at the same rate is referred to as a **chemical equilibrium**. Reactions undergoing chemical equilibrium never go to completion (100% yield); reactions that have high yields (lots of products) are called **product-favored** reactions and reactions with low yields (few products) are called **reactant-favored** reactions. The way you can tell that a reaction has reached equilibrium is that the concentrations of reactants and products have stopped changing (they are now constant). Reactions will reach equilibrium regardless of whether you start with only reactants, only products, or a mixture of both. The addition of a catalyst will decrease the time needed to reach equilibrium, but it will not affect the amount of reactants and products present in the equilibrium system.

The Dynamic Nature of Equilibrium (14.1 & 14.2)

When a chemical reaction has a small E_a(forward) value and a large E_a(reverse) value, the reaction is product-favored because many reactant molecules will have enough energy to react (making products), but very few of the product molecules will have enough energy to make reactants. Similarly, a reaction with a large E_a(forward) value and a small E_a(reverse) value will be reactant-favored. When a reaction has similar E_a values, both reactions are likely to be occurring at comparable rates (similar k_f and k_r values).

Since the reaction of hydrogen and iodine listed below involves an elementary (one-step) mechanism, you can predict the forward and reverse rates from the equation—$Rate_f = k_f[H_2][I_2]$ and $Rate_r = k_r[HI]^2$. The rate constants for this reaction are $k_f = 0.0306$ M^{-1} s^{-1} and $k_r = 0.0857$ M^{-1} s^{-1} (the reaction is slightly reactant-favored because $k_r > k_f$). The table below shows how the concentrations change over time.

$$H_2(g) + I_2(g) \rightleftharpoons 2\,HI(g)$$

t (s)	$[H_2]$ (M)	$[I_2]$ (M)	$[HI]$ (M)
0	1.000	2.000	0.000
5	0.821	1.821	0.358
10	0.680	1.680	0.640
15	0.680	1.680	0.640

When the reaction begins ($t = 0$), the rates are $Rate_f = (0.0306$ M^{-1} $s^{-1})[1.000$ M$][2.000$ M$] = 0.0612$ M/s and $Rate_r = (0.0857$ M^{-1} $s^{-1})[0.000$ M$]^2 = 0.0000$ M/s. Since the forward rate is greater than the reverse rate, some of the H_2 and I_2 will be converted into HI (which can be seen in the data at $t = 5$ s). Similar calculations show the rates at $t = 5$ s to be $Rate_f = 0.0457$ M/s and $Rate_r = 0.0110$ M/s, and more H_2 and I_2 will be converted into HI. At $t = 10$ s, the rates are $Rate_f = 0.0350$ M/s and $Rate_r = 0.0351$ M/s. Since these two rates are equal, the amount of HI being produced from H_2 and I_2 is exactly balanced by the amount of HI reacting to make H_2 and I_2, and the concentrations of H_2, I_2, and HI do not change. Although the concentrations are no longer changing and the reaction appears to have stopped at the macroscale, the reaction is still occurring at the nanoscale (since the two rates are <u>not</u> zero). Chemists refer to this as a **dynamic equilibrium** to emphasize the fact that even though it appears as though the forward and reverse reactions have stopped, they are still occurring at the nanoscale level.

Writing Equilibrium Constant Expressions (14.2)

When a chemical reaction reaches equilibrium, the forward and reverse rates must be equal ($Rate_f = Rate_r$). From the example above, $k_f[H_2][I_2] = k_r[HI]^2$. The rate constants do not change (as long as the temperature is held constant), but concentrations can change. So, chemists tend to group the constant objects on one side of the equation (the k values) and the variables that can change (concentrations) on the other side.

$$k_f[H_2][I_2] = k_r[HI]^2, \quad \frac{k_f}{k_r} = \frac{[HI]^2}{[H_2][I_2]} = K_c$$

The ratio of k_f/k_r is called the **equilibrium constant** (K_c), and the mathematical relationship between the reactant and product concentrations is called the **equilibrium constant expression**. Although equilibrium constant expressions can have units, most chemists and chemistry textbooks drop the units on equilibrium constants. In this study guide, units will be used for the equilibrium constant expression only when you are performing calculations using these values. Plugging in the values for the rate constants or the equilibrium concentrations will yield a numerical value for K_c.

$$K_c = \frac{k_f}{k_r} = \frac{0.0306 \text{ M}^{-1} \text{ s}^{-1}}{0.0857 \text{ M}^{-1} \text{ s}^{-1}} = 0.357, \quad K_c = \frac{[HI]^2}{[H_2][I_2]} = \frac{(0.640 \text{ M})^2}{(0.680 \text{ M})(1.680 \text{ M})} = 0.359$$

Note that the exponents in the equilibrium constant expression match the balanced chemical equation, with products appearing in the numerator and reactants appearing in the denominator. In general, the expression of the equilibrium constant for the reaction $a \text{ A} + b \text{ B} \rightarrow c \text{ C} + d \text{ D}$ is:

$$K_c = \frac{[C]^c[D]^d}{[A]^a[B]^b}$$

The 'c' in the K_c tells you that the equilibrium constant expression is written in terms of reactant and product concentrations. When dealing with heterogeneous chemical reactions (involving reactants and products with two or more states of matter), the concentrations of solids and liquids are left out of the equilibrium constant expression because these values are constants. Only gases and aqueous compounds appear in equilibrium constant expressions.

Example: Write the equilibrium constant expression for the following reactions.

(a) $3 \text{ N}_2\text{H}_4(g) \rightleftharpoons 4 \text{ NH}_3(g) + \text{N}_2(g)$, (b) $\text{H}_2\text{PO}_4^-(aq) + \text{H}_2\text{O}(\ell) \rightleftharpoons \text{H}_3\text{O}^+(aq) + \text{HPO}_4^{2-}(aq)$,

(c) $\text{Al}^{3+}(aq) + 3 \text{ OH}^-(aq) \rightleftharpoons \text{Al(OH)}_3(s)$, (d) $\text{Li}_2\text{O}(s) + 2 \text{ H}_2\text{O}(\ell) \rightleftharpoons 2 \text{ Li}^+(aq) + 2 \text{ OH}^-(aq)$

Solution: The equilibrium constant expressions are written with products on the top, reactants on the bottom, the exponents of each concentration matching its coefficient in the equation, and solid and liquid compounds left out of the expression. Whenever all chemicals on one side of the equation are solids or liquids, their concentrations are replaced by a value of 1 in the equilibrium constant expression.

(a) $K_c = \dfrac{[NH_3]^4[N_2]}{[N_2H_4]^3}$, (b) $K_c = \dfrac{[H_3O^+][HPO_4^{2-}]}{[H_2PO_4^-]}$, (c) $K_c = \dfrac{1}{[Al^{3+}][OH^-]^3}$, (d)

Equilibrium Constant Expressions for Related Reactions (14.2)

You can adapt the concept of Hess's law (used for determining $\Delta H°$ values when combining balanced equations together) to the numerical values of the equilibrium constants. (1) If you multiply a balanced chemical equation by a number, then you must raise K_c to the same power (exponent); (2) If you write a balanced chemical equation backwards, then the new value of K_c is the reciprocal of the old value; and (3) If you add two balanced chemical equations together, then you must multiply their K_c values together.

124

Example: Using the equilibrium equations listed below, calculate a value of K_c for the following reactions.

(a) $H_2O(g) \rightleftharpoons H_2(g) + \frac{1}{2} O_2(g)$, (b) $N_2(g) + 2 H_2O(g) \rightleftharpoons 2 NO(g) + 2 H_2(g)$

$$2 H_2(g) + O_2(g) \rightleftharpoons 2 H_2O(g) \qquad\qquad K_c = 3.3 \times 10^{81}$$

$$N_2(g) + O_2(g) \rightleftharpoons 2 NO(g) \qquad\qquad K_c = 4.5 \times 10^{-31}$$

Solution:

(a) This equation is simply the first equation written backward and divided by two. So, the new K_c value will be the reciprocal of the first K_c value (because the equation is written backward) raised to the 1/2 power (because the equation was multiplied by 1/2).

$$K_c = \frac{1}{K_{c_1}^{1/2}} = \frac{1}{(3.3 \times 10^{81})^{1/2}} = 1.7 \times 10^{-41}$$

(b) This equation is simply the second equation added to the reverse of the first equation (with the $O_2(g)$ canceling out). So, the new K_c value will be the second K_c value times the reciprocal of the first K_c value.

$$K_c = K_{c_2} \times \frac{1}{K_{c_1}} = 4.5 \times 10^{-31} \times \frac{1}{3.3 \times 10^{81}} = 1.4 \times 10^{-112}$$

Converting Between K_c and K_p Values (14.2)

For equilibrium reactions involving reactants or products in the gas phase, you can write an equilibrium constant expression using partial pressures instead of concentrations. The equilibrium constant for this reaction is called K_P ('P' for partial pressures), and the equilibrium constant expression for the general reaction $a A + b B \rightarrow c C + d D$ is written below. Because the partial pressure of a gas is proportional to its concentration ($[X] = P_X RT$), K_P and K_c are related by the equation listed below, where R is the ideal gas constant, T is the Kelvin temperature, and Δn is the moles of gaseous products minus the moles of gaseous reactants (which equals $c + d - a - b$, assuming all of the chemicals in the equation are gases).

$$K_P = \frac{P_C^c P_D^d}{P_A^a P_B^b} = K_c \times (RT)^{\Delta n}$$

Example: Write the K_P expression for the following reactions and calculate a value for K_P from K_c.

(a) $H_2O(g) \rightleftharpoons H_2(g) + \frac{1}{2} O_2(g)$, (b) $N_2(g) + 2 H_2O(g) \rightleftharpoons 2 NO(g) + 2 H_2(g)$

Solution: Use the K_c values that were calculated in the last example. It makes sense to list the units on the equilibrium constants (listed in italics) so you can see that the units from $(RT)^{\Delta n}$ convert units of M into units of atm (or vice versa). Since M = mol/L, the units of R can be written as L atm/mol K or atm/M K.

(a) $K_P = \frac{P_{H_2} P_{O_2}^{1/2}}{P_{H_2O}} = K_c(RT)^{1/2} = (1.7 \times 10^{-41} \ M^{1/2})\left(0.082057 \ \frac{atm}{M \ K} \times 298.15 \ K\right)^{1/2} = 8.4 \times 10^{-41} \ atm^{1/2}$.

(b) $K_P = \frac{P_{NO}^2 P_{H_2}^2}{P_{N_2} P_{H_2O}^2} = K_c(RT)^{+1} = (1.4 \times 10^{-112} \ M)\left(0.082057 \ \frac{atm}{M \ K} \times 298.15 \ K\right)^{1} = 3.3 \times 10^{-111} \ atm$.

So, the K_P values for these reactions are 8.4×10^{-41} and 3.3×10^{-111} (units removed).

Determining K_c Values from Initial and Equilibrium Concentrations (14.3)

In order to determine values for equilibrium constants, chemists usually perform an experiment where the initial concentrations are known, and one of the **equilibrium concentrations** is measured. Equilibrium calculations are usually solved using reaction tables called *ICE tables*. *ICE* tables have a column for each chemical that appears in the equilibrium constant (no solids or liquids) and three rows—one for the initial concentrations (I), one for the change in concentrations (C), and one for the equilibrium concentrations (E). The C row is written in terms of unknowns (x) with coefficients that match the balanced equation, and a negative sign when the chemical is reacting and a positive sign if the chemical is being made. The E row is the sum of other two rows ($I + C = E$).

Example: When a 1.500 mol of nitrogen gas and 2.250 mol of hydrogen gas are added to a 1.500 L flask, they react according to the equation below to make ammonia. The equilibrium concentration of ammonia was measured to be 0.066 M. What is the value of the equilibrium constant (K_c) for this reaction?

$$N_2(g) + 3 H_2(g) \rightleftharpoons 2 NH_3(g)$$

Solution: The initial concentrations will be $[N_2] = 1.500$ mol/1.500 L = 1.000 M, $[H_2] = 2.250$ mol/1.500 L = 1.500 M, and $[NH_3] = 0.000$ M. Since there is no NH_3 under the initial conditions, some NH_3 must be made and some N_2 and H_2 must react, so the signs for the C row are positive for NH_3 and negative for N_2 and H_2. The coefficients in front of the unknown (x) come from the balanced equation—for every x molecules of N_2 that react, $3x$ molecules of H_2 will react, and $2x$ molecules of NH_3 will be produced.

	$N_2(g)$ +	$3 H_2(g)$ $\rightleftharpoons$	$2 NH_3(g)$
I	1.000 M	1.500 M	0.000 M
C	$-x$	$-3x$	$+2x$
E	1.000 M $- x$	1.500 M $- 3x$	$2x$

At equilibrium, you know that $[NH_3] = 0.066$ M $= 2x$. Solve for x, and use it to find the equilibrium values of $[N_2]$ and $[H_2]$ and plug those into the equilibrium constant expression to find a value for K_c.

$$[NH_3] = 2x = 0.066 \text{ M}, \ x = 0.033 \text{ M}$$
$$[N_2] = 1.000 \text{ M} - (0.033 \text{ M}) = 0.967 \text{ M}, \ [H_2] = 1.500 \text{ M} - 3(0.033 \text{ M}) = 1.401 \text{ M}$$
$$K_c = \frac{[NH_3]^2}{[N_2][H_2]^3} = \frac{[0.066 \text{ M}]^2}{[0.967 \text{ M}][1.401 \text{ M}]^3} = 0.0016 \ M^{-2} = 0.0016$$

Using K_c Values to Predict the Favored Direction of a Reaction (14.4)

You can use the numerical value of K_c to predict whether a reaction is reactant- or product-favored. In product-favored reactions, the concentrations of products are large compared to those of the reactants, and in reactant-favored reactions, the concentrations of products are small compared to those of the reactants. Since K_c (or K_P) is calculated by dividing the concentrations (partial pressures) of the products by those of the reactants, product-favored reactions have very large K values (large numbers/small numbers = 10^{+x}) while reactant-favored reactions have very small K values (small numbers/large numbers = 10^{-x}). When the K values are close to 1 (in the range of about 0.01-100), significant amounts of both products and reactants will be present.

Calculating Equilibrium Concentrations (14.5)

In order to determine equilibrium concentrations, chemists usually use *ICE* tables to perform calculations using the initial concentrations and the numerical value of the equilibrium constant. Although some of the more complex problems could involve calculations with exponents greater than 2 (x^3, x^4, etc.), most chemistry teachers and textbooks focus on simpler calculations with exponents of two or lower. These calculations can be solved using the quadratic formula; however, some of these calculations can be made easier by making assumptions (described more fully in Chapters 16 and 17) or by using the method of *perfect squares*.

<u>Example:</u> When a 225 mol of phosphorus trichloride gas and 225 mol of water gas are added to a 10.0 L steel tank, they will react according to the equation below. What are the equilibrium concentrations of the chemicals in this reaction ($K_c = 15.2$)?

$$PCl_3(g) + H_2O(g) \rightleftarrows PCl_3O(g) + H_2(g)$$

<u>Solution:</u> The initial concentrations are $[PCl_3] = [H_2O] = 225$ mol/10.0 L = 22.5 M, and $[PCl_3O] = [H_2] = 0.0$ M. Since there is no PCl_3O or H_2 under the initial conditions, they will be made and some PCl_3 and H_2O will react, giving positive signs for the products and negative signs for the reactants.

	$PCl_3(g)$	+	$H_2O(g)$	$\rightleftarrows$	$PCl_3O(g)$	+	$H_2(g)$
I	22.5 M		22.5.0 M		0.0 M		0.0 M
C	$-x$		$-x$		$+x$		$+x$
E	$22.5\,M - x$		$22.5\,M - x$		x		x

Since you don't know the equilibrium concentration of any chemical in this reaction, you must solve for x and use it to find the equilibrium concentrations.

$$K_c = \frac{[PCl_3O][H_2]}{[PCl_3][H_2O]} = \frac{[x][x]}{[22.5\,M - x][22.5\,M - x]} = \frac{[x]^2}{[22.5\,M - x]^2} = 15.2$$

$$\frac{[x]}{[22.5\,M - x]} = \sqrt{15.2} = 3.90, \quad x = 3.90[22.5\,M - x] = 87.7\,M - 3.90x$$

$$x + 3.90x = 4.90x = 87.7\,M, \quad x = \frac{87.7\,M}{4.90} = 17.9\,M = [PCl_3O] = [H_2]$$

$$[PCl_3] = [H_2O] = 22.5\,M - x = 22.5\,M - (17.9\,M) = 4.6\,M$$

You can test to make sure your calculations are correct by using these values to calculate K_c ($= 15.2$).

$$K_c = \frac{[PCl_3O][H_2]}{[PCl_3][H_2O]} = \frac{[17.9\,M][17.9\,M]}{[4.6\,M][4.6\,M]} = 15.1$$

Using Reaction Quotients to Predict the Direct of a Reaction (14.5)

The **reaction quotient** (Q) for an equilibrium reaction is simply the equilibrium constant expression (K) when the reaction is not at equilibrium. Since there are many possible sets of concentrations when the reaction is not at equilibrium, Q values are not constant. When the Q value is smaller than K, the reactants will be converted into products and the Q value will continue to increase until it equals the K value; when the Q value is larger than K, the products will be converted into reactants and the Q value will continue to decrease until it equals the K value. When $Q = K$, the reaction is at equilibrium.

Using Le Chatelier's Principles to Predict the Direction of a Reaction (14.6)

Le Chatelier's principle states that when a system at equilibrium is disturbed, the chemical system will react to counteract (or minimize) this change, **shifting the equilibrium** until the system is again at equilibrium. Le Chatelier's Principle is a consequence of dynamic equilibria—if reactions stopped after reaching equilibrium, they would not react to these changes. When the equilibrium system reacts to these changes, either some reactants are converted into products (shifting the reaction to the right) or some products are converted into reactants (shifting the reaction to the left).

When the concentrations of one or more of the chemicals in the system are changed, the forward or reverse rates will also change. Since these rates are no longer equal, the system is no longer at equilibrium and the reaction will be shifted in the direction with the larger rate.

Example: How will the reaction below be shifted if the following changes are made to the system? (a) additional C_2H_6 is added, (b) some F_2 is removed, (c) some CH_3F is added

$$C_2H_6(g) + F_2(g) \rightleftharpoons 2 CH_3F(g) \qquad\qquad K_c = \frac{[CH_3F]^2}{[C_2H_6][F_2]}$$

Solution: This is an elementary (one-step) reaction, so $Rate_f = k_f[C_2H_6][F_2]$ and $Rate_r = k_r[CH_3F]^2$.
(a) When C_2H_6 is added, $[C_2H_6]$ increases and that makes $Rate_f$ increase (but $Rate_r$ stays the same since k_r and $[CH_3F]$ have not changed). Since $Rate_f > Rate_r$, more CH_3F is being made (and more C_2H_6 and F_2 are being used up) in the forward reaction than is reacting in the reverse reaction and the reaction is shifted to the right. This shift in reaction will decrease $[C_2H_6]$ and $[F_2]$ (making $Rate_f$ decrease) and increase $[CH_3F]$ (making $Rate_r$ increase) until $Rate_f = Rate_r$ again.
(b) When F_2 is removed, $[F_2]$ decreases and $Rate_f$ decreases ($Rate_r$ stays the same since k_r and $[CH_3F]$ have not changed). Since $Rate_f < Rate_r$, more C_2H_6 and F_2 are being made (and more CH_3F is being used up) in the reverse reaction than is reacting in the forward reaction and the reaction is shifted to the left. This shift in reaction will increase $[C_2H_6]$ and $[F_2]$ and decrease $[CH_3F]$ until $Rate_f = Rate_r$ again.
(c) When CH_3F is added, $[CH_3F]$ increases and $Rate_r$ increases ($Rate_f$ stays the same since k_f, $[C_2H_6]$, and $[F_2]$ haven't changed). Since $Rate_r > Rate_f$, more C_2H_6 and F_2 are being made (and more CH_3F is being used up) in the reverse reaction than is reacting in the forward reaction and the reaction is shifted to the left. This shift in reaction will increase $[C_2H_6]$ and $[F_2]$ and decrease $[CH_3F]$ until $Rate_f = Rate_r$ again.

When the volume of an equilibrium system involving gaseous compounds is changed, the number of particle collisions also changes (Boyle's law—as volume decreases, the number of collisions increases and vice versa). The same effect on the number of collisions occurs when the volume of an aqueous solution is changed (by evaporating or adding water). When the volume of the system is increased, there are fewer collisions (and therefore fewer reactions in both directions), and the system will shift in the direction with more gaseous or aqueous products. When the volume is decreased, there are more collisions and reactions in both directions, and the system will shift in the direction with fewer gaseous or aqueous products.

Example: How will the following reactions shift if the volume is increased?
(a) $N_2(g) + 3 H_2(g) \rightleftharpoons 2 NH_3(g)$, (b) $Fe(CN)_6^{3-}(aq) \rightleftharpoons Fe^{3+}(aq) + 6 CN^-(aq)$.

Solution:
(a) When the volume is increased, the number of collisions decreases. Since the forward reaction needs four molecules to collide (in at least three subsequent bimolecular reactions) and the reverse reactions needs only two molecules to collide (in one bimolecular reaction), when the number of collisions decreases the forward reaction is affected more than the reverse reaction (three vs. one bimolecular collisions). So, even though both reactions have slowed, the forward rate has decreased much more than the reverse rate ($Rate_f < Rate_r$). This results in a shift of the reaction to left, producing more N_2 and H_2 and less NH_3.
(b) When the volume is increased (diluted by the addition of water), the number of collisions between the aqueous ions decreases. The forward reaction needs one ion to react (one unimolecular reaction) but the reverse reactions needs seven ions to collide (six bimolecular reactions). So, decreasing the number of collisions causes the reverse reaction to decrease much more than the forward rate does, and $Rate_r < Rate_f$. This results in a shift of the reaction to right, producing more Fe^{3+} and CN^- ions and less $Fe(CN)_6^{3-}$ ions.

Changing the temperature of an equilibrium system has opposite effects on endothermic and exothermic reactions. This is because of the relative values of E_a(forward) and E_a(reverse)—E_a(forward) $> E_a$(reverse) for endothermic reactions and E_a(forward) $< E_a$(reverse) for exothermic reactions. When the temperature is low and the molecules have low kinetic energies, the reaction direction with the lower E_a value is favored (reverse for endothermic reactions; forward for exothermic reactions). When T is increased and the molecules have higher kinetic energies, both reactions become more likely but the reaction direction with the higher E_a value (forward for endothermic reactions; reverse for exothermic reactions) is increased more than the reaction direction with the lower E_a value. The temperature effect on equilibrium reactions is seen in the K_c values—as the temperature is increased, K_c values increase for endothermic reactions and decrease for exothermic reactions, and vice versa for decreases in temperature.

Example: How will the following reactions shift if the temperature is increased?

(a) $CH_4(g) + 2\,O_2(g) \rightleftharpoons CO_2(g) + 2\,H_2O(g)$, $\Delta H° = -802$ kJ; (b) $Br_2(g) \rightleftharpoons 2\,Br\cdot(g)$, $\Delta H° = +193$ kJ

Solution: Besides the E_a argument described above, you can also predict the direction of the reaction shift using the idea that heat is absorbed as a reactant in endothermic reactions and released as a product in exothermic reactions. Then, you can treat 'heat' as another reactant or product that is added to the system when the temperature is increased and removed when the temperature is decreased.

$$CH_4(g) + 2\,O_2(g) \rightleftharpoons CO_2(g) + 2\,H_2O(g) + \text{heat}$$

$$Br_2(g) + \text{heat} \rightleftharpoons 2\,Br\cdot(g)$$

(a) Increasing the temperature adds heat to the system. Since this reaction is exothermic, heat is a product. Adding a product like heat will shift the reaction to the left, because CO_2 and H_2O will use some of the added heat to react, producing CH_4 and O_2.

(b) Since this reaction is endothermic, heat is a reactant. Adding a reactant like heat will shift the reaction to the right, because Br_2 will use some of the added heat to react, producing $Br\cdot$ atoms.

Using Enthalpy and Entropy Changes to Predict the Favored Direction of a Reaction (14.7)

Reactions with negative enthalpy values ($\Delta H° < 0$, exothermic) tend to be product-favored and reactions with positive enthalpy values ($\Delta H° > 0$, endothermic) tend to be reactant-favored. This is because the forward direction of an exothermic reaction and the reverse direction of an endothermic reaction have lower activation energies (and thus happen more often). The **entropy** change ($\Delta S°$, which will be described in greater detail in Chapter 18) also affects whether a reaction is reactant- or product-favored.

Entropy is a quantitative measure of energy dispersal. When energy is dispersed over a larger volume or over a larger number of molecules, the entropy change is positive ($\Delta S° > 0$) and the reaction tends to be product-favored; when the energy is forced into a smaller volume or over a smaller number of molecules, the entropy change is negative ($\Delta S° < 0$) and the reaction tends to be reactant-favored. In general, the effect of $\Delta H°$ is most important at lower temperatures and the effect of $\Delta S°$ becomes more important at higher temperatures.

Predicting Optimal Conditions for the Formation of Products (14.8)

Reactions are product-favored when the K values are large (10^{+x}); reactions with negative $\Delta H°$ and positive $\Delta S°$ are also product-favored. When the enthalpy and entropy effects contradict each other (both positive or both negative), the reaction should be carried out at low temperatures when enthalpy favors the reaction, but should be performed at high temperatures when entropy favors the reaction. In general, reactions are quicker when the reactants are present at high temperatures and concentrations, when the reactants are in the gas or aqueous phase or are solids or liquids with large surface areas, and when a catalyst is added.

The textbook provides a specific example of how these general rules can be applied to an actual chemical system for the production of ammonia. The industrial process for making ammonia from its elements is called the **Haber-Bosch process** (or sometimes the *Haber process*). Since this reaction has negative $\Delta H°$ and $\Delta S°$ values, this reaction is more product-favored at low temperatures. However, the reaction is still heated to 450°C because the reaction is too slow at lower temperatures, even though it is more favored.

Chapter Review — Key Terms

The key terms that were introduced in this chapter are listed below, along with the section in which they were introduced. You should understand these terms and be able to apply them in appropriate situations.

chemical equilibrium (14.1)   equilibrium constant (14.2)   product-favored (14.1)
dynamic equilibrium (14.1)   equilibrium constant expression (14.2)   reactant-favored (14.1)
entropy (14.7)   Haber-Bosch process (14.8)   reaction quotient (14.5)
equilibrium concentration (14.3) Le Chatelier's principle (14.6)   shifting an equilibrium (14.6)

Practice Test

After you have finished studying the chapter and the homework problems, the following questions can serve as a test to determine how well you have learned the chapter objectives.

1. Dinitrogen tetraoxide decomposes according to the equation $N_2O_4(g) \rightleftarrows 2\,NO_2(g)$.

t (s)	$[N_2O_4]$ (M)	$[NO_2]$ (M)
0	5.000	0.000
9	4.962	0.076
18	4.923	0.154
27	4.915	0.170
36	4.915	0.170

$T = 25°C$
$k_f = 0.013 \text{ s}^{-1}$
$k_r = 2.2 \text{ M}^{-1}\,\text{s}^{-1}$

   (a) When has this system reached equilibrium? Explain.
   (b) Assuming this is an elementary (one-step) reaction, calculate the forward and reverse rates at equilibrium. How do these values support the concept of a dynamic equilibrium?
   (c) Write the equilibrium constant expression for this reaction and calculate a value for K_c.
   (d) Calculate the reaction quotient at $t = 18$ s. How does this value compare to K_c? In which direction will the reaction shift to reach equilibrium?
   (e) Is this reaction reactant-favored or product-favored? Explain your answer.
   (f) Write the equilibrium constant expression for K_P and calculate its value at 25°C.

2. Ammonia and hydrofluoric acid will react when placed in water according to the following equation:

$$NH_3(aq) + HF(aq) \rightleftarrows NH_4^+(aq) + F^-(aq)$$

   (a) What is the value of K_c for this reaction? Write the equilibrium constant expression for K_c.
   (b) Why doesn't it make sense to write a K_P expression for this reaction?
   (c) Is this reactant-favored or product-favored? Explain your answer.

$$NH_3(aq) + H_2O(\ell) \rightleftarrows NH_4^+(aq) + OH^-(aq) \qquad K_c = 1.8 \times 10^{-5}$$

$$HF(aq) + H_2O(\ell) \rightleftarrows H_3O^+(aq) + F^-(aq) \qquad K_c = 7.2 \times 10^{-4}$$

$$2\,H_2O(\ell) \rightleftarrows H_3O^+(aq) + OH^-(aq) \qquad K_c = 1.0 \times 10^{-14}$$

3. When 3.00 mol of gaseous water and 4.00 mol of solid carbon are added to a 0.500 L steel tank at 300°C, they react according to the following equation to make carbon dioxide and hydrogen. The equilibrium concentration of hydrogen was measured to be 0.30 M.

$$2\,H_2O(g) + C(s) \rightleftarrows CO_2(g) + 2\,H_2(g)$$

   (a) Write the K_c expression for this reaction. Why is carbon omitted from this equation?
   (b) What are the equilibrium concentrations of each chemical in the system?
   (c) What is the value of K_c for this reaction?
   (d) Is this reactant-favored or product-favored? Explain your answer.
   (e) Write the equilibrium constant expression for K_P and calculate its value.

4.   When 18.0 mol of solid cobalt(II) sulfate and 18.0 mol of carbon dioxide gas are added to a 1.00 L steel tank, they react according to the following equation to make cobalt(II) carbonate and sulfur trioxide. The equilibrium constant is $K_c = 61.2$ for this reaction at 2500°C.

$$CoCO_3(s) + SO_3(g) \rightleftharpoons CoSO_4(s) + CO_2(g)$$

(a)   Write the K_c expression for this reaction.
(b)   What are the equilibrium concentrations of each chemical in the system?
(c)   Is this reactant-favored or product-favored? Explain.
(d)   Write the equilibrium constant expression for K_P and calculate its value.

5.   When 2.00 mol of phosphorus pentachloride is mixed with 3.00 mol of phosphorus trichloride and 1.00 mol of chlorine gas in a 1.00 L flask, they react according to the following equation. The equilibrium constant is $K_c = 1.21$ for this reaction at 250°C.

$$PCl_5(g) \rightleftharpoons PCl_3(g) + Cl_2(g)$$

(a)   Write the K_c expression for this reaction.
(b)   Calculate the reaction quotient at the beginning of this reaction. In which direction will the reaction shift to reach equilibrium?
(c)   What are the equilibrium concentrations of each chemical in the system?
(d)   Is this reactant-favored or product-favored? Explain.

6.   The reaction of hydrogen and bromine can be monitored by sight since $Br_2(g)$ is rusty brown, but $H_2(g)$ and $HBr(g)$ are colorless. Predict how the following changes will shift this exothermic reaction.

$$H_2(g) + Br_2(g) \rightleftharpoons 2 HBr(g)$$

(a)   Adding hydrogen gas   (c)   Decreasing the system's volume
(b)   Adding hydrogen bromide gas   (d)   Increasing the system's temperature

7.   The reaction of cobalt(II) and chloride ions to make the tetrachlorocobaltate(II) ion can be monitored by sight since $Co(H_2O)_6^{2+}(aq)$ is red, $CoCl_4^{2-}(aq)$ is blue, and $H_2O(\ell)$ and $Cl^-(aq)$ are colorless.

$$Co(H_2O)_6^{2+}(aq) + 4 Cl^-(aq) \rightleftharpoons CoCl_4^{2-}(aq) + 6 H_2O(\ell)$$

(a)   How will the reaction shift if chloride ions are added from the system?
(b)   How will the reaction shift if liquid water is added to the system?
(c)   When the a test tube containing a system at equilibrium is placed in boiling water, the solution changes from red to purple. Is this reaction endothermic or exothermic?

8.   Answer the following questions for the equilibrium systems described in questions 6 and 7.
(a)   What happens to the K_c values when the concentration of a reactant is increased?
(b)   What happens to the K_c values when the volume of the system is decreased?
(c)   What happens to the K_c values when the temperature of the system is increased?
(d)   Predict whether the $\Delta S°$ values for these reactions are positive, negative, or close to zero.
(e)   Predict whether these reactions will be product-favored or reactant-favored at low temperatures and at high temperatures.

131

Chapter 15
The Chemistry of Solutes and Solutions

The goal of this chapter is to introduce the chemical and physical properties of aqueous solutions, which are homogeneous mixtures made by dissolving a small amount of a chemical (the *solute*) in a larger amount of liquid water (the *solvent*). One of these properties is the **solubility** of the solute in water, and the chapter discusses the factors that influence how much of a solute can dissolve in water. This chapter also introduces colligative properties including vapor pressure lowering, freezing point lowering, boiling point elevation, and osmotic pressure. Finally, this chapter describes the sources of fresh water and how municipal drinking water is purified and softened.

Dissolving Solids, Liquids, and Gases in a Solvent (15.1)

When a solid, liquid, or gaseous sample dissolves in water or any other solvent, the molecules or ions of the solute and the solvent molecules interact with each other, making a homogeneous mixture. Solid or gaseous compounds that dissolve in water are *soluble* ('dissolvable') in water; liquid compounds that dissolve in water are **miscible** ('mixable'). When a solute does not dissolve in water, the particles in the solute do not interact much with the water molecules and the two substances stay separate from each other (i.e., a heterogeneous mixture). Solid or gaseous compounds that do not dissolve in water are *insoluble* in water, and liquid compounds that do not dissolve in water are **immiscible**.

Predicting Solubility Based on Solute and Solvent Properties (15.1)

Many chemistry teachers and textbooks use the concept of 'like dissolves like' to determine whether a particular solute molecule (or ions) will dissolve in a particular solvent. The general rule for 'like dissolves like' is that polar molecules are soluble in polar solvents and insoluble in nonpolar solvents, and nonpolar molecules are soluble in nonpolar solvents and insoluble in polar solvents. Unfortunately, this general rule is not always true because there are many exceptions to the rules listed above.

The molecules in nonpolar liquids are held together by London forces (which are rather weak), and these nonpolar liquids will also be attracted to solute molecules or ions by London forces. If the solute is nonpolar, it will be attracted to the solvent and other solute molecules by London forces. Since neither of these molecules have a particular preference (stronger attraction) for their own molecules or for the other molecule, the solution will mix. If the solute molecules are polar, however, they will have much stronger attractions to their own molecules (dipole-dipole attractions or hydrogen bonds) than they would to the nonpolar solvent molecules (London forces). Therefore, you should expect polar molecules to be insoluble in nonpolar liquids.

The molecules in polar liquids are held together by dipole-dipole attractions or hydrogen bonds, which are rather strong. If the solute is polar, it will be attracted to solvent and other solute molecules by dipole-dipole attractions or hydrogen bonds. Since neither molecule has a particular preference (stronger attraction) for their own or for the other molecule, the solution will mix. However, if the solute molecules are nonpolar they will be attracted to each other and to the solvent molecules by much weaker attractions (London forces). Therefore, you should expect nonpolar molecules to be insoluble in polar liquids because the polar solvent molecules are more attracted to each other than they are to the solute molecules.

Example: Would you predict solid potassium sulfate (K_2SO_4) to be more soluble in the polar liquid water (H_2O) or in the nonpolar liquid hexane (C_6H_{14})?

Solution: K_2SO_4 contains K^+ and SO_4^{2-} ions. Since the attractions between ions are very strong, this solid would be more likely to dissolve in a liquid with strong intermolecular forces like liquid water, and it is unlikely to dissolve in a nonpolar liquid with rather weak intermolecular forces like hexane.

Example: Would you predict liquid 1-hexanol ($C_6H_{13}OH$) to be more miscible in the polar liquid water (H_2O) or in the nonpolar liquid hexane (C_6H_{14})?

Solution: 1-hexanol is a polar molecule that has an O-H bond in it (making it an alcohol) that can undergo hydrogen bonding. Since 1-hexanol is polar and has hydrogen-bonding forces, you might expect it to be more miscible in the polar liquid water. However, it is only slightly soluble in water and mixes much more completely with the nonpolar liquid hexane. This does not follow the 'like dissolves like' rule.

Although 1-hexanol has an O-H bond that can undergo hydrogen bonding with water, it has a very large nonpolar region (C_6H_{13}-) that is attracted to the hexane and water molecules by London forces. In order to dissolve in water, many strong hydrogen bonds between water molecules must be broken to let the large molecule in water, and they will be compensated not by strong dipole-dipole attractions or hydrogen bonds, but by weaker London forces between the H_2O molecules and the long nonpolar region of 1-hexanol. In order to dissolve in hexane, weak London forces between hexane molecules must be broken to let the large molecule in hexane, and they will be compensated by other London forces between the two molecules.

Interpreting Solubility in Terms of Enthalpy and Entropy Changes (15.2)

The qualitative description above regarding the attractions of solvent and solute molecules can be applied quantitatively to changes in enthalpies ($\Delta H°$). The **enthalpy of solution** ($\Delta H°_{soln}$) consists of three enthalpy factors—the enthalpy of separating solvent molecules (always endothermic), the enthalpy of separating solute molecules (always endothermic), and the enthalpy of mixing solvent and solute molecules (always exothermic). If the attractions of the solvent molecules to each other and the attractions of the solute molecules to each other (the endothermic terms) are larger than the attractions of the solvent and solute molecules (the exothermic term), then $\Delta H°_{soln}$ is positive (endothermic). If the attractions of the solvent molecules to each other and the attractions of the solute molecules to each other are smaller than the attractions of the solvent and solute molecules, then $\Delta H°_{soln}$ is negative (exothermic).

When solid or liquid solutes are dissolved in liquid solvents, the solvent and solute molecules become more dispersed, and this results in a larger dispersal of energy in the solution compared to the solvent and solute (positive $\Delta S°_{soln}$). When gaseous solutes dissolve in liquid solvents, the gas molecules are forced into the smaller liquid volume and this leads to negative $\Delta S°_{soln}$ values. Therefore, the entropy effect favors the dissolution of solids and liquids in liquid solvents but does not favor dissolving gases in liquid solvents.

Unsaturated, Saturated, and Supersaturated Solutions (15.3)

The solubility of a solid compound in a liquid solvent is an equilibrium process, in which the dissolution of the solid compound into the aqueous phase is matched by the precipitation of the aqueous compound into the solid phase. A **saturated solution** is a solution that is at equilibrium with the solid and contains the *equilibrium concentration* of the compound dissolved in the aqueous solution. Solutions containing more or less than the equilibrium concentration of the compound dissolved in the aqueous solution are called **supersaturated solutions** and **unsaturated solutions**, respectively. Unsaturated solutions are common and usually occur when the solid has just started dissolving (when $Q \neq K_c$) or if there is not enough solid present to make a saturated solution. Supersaturated solutions are much less common, and are usually made by taking advantage of the fact that most solids become more soluble as the temperature is increased. So, the solution is heated to higher temperatures and more of the solid dissolves. Then, it is slowly and carefully cooled to prevent precipitation of the dissolved solid. Supersaturated solutions are unstable, and the excess solid will eventually precipitate out of the supersaturated solution. This process can be accelerated by adding a small crystal of the solid to the supersaturated solution or by stirring the solution.

Dissolving Ionic Compounds in Water (15.3)

Ionic compounds have very strong intermolecular forces (ion-ion attractions) holding them together in the solid. In order for ionic compounds to dissolve in a liquid solvent, the interactions between the individual

133

ions and the solvent molecules also need to be very strong. So, ionic compounds will only dissolve in very polar liquids like water. When these compounds dissolve in water, the negative ends of water (O atoms) are attracted to the cations and the positive ends of water (H atoms) are attracted to the anions. The process of water molecules arranging themselves to point their negative ends around the positive ions and their positive ends around the negative ions is called **hydration**.

The Effect of Temperature on the Solubility of Ionic Compounds (15.4)

In general, the solubilities of ionic compounds increase as the temperature is increased. An increase in temperature results in the ions in the solid phase having more energy. When the ions in the solid phase absorb energy, more of these ions have enough energy to escape the attractions holding them in the solid and more of the solid particles dissolve into the aqueous phase.

<u>Example</u>: Use $\Delta H°_{soln}$ and $\Delta S°_{soln}$ to explain why ionic compounds are more soluble at high temperatures.

<u>Solution</u>: Depending on the relative strengths of the attractions of the ions and the water molecules to themselves and to each other, the $\Delta H°_{soln}$ value may be positive or negative. When the ions in the solid dissolve in the aqueous solution, they are more dispersed in space and allow a greater dispersal of energy (positive $\Delta S°_{soln}$ value). The positive $\Delta S°_{soln}$ value suggests that this process will become more product-favored as the temperature increases.

The Effect of Temperature and Pressure on the Solubility of Gases (15.4-15.5)

In general, the solubilities of gases decrease as the temperature is increased. An increase in temperature results in the solute molecules in the aqueous phase having more energy. When the dissolved gas molecules absorb energy, more of these molecules have enough energy to escape the attractions holding them in the aqueous phase and more of the gas molecules escape from the aqueous solution. The $\Delta S°_{soln}$ value for dissolving a gas in an aqueous solution is negative because these gas molecules were forced into a much smaller (liquid) volume. This negative $\Delta S°_{soln}$ value makes the dissolution of gases in liquid solvents more product-favored at low temperature and less product-favored at high temperatures.

The solubility of a gas increases when the partial pressure of that gas over the solution is increased. Increasing the partial pressure of the gas over the aqueous phase results in more collisions of the gas molecules with the liquid's surface. This leads to more gas molecules becoming trapped in the solvent, increasing its solubility. The solubility of the gas increases only if <u>its</u> partial pressure increases (e.g., increasing the total pressure of the system by adding nitrogen gas will not make any gas other than nitrogen more soluble). The mathematical relationship between the gas solubility and its partial pressure is called **Henry's law**—$[X] = k_H P_X$, where $[X]$ is the molar solubility (concentration) of gas X, k_H is the *Henry's law constant* for the gas, and P_X is the partial pressure of the gas.

Converting Between Solution Concentration Units (15.6)

Chemists use many different units for solution concentrations. The first five units described in this chapter (mass fraction, weight percent, parts per million, parts per billion, and parts per trillion) are clearly related to each other. The **mass fraction** is the proportion (fraction) of a solution's mass due to the solute:

$$mass\ fraction\ X = \frac{m_X}{m_{total}}$$

The proportionality factors for the mass fraction (1) and the **weight percent** value, the **parts per million** (ppm) value, the **parts per billion** (ppb) value, and the **parts per trillion** (ppt) value are listed below.

$$1 = 100\% = 10^6\ ppm = 10^9\ ppb = 10^{12}\ ppt$$

Example: Determine the mass fraction, weight percent, ppm, ppb, and ppt values for a solution containing 2.9×10^{-5} mol of oxygen gas (O_2) dissolved in 100.0 mL of water.

Solution: The mass fraction is calculated using the mass fraction equation. The other concentration values are determined by using proportionality factors listed above.

$$m_{O_2} = 2.9 \times 10^{-5} \text{ mol } O_2 \times \frac{32.00 \text{ g } O_2}{\text{mol } O_2} = 0.0093 \text{ g } O_2$$

$$m_{H_2O} = 100. \text{ mL} \times 1.00 \frac{\text{g } H_2O}{\text{mL}} = 100. \text{ g } H_2O$$

$$mass\ fraction\ O_2 = \frac{m_{O_2}}{m_{total}} = \frac{0.0093 \text{ g } O_2}{100. \text{ g } H_2O + 0.0093 \text{ g } O_2} = 9.3 \times 10^{-5}$$

$$9.3 \times 10^{-5} \times \frac{100\%}{1} = 9.3 \times 10^{-3}\% \times \frac{10^6 \text{ ppm}}{100\%} = 93 \text{ ppm}$$

$$93 \text{ ppm} \times \frac{10^9 \text{ ppb}}{10^6 \text{ ppm}} = 9.3 \times 10^4 \text{ ppb} \times \frac{10^{12} \text{ ppt}}{10^9 \text{ ppb}} = 9.3 \times 10^7 \text{ ppt}$$

The other two concentration units used in this chapter are molarity [X] (previously described in chapter 5) and molality. Molarities are used extensively in chemistry; however, they are not very useful when the temperature is allowed to change because solution molarities change when T changes. **Molality**, which is related to molarity, is used when T changes because it is temperature independent. While [X] is defined as the moles of solute per liter of solution (solvent plus solute), molality (m) is defined as the moles of solute per kilogram of solvent (no solute). The unit 'molal' (m) is synonymous with the units 'mol/kg'.

$$[X] = \frac{n_X}{V_{soln}} \left(\text{units of } \frac{\text{mol}}{\text{L}} = M \right); \quad m = \frac{n_X}{m_{solvent}} \left(\text{units of } \frac{\text{mol}}{\text{kg}} = m \right)$$

Vapor Pressure Lowering and Raoult's Law (15.7)

The vapor pressure of a pure liquid is constant as long as its temperature is held constant. However, the vapor pressure of a solution depends on its concentration—the more nonvolatile molecules or ions dissolved in the liquid, the lower its vapor pressure. This is because some of the nonvolatile species sit at the liquid's surface taking up space at the surface and preventing some of the solvent molecules from reaching the liquid's surface where they can evaporate. This results in the evaporation of fewer solvent molecules and a decreased vapor pressure. The mathematical relationship between the solution's vapor pressure (P_{soln}), the vapor pressure of the pure solvent ($P°$) and its concentration is called **Raoult's law**—$P_{soln} = X_{solvent}P°$, where $X_{solvent}$ is the mole fraction of the volatile solvent.

Calculating Colligative Properties Using Molality Values (15.7)

Colligative properties are properties of a solution that depend only on the <u>concentration</u> of the solute dissolved in the solution and not the chemical identity of these particles. The three colligative properties described in the textbook are boiling point elevation, freezing point lowering, and osmotic pressure.

Because the addition of a nonvolatile solute decreases the vapor pressure of the liquid solvent in a solution, the solution must be heated to a higher temperature for the liquid's vapor pressure to equal the atmospheric pressure and for the liquid to boil. This increase in the boiling point, called **boiling point elevation**, is proportional to the molality of the solute, where K_b is the *boiling point elevation constant* for the solvent.

$$\Delta bp = bp(\text{solution}) - bp(\text{solvent}) = K_b m_{solute}$$

The presence of the solute in the solution also lowers the melting point of the solution compared to the pure solvent. This process is called **freezing point lowering**, but is also referred to as *melting point lowering*, *freezing point depression* or *melting point depression*. The freezing point of the solution is lowered because the solute molecules interfere with the formation of solid solvent crystals as the liquid starts to freeze. Therefore, the solution must be cooled even lower than the pure solvent's freezing point to force the solvent molecules to crystallize. Note that the solid crystals made when a solution freezes contain only the solvent molecules and not the solute particles. This decrease in the freezing point is proportional to the molality of the solute, where K_f is the *freezing point elevation constant* for the solvent.

$$\Delta fp = fp(\text{solution}) - fp\,(\text{solvent}) = -K_f\, m_{solute}$$

Example: What are the boiling point and freezing point for a solution containing 80.0 g of glucose ($C_6H_{12}O_6$) dissolved in 100.0 g of water? $K_b(H_2O) = 0.52°C/m$, $K_f(H_2O) = 1.86°C/m$.

Solution: In order to calculate the boiling point and melting point for this solution, you must first calculate the molality of the solution.

$$n_{C_6H_{12}O_6} = 80.0 \text{ g } C_6H_{12}O_6 \times \frac{\text{mol } C_6H_{12}O_6}{180.16 \text{ g } C_6H_{12}O_6} = 0.44405 \text{ mol } C_6H_{12}O_6 \text{ (unrounded)}$$

$$m_{C_6H_{12}O_6} = \frac{n_{C_6H_{12}O_6}}{m_{H_2O}} = \frac{0.44405 \text{ mol } C_6H_{12}O_6}{100.0 \text{ g } H_2O \times \frac{1 \text{ kg}}{10^3 \text{ g}}} = 4.44 \text{ m}$$

$$bp(\text{solution}) - bp\,(\text{solvent}) = K_b m_{solute} = \left(0.52\, \frac{C}{m}\right)(4.44 \text{ m}) = 2.3 \text{ C}$$

$$bp(\text{solution}) = bp\,(\text{solvent}) + 2.3 \text{ C} = 100.0 \text{ C} + 2.3 \text{ C} = 102.3 \text{ C}$$

$$fp(\text{solution}) - fp\,(\text{solvent}) = -K_f m_{solute} = -\left(1.86\, \frac{C}{m}\right)(4.44 \text{ m}) = -8.26 \text{ C}$$

$$fp(\text{solution}) = fp\,(\text{solvent}) - 8.26 \text{ C} = 0.0 \text{ C} - 8.26 \text{ C} = -8.3 \text{ C}$$

The Colligative Properties of Electrolyte Solutions (15.7)

When measuring the boiling point elevation or freezing point lowering of ionic compounds dissolved in water, chemists noted that the changes in the colligative properties of dilute solutions of ionic compounds like KNO_3 or NH_4Cl had twice the expected effect for its molality, and ionic compounds like $MgCl_2$ or Na_2CO_3 had three times the expected effect for its molality. These increased effects can be explained by recognizing that ionic compounds dissolve into two or more ions when placed in water. So, a solution that is 1.0 m in KNO_3 has 1.0 m K^+ ions and 1.0 m NO_3^- ions (for a total of 2.0 m ions in solution). To correct for the formation of multiple ions in electrolyte solutions, the *van't Hoff factor* (i_{solute}), which represents the number of ions produced from the formula unit of the electrolyte, is added to the ΔT equations:

$$\Delta bp = K_b m_{solute} i_{solute} \text{ and } \Delta fp = -K_f m_{solute} i_{solute}$$

Osmosis, Osmotic Pressure, and Reverse Osmosis (15.7)

Osmosis is the movement of solvent molecules through a **semipermeable membrane** (a membrane that allows small molecules like water to pass through it, but prevents the passage of larger molecules or ions). When the concentrations of solute particles dissolved in the solutions on either side of the membrane are different, the rate at which the water molecules travel through either side of the membrane is different. The solution with a lower concentration of solute particles has a higher concentration of water molecules, so there will be more water molecules moving through the membrane from the less concentrated (solute) solution, entering the more concentrated (solute) solution, than in the reverse direction.

The net migration of solvent molecules from a pure sample of the solvent into a solution can be stopped by applying a pressure to the solution (called the **osmotic pressure**). This pressure will force more solvent molecules from the solution to pass through the membrane into the pure solvent until the rates of solvent

©2011 Cengage Learning. All Rights Reserved. May not be scanned, copied or duplicated, or posted to a publicly accessible website, in whole or in part.

migration through both sides of the membrane are equal. The osmotic pressure of a solution can be calculated from the following equation, where Π_X is the osmotic pressure.

$$\Pi_X = [X]RTi_{solute}$$

If a pressure greater than the osmotic pressure is applied to a solution, you can force solvent molecules to leave a solution through the semipermeable membrane, resulting in a sample of pure solvent and a more concentrated solution. This process, called **reverse osmosis**, is often used to make pure drinking water from salty (sea) water.

Types and Properties of Colloids (15.8)

Colloids are mixtures in which relatively large particles (the **dispersed phase**) are uniformly distributed within a solvent (the **continuous phase**). These solutions can be considered as homogeneous mixtures of the dispersed and continuous phases at the macroscale, but are actually heterogeneous mixtures at the nanoscale level. A *true solution* is a colloid where the particles in the dispersed phase are less than 2 nm in size; in a *colloidal dispersion* the particles are 2-2000 nm in size, and in a *suspension* the particles are larger than 2000 nm. The small particles within a colloid tend to disperse light traveling through it (the **Tyndall effect**). This is why colloids (like fog, milk, or whipped cream) appear to be opaque.

Surfactants (15.9)

Surfactants (also called *surface agents*) are polar or ionic compounds that have a polar/ionic (*hydrophilic*) end and a nonpolar hydrocarbon (*hydrophobic*) end. Synthetic (man-made) surfactants are referred to as **detergents**. Sodium stearate, $CH_3(CH_2)_{16}CO_2^-Na^+$, is an example of a natural surfactant while sodium lauryl sulfate, $CH_3(CH_2)_{11}OSO_3^-Na^+$, is an example of a synthetic detergent.

<u>Example</u>: Draw a picture showing how soap molecules dissolve in water. Use the symbol ⌇⌇● for soap (the squiggly line is the hydrophobic tail and the circle is the hydrophilic, ionic, head).

<u>Solution</u>: In general, the soap molecules will orient themselves so that their hydrophilic (ionic) ends are in water and their nonpolar ends do not interact with water molecules. One way for the soap molecules to do this is by sitting on the surface of water with their ionic ends in the water and their nonpolar tails hanging out of the water (called a *monolayer*). Another way is for the soap molecules to cluster together in a sphere with their ionic ends pointing out into the water and their nonpolar tails clustering together in a nonpolar region (called a *micelle*).

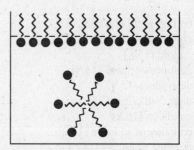

Surfactants and detergents can be used to remove nonpolar substances like grease, oil, or dirt from your hands (or your clothes). Consider what happens when you have motor oil or vegetable oil on your hands. When you wash your hands with water, the oil stays on your hands because the water molecules are more attracted to each other than they are to the nonpolar oil molecules and the water molecules go down the drain without the oil molecules. When the soap is added to the oil on your hands and agitated, the nonpolar tails of the soap molecules embed themselves into the oil droplets creating a soap-oil micelle with the ionic heads of the soap molecules sticking out of the micelle. When your rinse your hands with water, the micelles are attracted to the water molecules by the ionic ends sticking out of the micelle. This makes the micelle miscible in water and the whole mixture goes down the drain.

Earth's Water Supply and Sources of Fresh Water (15.10)

The vast majority of water on the earth appears in the oceans, which is too salty to be potable (drinkable) to humans or other animals. Of the remaining 2.5% of water that is fresh (not salty), most of that (1.72%) is tied up in solid glaciers at the poles of the earth. Only 0.77% of earth's water supplies appear in sources that are accessible as drinkable water (groundwater, lakes, rivers, etc.). The huge supply of salty ocean water makes the prospect of using reverse osmosis to *desalinate* (removing salt) ocean water an attractive alternative for making potable water.

Purifying Municipal Drinking Water Supplies (15.10)

Even samples of fresh water must be purified to remove possibly dangerous or distasteful materials. The general steps for the purification of drinking water include coarse filtration to remove large objects from the water, forcing sediments to settle out of the water using $Al_2(SO_4)_3$ and $Ca(OH)_2$, fine filtration using a sand filter, aeration to allow dissolved organic compounds to be oxidized (producing CO_2, H_2O, etc.), disinfection using chlorine or ozone to kill microorganisms in the water, and finally storage and delivery of the water to the consumer.

Causes and Neutralization of Hard Water (15.10)

Water containing relative high amounts of Ca^{2+}, Fe^{2+}, Mg^{2+}, or Mn^{2+} ions is usually called *hard water*. Hard water is generally considered a problem because it lead to the formation of insoluble ionic salts in hot-water heaters, pipes, and plumbing fixtures (sinks, toilets, showers, etc.), and in the presence of soaps (known as *soap scum*). Hard water can be 'neutralized' (made into *soft water*) by replacing these ions by ions that do not make precipitates (H_3O^+ ions or alkali metal cations). In the *lime-soda process*, $Ca(OH)_2$ and Na_2CO_3 are added to the hard water, resulting in the precipitation of carbonate or hydroxide salts of the hard-water cations, leaving Na^+ ions in their place. In the *ion exchange process*, the hard water is sent through a column containing a *cation-exchange resin*, which trades the hard-water cations for H_3O^+ ions or alkali metal cations.

Chapter Review — Key Terms

The key terms that were introduced in this chapter are listed below, along with the section in which they were introduced. You should understand these terms and be able to apply them in appropriate situations.

boiling point elevation (15.7)	hypertonic (15.7)	parts per trillion (15.6)
colligative properties (15.7)	hypotonic (15.7)	Raoult's law (15.7)
colloid (15.8)	immiscible (15.1)	reverse osmosis (15.7)
continuous phase (15.8)	isotonic (15.7)	saturated solution (15.3)
detergents (15.9)	mass fraction (15.6)	semipermeable membrane (15.7)
dispersed phase (15.8)	miscible (15.1)	solubility (15.3)
emulsion (15.8)	molality (15.6)	supersaturated solution (15.3)
enthalpy of solution (15.2)	osmosis (15.7)	surfactants (15.9)
freezing point lowering (15.7)	osmotic pressure (15.7)	Tyndall effect (15.8)
Henry's law (15.5)	parts per billion (15.6)	unsaturated solution (15.3)
hydration (15.3)	parts per million (15.6)	weight percent (15.6)

Practice Test

After you have finished studying the chapter and the homework problems, the following questions can serve as a test to determine how well you have learned the chapter objectives.

1. When a small amount of the dark purple solid iodine (I_2) is added to a sample of liquid water (H_2O), the liquid remains colorless and the solid sinks to the bottom. When a small amount of iodine is added to a sample of liquid hexane (C_6H_{14}), the liquid becomes bright purple.
   (a) Is iodine soluble in water? Explain this observation in terms of intermolecular forces.
   (b) Is iodine soluble in hexane? Explain this observation in terms of intermolecular forces.

2. When a small amount of the green solid nickel(II) chloride ($NiCl_2$) is added to a sample of liquid water (H_2O), the liquid becomes bright green. When a small amount of $NiCl_2$ is added to a sample of liquid hexane (C_6H_{14}), the liquid remains colorless and the solid sinks to the bottom.
   (a) Is $NiCl_2$ soluble in water? Explain this observation in terms of intermolecular forces.
   (b) Is $NiCl_2$ soluble in hexane? Explain this observation in terms of intermolecular forces.

3. Water (H_2O, a polar liquid) and benzene (C_6H_6, a nonpolar liquid) are immiscible.
   (a) Is $\Delta S°_{soln}$ positive or negative? Explain your answer.
   (b) Are the H_2O and C_6H_6 molecules attracted to each other, do they repel each other, or are there no attractions or repulsions between the molecules? Explain your answer.
   (c) Is $\Delta H°_{soln}$ positive or negative? Explain your answer.
   (d) Which of these two factors ($\Delta S°_{soln}$ or $\Delta H°_{soln}$) is responsible for the fact that these two liquids are immiscible?

4. Methane (CH_4, a nonpolar gas) and ammonia (NH_3, a polar gas) have similar sizes and shapes, and are isoelectronic with each other.
   (a) Is $\Delta S°_{soln}$ for dissolving these gases in water positive or negative? Explain your answer.
   (b) Is $\Delta H°_{soln}$ for dissolving these gases in water positive or negative? Explain your answer.
   (c) Predict whether these gases would be soluble at low temperatures and at high temperatures.

5. When dissolving a solid compound (like salt or sugar) in water, why is it necessary to have some undissolved solid at the bottom of the container to be sure that the solution is saturated?

6. A recent study reports that an exposure to 3,404 ppm of hydrogen cyanide (HCN) would prove fatal to most humans within one minute. What is the mass fraction and mass percent of this value?

7. The solubility of gaseous carbon dioxide in water is 0.169 g CO_2 per 100.0 g of H_2O when the partial pressure of CO_2 is 1.00 atm.
   (a) What is the value of the Henry's law constant for CO_2 (in units of M/mm Hg)?
   (b) What is the solubility of carbon dioxide gas in water if the partial pressure of CO_2 is equal to 8,400 mm Hg?
   (c) What partial pressure of carbon dioxide (in atm) is needed to get a CO_2 solubility of 1.00 M?

8. A 12-ounce (355 mL) can of soda has been allowed to go flat, so that there is no $CO_2(g)$ dissolved in it. The nutritional information label on the can says that the soda contains 44.0 g of sugar (fructose, $C_6H_{12}O_6$, a nonelectrolyte). Useful information: d(flat soda) = 1.025 g/mL, $P°(H_2O)$ at 85°C = 433.6 mm Hg, $K_b(H_2O)$ = 0.52°C/m, $K_f(H_2O)$ = 1.86°C/m, $fp°(H_2O)$ = 0.0°C, $bp°(H_2O)$ = 100.0°C.
   (a) Determine the mass fraction and weight percent of sugar in the soda.
   (b) Determine the molarity and molality of the sugar in the soda.
   (c) Estimate the vapor pressure of this solution at 85°C.
   (d) Estimate the boiling point and freezing point of this solution.
   (e) Estimate the osmotic pressure (in atm) for this solution at 37.0°C.

9. When elemental sulfur (a nonelectrolyte) is dissolved in cyclohexane (C_6H_{12}), the melting point of cyclohexane is lowered. A solution containing 152.00 g of cyclohexane and 2.05 g of sulfur has a freezing point of 5.44°C. Useful information: $K_f(C_6H_{12})$ = 20.2°C/m, $mp°(C_6H_{12})$ = 6.50°C.
   (a) Determine the freezing point lowering of the solution.
   (b) Determine the molality of the solution.
   (c) Determine the number of moles of sulfur present in the solution.
   (d) Estimate the molar mass of sulfur. What is its molecular formula?

10. Rank the following solutions and water in order of increasing boiling points. What is the order of their freezing points?
    (a) Pure water, a nonelectrolyte
    (b) 0.75 m Li_2SO_4, an electrolyte
    (c) 1.0 m CH_2O, a nonelectrolyte
    (d) 1.0 m $NaCH_3COO$, an electrolyte
    (e) 1.5 m $Al(NO_3)_3$, an electrolyte
    (f) 2.25 m $C_{12}H_{22}O_{11}$, a nonelectrolyte

139

11. Ethanol (C_2H_5OH) is a polar solvent that can dissolve ionic compounds like NH_4F. A 0.0052 M solution of NH_4F dissolved in C_2H_5OH has a density of 0.789 g/mL.
    (a) Determine the mass fraction and ppm value for NH_4F in this solution.
    (b) Determine the molality of NH_4F in this solution.
    (c) Estimate the osmotic pressure (in atm) for this solution at 22°C.
    (d) The osmotic pressure of this solution was measured to be 0.23 atm. Explain why the actual value is different than the theoretical value. (*Hint:* Focus on i_{solute}).

12. Explain how the process of reverse osmosis can produce pure water from salty ocean water.

13. Soap molecules will also dissolve in nonpolar liquids like heptane (C_7H_{16}).
    (a) Draw a molecular picture showing how soap molecules dissolve in heptane (C_7H_{16}). Use the 'squiggly line/circle' symbol described in the chapter.
    (b) Explain why the soap molecules behave this way in heptane.

14. Water containing cations and anions as impurities can purified by using cation-exchange resin that replaces all cations in the solution with H_3O^+(aq) ions and then using an anion-exchange resin that replaces all anions with OH^-(aq) ions.
    (a) What ions would be present in the solution when a sample of 0.32 M Na_3AsO_4 is sent through these two resins?
    (b) Water purified by these two resins is referred to as *deionized water*. Explain why this water can be viewed as 'deionized'.
    (c) When a sample of impure water is sent through these two resins, is the solution guaranteed to be pure water? Explain your answer.

Chapter 16
Acids and Bases

The goal of this chapter is to describe the properties of acids and bases (including their equilibrium reactions in aqueous solutions) and **acid-base reactions** more fully. The chapter introduces the concept of conjugate acids and bases and the autoionization of water, and uses them to explain the concept of pH, which is widely used when working with acids and bases. This chapter also describes how a chemical's molecular structure affects its acidity or basicity. Finally, the chapter introduces Lewis acids and bases, and describes the uses of acids and bases in the household.

Water's Role in Aqueous Acid-Base Reactions (16.1)

In Chapter 5 of the textbook, the *Arrhenius definition of acids* (compounds that produce H_3O^+ ions) *and bases* (compounds that produce OH^- ions) was introduced. However, this definition is limiting because it does not describe how these ions are made, and few bases (other than alkali metal or alkaline earth metal hydroxide salts) release OH^- ions into the solution. Another definition, the *Bronsted-Lowry definition of acids and bases*, explains how these ions are made in aqueous solutions. A **Bronsted-Lowry acid** donates a proton (H^+ ion) to a water molecule, producing an H_3O^+ ion, while a **Bronsted-Lowry base** accepts (takes) a proton from a water molecule, producing an OH^- ion. The general formula of an acid is HA (it must have an H atom that can be lost as an H^+ ion), and the general formula for a base is B.

$$HA(aq) + H_2O(\ell) \rightleftharpoons H_3O^+(aq) + A^-(aq)$$

$$B(aq) + H_2O(\ell) \rightleftharpoons BH^+(aq) + OH^-(aq)$$

Conjugate Acids and Bases (16.1)

Since the Bronsted-Lowry acid-base reactions discussed above are equilibrium reactions, the products can react to form the reactants. So, these reactions show two different acid-base reactions with two different acids and bases. When an acid (HA) loses its proton to a base, it produces a base (A^-, called the *conjugate base*) that can accept a proton. Similarly, when a base (B) accepts a proton from an acid, it produces an acid (BH^+, called the *conjugate acid*) that can donate a proton. The relationships between these **conjugate acid-base pairs** are: $CB = A - H^+$ and $CA = B + H^+$, where CB is the conjugate base formula, A is the acid formula, CA is the conjugate acid formula, and B is the base formula.

Example: What are the conjugate acids and conjugate bases for the following ions or molecules?
(a) $H_2O(\ell)$, (b) $HCO_3^-(aq)$, (c) $NH_3(aq)$, (d) $HPO_4^{2-}(aq)$

Solution: To determine the conjugate acids for these species (when they act as a base), simply add an H^+ ion to their formulas; to determine the conjugate bases for these species (when they act as an acid), remove an H^+ ion from their formulas. These products will all be aqueous.
(a) The conjugate acid of H_2O is $H_2O + H^+ = H_3O^+(aq)$; the conjugate base of H_2O is $H_2O - H^+ = OH^-(aq)$.
(b) The conjugate acid of HCO_3^- is $HCO_3^- + H^+ = H_2CO_3(aq)$; the conjugate base of HCO_3^- is $HCO_3^- - H^+ = CO_3^{2-}(aq)$. Remember that you must not only add (or remove) an H atom, but also a positive charge.
(c) The conjugate acid of NH_3 is $NH_3 + H^+ = NH_4^+(aq)$; the conjugate base of NH_3 is $NH_3 - H^+ = NH_2^-(aq)$.
(d) The conjugate acid of HPO_4^{2-} is $HPO_4^{2-} + H^+ = H_2PO_4^-(aq)$; the conjugate base of HPO_4^{2-} is $HPO_4^{2-} - H^+ = PO_4^{3-}(aq)$. Chemicals that can gain or lose an H^+ ion are called **amphiprotic** or **amphoteric**.

In general, the stronger an acid is (the more it is willing to lose an H^+ ion), the weaker its conjugate base (the less willing it is to gain the H^+ ion back), and the stronger a base is (the more it is willing to gain an H^+ ion), the weaker its conjugate acid (the less willing it is to lose the gained H^+ ion). Figure 16.2 in the textbook (p. 761) contains a list of the relative strengths of several acids and their conjugate bases.

Strong acids like HCl, HBr, HI, $HClO_4$, HNO_3, and H_2SO_4 ionize completely into H_3O^+ and A^- ions in solution. Therefore, their conjugate base ions (Cl^-, Br^-, I^-, ClO_4^-, NO_3^-, and HSO_4^-) are exceedingly weak bases (nonbases) since they have no interest in the H^+ ions. Similarly, *strong bases* like NaOH, KOH, H^-, and CH_3^- react completely to form BH^+ and OH^- ions in solution. Therefore, the conjugate acids of these bases ($NaOH_2^+$ = Na^+ + H_2O, KOH_2^+ = K^+ + H_2O, H_2, and CH_4) are exceedingly weak acids (nonacids) since they have no interest in losing the H^+ ions they have just gained.

Acid-Base Properties of Amines and Carboxylic Acids (16.2)

The H atom of the carboxylic acid group, R-C(=O)OH, is acidic and can be donated to water in an aqueous solution. When the H^+ ion is lost, the carboxylic acid becomes a carboxylate ion, R-C(=O)O$^-$. **Amines**, organic compounds containing N atoms, act as bases because the N atom in these compounds can accept an H^+ ion from water, forming a positively-charged compound (the R groups in the amine listed below can be the same or different from each other and can be hydrocarbon groups or H atoms).

$$RCOOH(aq) + H_2O(\ell) \rightleftharpoons H_3O^+(aq) + RCOO^-(aq)$$

$$R_3N(aq) + H_2O(\ell) \rightleftharpoons R_3NH^+(aq) + OH^-(aq)$$

The Autoionization of Water (16.3)

Even samples of pure water have H_3O^+ and OH^- ions in them because one water molecule can accept an H^+ ion from another water molecule. This reaction is called the **autoionization** of water, since water is ionizing itself. The equilibrium constant for this reaction, called the **ionization constant for water**, is $K_w = [H_3O^+][OH^-]$, and equals 1.0×10^{-14} at 25°C.

$$2 H_2O(\ell) \rightleftharpoons H_3O^+(aq) + OH^-(aq)$$

Example: What is the [OH$^-$] concentration for a solution of 0.053 M HBr? What is the [H_3O^+] concentration for a solution of 0.00075 M KOH?

Solution: A solution of 0.053 M HBr (a strong acid) has [H_3O^+] = [Br^-] = 0.053 M, and a solution of 0.00075 M KOH (a strong base) has [K^+] = [OH^-] = 0.0075 M. Use the K_w expression and the [H_3O^+] value to determine the [OH$^-$] concentration in the HBr solution, and K_w and the [OH$^-$] value to find [H_3O^+] in the KOH solution.

$$K_w = [H_3O^+][OH^-], [OH^-] = \frac{K_w}{[H_3O^+]} = \frac{1.0 \times 10^{-14} \, M^2}{0.053 \, M} = 1.9 \times 10^{-13} \, M$$

$$K_w = [H_3O^+][OH^-], [H_3O^+] = \frac{K_w}{[OH^-]} = \frac{1.0 \times 10^{-14} \, M^2}{0.0075 \, M} = 1.3 \times 10^{-12} \, M$$

Determining Whether a Solution is Acidic, Basic, or Neutral (16.4)

As you can see above, even solutions of strong acids contain OH$^-$ ions in them and solutions of strong bases contain H_3O^+ ions. In order to determine whether a solution is acidic, basic, or neutral, you must look at the relative concentrations of these two ions. If [H_3O^+] > [OH$^-$], [H_3O^+] > 1.0×10^{-7} M, or [OH$^-$] < 1.0×10^{-7} M, it is an **acidic solution**; if [H_3O^+] < [OH$^-$], [H_3O^+] < 1.0×10^{-7} M, or [OH$^-$] > 1.0×10^{-7} M, it is a **basic solution**; and if [H_3O^+] = [OH$^-$] = 1.0×10^{-7} M, it is a **neutral solution**.

The acidity or basicity of a solution can also be determined by the **pH** (or pOH) of the solution. The pH of a solution is the negative logarithm (base 10) of the H_3O^+ concentration. Similarly, the pOH of a solution is the negative logarithm of the OH$^-$ concentration. At 25°C, the pH-pOH relationship is pH + pOH =

14.00. Solutions with pH < 7.00 or pOH > 7.00 are acidic, solutions with pH > 7.00 or pOH < 7.00 are basic, and solutions with pH = pOH = 7.00 are neutral.

$$pH = -\log[H_3O^+], \text{ or } [H_3O^+] = 10^{-pH} \text{ M}; \quad pOH = -\log[OH^-], \text{ or } [OH^-] = 10^{-pOH} \text{ M}$$

Converting Between Ion Concentrations and pH or pOH Values (16.4)

You can use the K_w expression, the definitions of pH and pOH, and the pH-pOH relationship to convert between values of $[H_3O^+]$, $[OH^-]$, pH, and pOH.

Example: What are the values of $[H_3O^+]$, $[OH^-]$, and pOH for a solution with a pH of 3.84? Is this solution acidic or basic?

Solution: Use the pH formula to determine $[H_3O^+]$ and the pH-pOH relationship to find pOH. Once you have $[H_3O^+]$ and pOH, you can use the K_w and $[H_3O^+]$ or the definition of pOH to calculate $[OH^-]$.

$$[H_3O^+] = 10^{-pH} \text{ M} = 10^{-3.84} \text{ M} = 1.4 \times 10^{-4} \text{ M}$$
$$pH + pOH = 14.00, \quad pOH = 14.00 - pH = 14.00 - 3.84 = 10.16$$

$$K_w = [H_3O^+][OH^-], \quad [OH^-] = \frac{K_w}{[H_3O^+]} = \frac{1.0 \times 10^{-14} \, M^2}{1.4 \times 10^{-4} \text{ M}} = 6.9 \times 10^{-11} \text{ M}$$

Since pH < 7.00, pOH > 7.00, $[H_3O^+] > 1.0 \times 10^{-7}$ M, and $[OH^-] < 1.0 \times 10^{-7}$ M, this solution is acidic.

Using K_a and K_b Values to Estimate Acid and Base Strengths (16.5)

The **acid ionization constant expression** and the **base ionization constant expression** for the equilibrium reactions of acids and bases in water are listed below, where K_a is the **acid ionization constant** and K_b is the **base ionization constant**.

$$HA(aq) + H_2O(\ell) \rightleftharpoons H_3O^+(aq) + A^-(aq) \qquad K_a = \frac{[H_3O^+][A^-]}{[HA]}$$

$$B(aq) + H_2O(\ell) \rightleftharpoons BH^+(aq) + OH^-(aq) \qquad K_b = \frac{[BH^+][OH^-]}{[B]}$$

If you have K_a or K_b values for two or more acids or bases, you can compare the relative strengths of these acids and bases. Acids with larger K_a values are more product-favored (and therefore stronger acids) than acids with smaller K_a values; similarly, bases with larger K_b values are stronger bases than bases with smaller K_b values. You can also compare K_a and K_b values for the same chemical.

Example: The bicarbonate ion (HCO_3^-) is amphiprotic when dissolved in water.
(a) Write the balanced equations and the equilibrium constant expressions for $K_a(HCO_3^-)$ and $K_b(HCO_3^-)$.
(b) The K_a value for HCO_3^- is 4.8×10^{-11} and the K_b value for HCO_3^- is 2.3×10^{-8}. Predict whether a solution containing bicarbonate ions (HCO_3^-) would be acidic or basic.

Solution:
(a) For K_a equations, the acid and water are the reactants, and H_3O^+ and the conjugate base are products. For K_b equations, the base and water are reactants, and OH^- and the conjugate acid are products.

$$HCO_3^-(aq) + H_2O(\ell) \rightleftharpoons H_3O^+(aq) + CO_3^{2-}(aq) \qquad K_a = \frac{[H_3O^+][CO_3^{2-}]}{[HCO_3^-]}$$

$$HCO_3^-(aq) + H_2O(\ell) \rightleftharpoons H_2CO_3(aq) + OH^-(aq) \qquad K_b = \frac{[H_2CO_3][OH^-]}{[HCO_3^-]}$$

(b) Since $K_a < K_b$ for the HCO_3^- ion, it is a stronger base than it is an acid. So, it should make water basic (pH > 7.00).

$$HCO_3^-(aq) + H_2O(\ell) \rightleftharpoons H_2CO_3(aq) + OH^-(aq)$$

Writing Successive Ionizations for Polyprotic Acids (16.5)

Acids that have only one hydrogen atom that can be lost as an H^+ ion are called **monoprotic acids**. Acids that can lose more than one hydrogen atom as an H^+ ion are called **polyprotic acids**. The acid ionization expressions for polyprotic acids can be written for each of the successive ionization reactions. It is also possible to have polybasic ions that can gain more than one H^+ ion; these base reactions can also be written in subsequent steps.

Example: Citric acid, $C_6H_8O_7(aq)$, is a triprotic acid (i.e., it has 3 hydrogen atoms that can be lost as H^+ ions). Write the balanced equations and the equilibrium constant expressions for the three K_a reactions.

Solution: For each K_a reaction, water is a reactant and H_3O^+ is a product.

$$C_6H_8O_7(aq) + H_2O(\ell) \rightleftharpoons H_3O^+(aq) + C_6H_7O_7^-(aq) \qquad K_{a1} = \frac{[H_3O^+][C_6H_7O_7^-]}{[C_6H_8O_7]}$$

$$C_6H_7O_7^-(aq) + H_2O(\ell) \rightleftharpoons H_3O^+(aq) + C_6H_6O_7^{2-}(aq) \qquad K_{a2} = \frac{[H_3O^+][C_6H_6O_7^{2-}]}{[C_6H_7O_7^-]}$$

$$C_6H_6O_7^{2-}(aq) + H_2O(\ell) \rightleftharpoons H_3O^+(aq) + C_6H_5O_7^{3-}(aq) \qquad K_{a3} = \frac{[H_3O^+][C_6H_5O_7^{3-}]}{[C_6H_6O_7^{2-}]}$$

The Relationship Between Acid Strength and Molecular Structure (16.6)

In general, H atoms tend to be more acidic when they are attached to very electronegative atoms (including O, N, Cl, Br, or I). Acids in which the H atom is attached to an O atom are called **oxoacids**. Oxoacids tend to be more acidic when there are more strongly electronegative (electron-withdrawing) atoms in the acid molecule because they can pull electrons away from the H atom (leaving it positive, and better able to leave the molecule as an H^+ ion).

Example: Explain why HS^-, has a K_a value of 1×10^{-19}, HSO_3^- (an oxoacid) has a K_a value of 6.3×10^{-8}, and HSO_4^- (an oxoacid) has a K_a value of 1.1×10^{-2}.

Solution: HS^- is much less acidic (smaller K_a) than HSO_3^- or HSO_4^- because the H atom is attached to a less electronegative atom ($EN = 2.3$ for S and 3.4 for O). The reason HSO_3^- is less acidic than HSO_4^- is that HSO_4^- has more highly electronegative O atoms in it that pull electron density away from the H atom, making it more positive (acidic) in HSO_4^- than in HSO_3^-.

Zwitterions of Amino Acids (16.6)

Amino acids, the monomer units used to form proteins, have the general formula $NH_2CHRC(=O)OH$. Amino acids have an amine group that can behave as a base and a carboxylic acid group that can behave as an acid in them. In fact, amino acids do not exist in their neutral form but as **zwitterions**, neutral molecules that have a positive and a negative ionic end ($NH_3^+CHRCO_2^-$).

144

Example: Use the equilibrium equations listed below and Hess's law to calculate a value of K_c for the zwitterionic reaction of glycine (NH_2CH_2COOH).

$$NH_2CH_2COOH(aq) \rightleftharpoons NH_3^+CH_2COO^-(aq)$$

$$NH_2CH_2COOH(aq) + H_2O(\ell) \rightleftharpoons H_3O^+(aq) + NH_2CH_2COO^-(aq) \quad K_a = 1.4 \times 10^{-3}$$

$$NH_2CH_2COO^-(aq) + H_2O(\ell) \rightleftharpoons NH_3^+CH_2COO^-(aq) + OH^-(aq) \quad K_b = 6.0 \times 10^{-5}$$

$$2 H_2O(\ell) \rightleftharpoons H_3O^+(aq) + OH^-(aq) \quad K_w = 1.0 \times 10^{-14}$$

Solution: Perform a Hess's law calculation to get the target equation, then calculate K_c from these values. Use equation 1 as written to get the NH_2CH_2COOH as a reactant, use equation 2 as written to get the $NH_3^+CHRCO_2^-$ as a product, and use equation 3 written backwards to get rid of H_2O, H_3O^+, and OH^-.

$$\mathbf{NH_2CH_2COOH(aq)} + H_2O(\ell) \rightleftharpoons H_3O^+(aq) + NH_2CH_2COO^-(aq)$$

$$NH_2CH_2COO^-(aq) + H_2O(\ell) \rightleftharpoons \mathbf{NH_3^+CH_2COO^-(aq)} + OH^-(aq)$$

$$H_3O^+(aq) + OH^-(aq) \rightleftharpoons 2 H_2O(\ell)$$

$$K_c = K_a \times K_b \times \frac{1}{K_w} = (1.4 \times 10^{-3})(6.0 \times 10^{-5})\left(\frac{1}{1.0 \times 10^{-14}}\right) = 8.4 \times 10^6$$

The very large value for K_c shows that this reaction is product-favored and glycine exists as a zwitterion.

Calculating pH Values for Acid and Base Solutions (16.7)

When working with strong acids or bases that ionize completely, calculating the pH of a solution is very straightforward because the concentration of the H_3O^+ or OH^- ion is equal to the initial concentration.

Example: Calculate the pH for an aqueous solution of 0.90 M nitric acid, HNO_3 (a strong acid), and for an aqueous solution of 0.0042 M barium hydroxide, $Ba(OH)_2$ (a strong base).

Solution: Since HNO_3 is a strong acid, it will ionize completely according to the equation below. In this solution $[H_3O^+] = [NO_3^-] = 0.90$ M, and pH $= -\log[H_3O^+] = -\log[0.90] = 0.05$ (very acidic).

$$HNO_3(aq) + H_2O(\ell) \rightarrow H_3O^+(aq) + NO_3^-(aq)$$

Since $Ba(OH)_2$ is a strong base, it will ionize completely according to the equation below (note that this strong base is a hydroxide donor and not a proton acceptor). In this solution $[Ba^{2+}] = 0.0042$ M and $[OH^-] = 0.0084$ M. pOH $= -\log[OH^-] = -\log[0.0084] = 2.08$, and pH $= 14.00 - 2.08 = 11.92$ (very basic).

$$Ba(OH)_2(aq) \rightarrow Ba^{2+}(aq) + 2 OH^-(aq)$$

Chemists often use *ICE tables* (Chapter 14) to perform pH calculations involving weak acids or bases.

Example: Calculate the pH for an aqueous solution of 0.90 M nitrous acid, HNO_2 ($K_a = 4.5 \times 10^{-4}$).

Solution: Although you were not told that HNO_2 is a weak acid, it is not one of the six strong acids mentioned previously. The fact that you are given a K_a value tells you HNO_2 is a weak acid (so does its name). First you need to write the balanced chemical equation for the K_a value and create an *ICE* table. The initial concentrations are $[HNO_2] = 0.90$ M, and $[H_3O^+] = [NO_2^-] = 0.0$ M (although neutral water has $[H_3O^+] = 1.0 \times 10^{-7}$ M, this value is small enough to ignore for most pH calculations). Since there are no H_3O^+ or NO_2^- ions under the initial conditions, they will be made when some HNO_2 and H_2O react, giving positive signs for the products and negative signs for the reactants in the *C* row. Remember that since water is a liquid, it is left out of the *ICE* table and the equilibrium expression.

	HNO_2(aq)	+	$H_2O(\ell)$	$\rightleftharpoons$	H_3O^+(aq)	+	NO_2^-(aq)
I	0.90 M				0.0 M		0.0 M
C	$-x$				$+x$		$+x$
E	0.90 M $- x$				x		x

Since you don't know the equilibrium concentrations, you must solve for x and use it to find the pH. You can solve for x using the quadratic formula, but most chemists will simplify this calculation by making an assumption. Since the K_a value is very small, this reaction is reactant-favored and very little HNO_2 will react to form H_3O^+ and NO_2^- (i.e., x is a very small number). So, 0.90 M $- x \approx 0.90$ M

$$K_a = \frac{[H_3O^+][NO_2^-]}{[HNO_2]} = \frac{[x][x]}{[0.90 \text{ M} - x]} = \frac{x^2}{0.90 \text{ M} - x} \approx \frac{x^2}{0.90 \text{ M}}$$

$$\frac{x^2}{0.90 \text{ M}} = 4.5 \times 10^{-4} \text{ M}, \ x^2 = (0.90 \text{ M})(4.5 \times 10^{-4} \text{ M}) = 4.1 \times 10^{-4} \text{ M}^2$$

$$x = \sqrt{4.1 \times 10^{-4} \text{ M}^2} = 0.020 \text{ M}$$

Since $x = [H_3O^+]$, take the negative logarithm of x to calculate pH.

$$pH = -\log[H_3O^+] = -\log[0.020] = 1.70$$

Whenever you make an assumption, you must test to see if that assumption was valid. Chemists normally use the *5% rule* to test these assumptions. The 5% rule states that if the number you ignored (dropped) is smaller than 5% of the number you kept, then your assumption is valid. Since you compared x and 0.90 M:

$$5\% \text{ rule:} \quad \frac{x}{0.90 \text{ M}} = \frac{0.020 \text{ M}}{0.90 \text{ M}} \times 100\% = 2.2\% < 5\% \text{ (good)}$$

So, why is the pH of a 0.90 M solution of a weak acid less than the pH of a 0.90 M solution of a strong acid (pH = 1.70 for HNO_2 and 0.05 for HNO_3)? This is because the strong acid ionizes completely (so $[H_3O^+] = 0.90$ M) while the weak acid ionizes much less. You can calculate the *percent ionization* of a weak acid by dividing the total initial concentration of the weak acid by the concentration of the acid that ionizes and multiplying by 100%. In the case of 0.90 M HNO_2 shown above:

$$\% \text{ ionization:} \quad \frac{0.020 \text{ M}}{0.90 \text{ M}} \times 100\% = 2.2\%$$

The Hydrolysis of Salts in Aqueous Solutions (16.8)

When an ionic compound dissolves in water, the cation and anion from the salt can participate in acid-base reactions with water (called **hydrolysis** reactions). In general, cations tend to be acidic and anions tend to be basic, but there are exceptions. Cations that come from strong bases (Li^+, Na^+, K^+, Mg^{2+}, Ca^{2+}, and Ba^{2+}) and anions that come from strong acids (Cl^-, Br^-, I^-, NO_3^-, ClO_4^-, and to some extent SO_4^{2-}) do not undergo acid-base reactions with water and are therefore neutral cations and anions. Most other cations (transition metal element ions, Al^{3+}, NH_4^+, etc.) react with water, making it acidic. Similarly, most anions (F^-, CN^-, CH_3COO^-, CO_3^{2-}, PO_4^{3-}, etc.) will make water basic, although some amphiprotic anions (like $H_2PO_4^-$, HSO_3^-, and HSO_4^-) actually make it acidic.

In order to determine how the hydrolysis of an ionic salt affects the pH of the solution, first write a reaction for the ionic salt dissolving in water into separate cations and anions. Then, write a balanced equation for the acid-base reaction of the cation and another one for the acid-base reaction of the anion. If neither ion reacts with water, the solution is neutral; if only one of the ions reacts with water, that ion will determine whether the solution is acidic or basic; if the ions react with water, you will have to compare the K_a and K_b values to determine which effect is stronger (like example on pp. 143-144 of this book for the HCO_3^- ion).

Example: Use hydrolysis reactions to predict whether solutions of lithium nitrate ($LiNO_3$) and ammonium cyanide (NH_4CN) will form acidic, basic, or neutral solutions when dissolved in water.

Solution: For the $LiNO_3$ solution, write the balanced equation showing $LiNO_3$ dissolving into its ions. Then, write a balanced equation for the acid-base reaction of each ion with water. Since Li^+ is the cation of $LiOH$ (a strong base), you know that Li^+ has absolutely no interest in OH^- ions, so it will not react with water to steal and OH^- ion (leaving behind an H^+ ion), and will not make the water acidic. Similarly, since NO_3^- is the anion of HNO_3 (a strong acid), you know that NO_3^- has absolutely no interest in H^+ ions, and will not react with water to steal an H^+ ion (leaving behind an OH^- ion), and will not make the water basic. Since neither of these ions will undergo acid-base reactions with water, a solution containing these ions should be neutral.

$$LiNO_3(s) \rightarrow Li^+(aq) + NO_3^-(aq)$$

$$Li^+(aq) + H_2O(\ell) \rightarrow \text{N.R.}$$

$$NO_3^-(aq) + H_2O(\ell) \rightarrow \text{N.R.}$$

For the NH_4CN solution, write the balanced equation showing NH_4CN dissolving into its ions. Then, write a balanced equation for the acid-base reaction of each ion with water. Since NH_4^+ is not the cation of a strong base, it has some interest in losing an H^+ ion to water, resulting in an acidic solution. Similarly, since CN^- is not the anion of a strong acid, it has some interest in stealing an H^+ ion from water, leaving behind an OH^- ion behind, resulting in a basic solution.

$$NH_4CN(s) \rightarrow NH_4^+(aq) + CN^-(aq)$$

$$NH_4^+(aq) + H_2O(\ell) \rightleftharpoons H_3O^+(aq) + NH_3(aq)$$

$$CN^-(aq) + H_2O(\ell) \rightleftharpoons HCN(aq) + OH^-(aq)$$

Since the NH_4^+ ion makes the solution acidic and the CN^- ion makes it basic, you need to determine which effect is stronger. The easiest way to do this is to find the K values for these two reactions. The NH_4^+ equation is simply $K_a(NH_4^+) = 5.9 \times 10^{-10}$; the CN^- equation is $K_b(CN^-) = 3.0 \times 10^{-5}$ (both values were taken from Table 16.2 on p. 772 of the textbook). Since $K_b > K_a$, the second reaction is more important, and a solution of NH_4CN will be basic.

When making comparisons of K_a and K_b values for competing hydrolysis reactions, sometimes you have to calculate K_a or K_b values for the ions from their conjugate acids or bases. The general relationship between K_a for an acid and K_b for its conjugate base is $K_a(HA)K_b(A^-) = K_w = 1.0 \times 10^{-14}$. This relationship also holds for the K_b of a base and the K_a of its conjugate acid—$K_a(BH^+)K_b(B) = K_w$.

Lewis Acids and Bases (16.9)

Although H_3O^+ ions react with Bronsted-Lowry bases like OH^- or NH_3, neutralizing them, other chemicals also react with these bases. For example, Cu^{2+} ions will 'neutralize' OH^- ions by forming an insoluble precipitate of $Cu(OH)_2(s)$ (removing them OH^- ions the solution and making it less basic), and Cu^{2+} ions form the dark blue metal **complex ion** $Cu(NH_3)_4^{2+}(aq)$ (removing NH_3 molecules from the solution and making it less basic). Since Cu^{2+} is reacting with these bases and 'neutralizing' them, Cu^{2+} is acting like an acid. To account for the fact that some objects appear to be acting as acids without H_3O^+ ions (or H^+ transfers from one chemical to another), Gilbert Lewis proposed a different definition of acids and bases.

A **Lewis acid** is a chemical that can accept a lone pair of electrons from another object to form a chemical bond. The chemical that is donating the lone pair of electrons to the Lewis acid is the **Lewis base**. The bond formed between a Lewis acid and a Lewis base is called a **coordinate covalent bond**. Coordinate covalent bonds are the same as other chemical bonds; the only difference is that these bonds are made from one atom bringing two electrons to the bond (Lewis base) and one atom bringing none (Lewis acid).

Lewis acids must have an empty orbital in them (like H^+, metal cations, BF_3, or other electron-deficient chemicals) or they must have a double bond—double bonds represent electrons being shared by the central atom and an outer atom; if a Lewis base shares a pair of electrons with the central atom, it no longer needs the electrons shared by the outer atom and can give them back to the outer atom. Lewis bases, on the other hand, must have a lone pair of electrons that can be donated to a Lewis acid.

Example: Identify the Lewis acid and Lewis base in the following reaction.

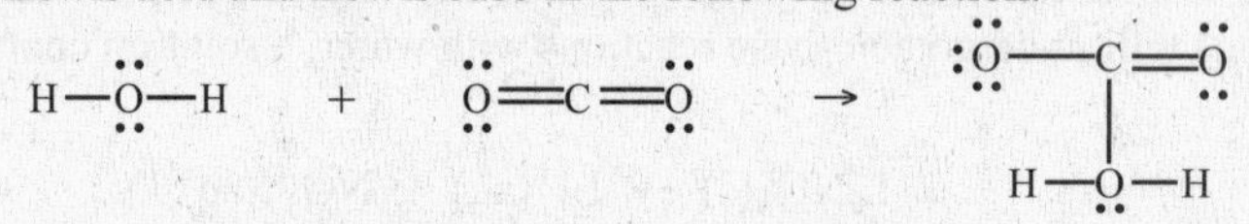

$$H - \overset{\cdot\cdot}{\underset{\cdot\cdot}{O}} - H \quad + \quad \overset{\cdot\cdot}{O} = C = \overset{\cdot\cdot}{O} \quad \rightarrow \quad \overset{\cdot\cdot}{\underset{\cdot\cdot}{:O}} - C = \overset{\cdot\cdot}{O}$$
$$|$$
$$H - \overset{\cdot\cdot}{O} - H$$

Solution: The easiest way to identify a Lewis base is to find the atom that has one less lone pair in the product side compared to the reactant side. In this case, the O from water started with two lone pairs but now has only one lone pair on it. So, this makes H_2O the Lewis base. Since the O atom in water formed a chemical bond with the C atom in CO_2, the CO_2 atom must be the electron-pair acceptor from H_2O (giving a pair of electrons it was sharing to one of its outer O atoms), making it the Lewis acid. Lewis acids can also be identified by looking to see if an atom has a larger number of electrons around it or more chemical bonds around it in the product side compared to the reactant side.

The Acidic Behavior of Hydrated Metal Ions (16.9)

Many metal cations, especially transition element cations and cations with large positive charges (+3 or higher), act as acids when placed in water. These metal cations are attracted to the negative O atom in water and form a bond in which the O atom shares one of its lone pairs with the metal cation. The metal cation pulls electrons from the M–O bond and the O–H bonds toward it, leaving the H atoms on this water molecule slightly more positive (acidic). Once an H^+ ion from this water molecule is transferred to another water molecule, the metal cation now has a hydroxide ion attached to it and has a smaller positive charge.

$$M\text{-}OH_2^{n+}(aq) + H_2O(\ell) \rightleftharpoons H_3O^+(aq) + M\text{-}OH^{(n-1)+}(aq)$$

Applying Acid-Base Principles to the Household (16.10)

Acid-base reactions are used extensively in your everyday life, and the textbook describes three examples of these reactions—antacids, chemical leavening agents, and drain cleaners. The human stomach is a very acidic environment (an approximately 0.10 M HCl solution), and most foods eaten by humans are on the acidic side of the pH scale (coffee, fruit juices, tomato sauces, etc.). When a person has eaten too much acidic food (or just too much food), they can experience indigestion that results from too much acid in their stomach. *Antacids* ('anti-acids') are chemicals that can neutralize some of the excess acid in the stomach. Antacids are bases that contain OH^-, HCO_3^-, or CO_3^{2-} ions that react with H_3O^+ ions to make water and carbon dioxide gas.

Chemical *leavening agents* (agents used to make dough rise) take advantage of the same reaction above, using H_3O^+ ions and either HCO_3^-, or CO_3^{2-} ions to produce carbon dioxide bubbles in baked goods that make them lighter and fluffier. The acids usually come from acidic foods like milk, buttermilk, or citrus juices, but the bicarbonate and/or carbonate ions must come from baking soda or baking powder (which also contains the solid acid KH_2PO_4).

Drain cleaners, on the other hand, tend to have very strong bases in them (like NaOH). These strong bases are useful in breaking up grease clogs (often made of triglyceride fats) by the process of saponification, in which the rather polar (and insoluble) triglycerides are broken into very soluble glycerol molecules and ionic sodium carboxylates. The strong bases in drain cleaners also decompose human hair in drain clogs.

Chapter Review — Key Terms

The key terms that were introduced in this chapter are listed below, along with the section in which they were introduced. You should understand these terms and be able to apply them in appropriate situations.

acid-base reaction (16.1)
acid ionization constant (16.5)
acid ionization constant expression (16.5)
acidic solution (16.3)
amines (16.2)
amphiprotic (16.1)
amphoteric (16.9)
autoionization (16.3)
base ionization constant (16.5)

base ionization constant expression (16.5)
basic solution (16.3)
Bronsted-Lowry acid (16.1)
Bronsted-Lowry base (16.1)
complex ion (16.9)
conjugate acid-base pair (16.1)
coordinate covalent bond (16.9)
hydrolysis (16.8)
ionization constant for water (16.3)

Lewis acid (16.9)
Lewis base (16.9)
monoprotic acids (16.5)
neutral solution (16.3)
oxoacids (16.6)
pH (16.4)
polyprotic acids (16.5)
zwitterion (16.6)

Practice Test

After you have finished studying the chapter and the homework problems, the following questions can serve as a test to determine how well you have learned the chapter objectives.

1. Identify the acid, the base, the conjugate acid, and the conjugate base in the following reactions.

    (a) $CH_3NH_2(aq) + H_2O(\ell) \rightleftharpoons CH_3NH_3^+(aq) + OH^-(aq)$

    (b) $H_2PO_4^-(aq) + CO_3^{2-}(aq) \rightleftharpoons HCO_3^-(aq) + HPO_4^{2-}(aq)$

    (c) $HCl(g) + NH_3(g) \rightleftharpoons NH_4Cl(s)$

    (d) $2\ HSO_3^-(aq) \rightleftharpoons SO_3^{2-}(aq) + H_2SO_3(aq)$

2. Consider the three nanoscale pictures that appear below:

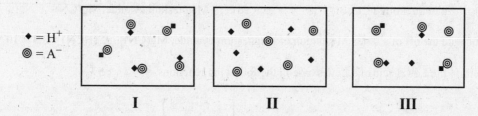

    ◆ = H^+
    ◎ = A^-

    I   II   III

    (a)   Classify these compounds as strong acids, weak acids, or nonacids. Explain your reasoning.
    (b)   Which of these solutions will have lowest pH value?
    (c)   Which of these solutions will react with the most OH^- ions?

3. Determine whether the solutions with the following conditions are acidic, basic, or neutral.
   (a) pOH = 2.43
   (b) $[OH^-] = 1.0 \times 10^{-7}$ M
   (c) $[H_3O^+] = 7.1 \times 10^{-5}$ M
   (d) pH = 7.00
   (e) $[H_3O^+] = 3.5 \times 10^{-12}$ M
   (f) pH = 3.50

4. You have a solution that contains a hydroxide concentration of $[OH^-] = 3.4 \times 10^{-5}$ M. What are the values of $[H_3O^+]$, pH, and pOH for this solution?

5. Hypochlorous acid (HOCl) has $K_a = 6.8 \times 10^{-8}$ and hypobromous acid (HOBr) has $K_a = 2.5 \times 10^{-9}$.
   (a) Write the balanced chemical equations for the acid-base reactions of these molecules with water and write the equilibrium expressions for these reactions.
   (b) Which chemical should have a lower pH value? Explain.
   (c) Calculate the pH of a 0.0015 M solution of each compound.
   (d) Calculate the percent ionization for each compound.
   (e) Explain why HOCl is a stronger acid than HOBr.

6. When the basic form of the amino acid alanine, $NH_2CH(CH_3)COO^-$, acts as a Bronsted-Lowry base with H_3O^+ ions, the H^+ ion can be added to the N atom or the O atom.
   (a) Write the balanced chemical equations for these two acid-base reactions.
   (b) The K_b value for the N atom is 7.4×10^{-5} and the K_b value for the O atom is 1.0×10^{-10}. Which of these two reactions is more likely to occur?

7. A 0.20 M solution of the hydrogen selenate ion ($HSeO_4^-$) has a pH of 1.26.
   (a) Write the balanced chemical equation for the acid-base reaction of this ion with water and write the equilibrium expression for this reaction.
   (b) Calculate a value for the acid ionization constant of $HSeO_4^-$.
   (c) Is $HSeO_4^-$ a weak acid or a strong acid? Explain your answer.

8. Fisher Scientific, a chemical supply company, sells concentrated solutions (14.8 M) of ammonia (NH_3, a weak base), but labels these bottles 'ammonium hydroxide' (NH_4OH).
   (a) Write the balanced chemical equation for the acid-base reaction of this ion with water and write the equilibrium expression for this reaction.
   (b) Find the K_c value for this reaction in the textbook. Given the value of K_c, would you expect this reaction be product-favored or reactant-favored?
   (c) Determine the pH of the concentrated ammonia solution.
   (d) Calculate the percent ionization of the concentrated ammonia solution.
   (e) Is this solution predominantly $NH_3(aq)$ or $NH_4OH(aq)$. Explain your answer.

9. Write the three-step hydrolysis reactions for the following ionic salts, and predict whether these solutions are acidic, basic, or neutral.
   (a) Aluminum nitrate, $Al(NO_3)_3$
   (b) Calcium bromide, $CaBr_2$
   (c) Iron(III) acetate, $Fe(CH_3COO)_3$
   (d) Magnesium cyanide, $Mg(CN)_2$

10. Calculate the pH of a 0.038 M solution of magnesium cyanide, $Mg(CN)_2$. $K_a(HCN) = 3.3 \times 10^{-10}$

11. Identify the Lewis acid and Lewis base in the following reactions.

   (a)

   (b) $4 \ :C\equiv O: \ + \ Ni \ \rightarrow \ Ni \!-\!\!\left(\!C\equiv O: \right)_4$

150
©2011 Cengage Learning. All Rights Reserved. May not be scanned, copied or duplicated, or posted to a publicly accessible website, in whole or in part.

Chapter 17
Additional Aqueous Equilibria

The goal of this chapter is to describe three additional aqueous equilibria reactions. These reactions are acid-base buffer solutions, the solubility of ionic compounds in water, and the formation of complex ions in water. This chapter also discusses acid-base titration curves and the formation and effects of acid rain.

Acid-Base Buffer Solutions and Buffer Capacity (17.1)

Acid-base **buffer solutions** are very important, especially in biological systems. **Buffers** are defined as solutions that resist changes in pH upon the addition of small amounts of acid (H_3O^+) or base (OH^-). Blood is an example of a buffered solution whose pH is about 7.40. When the pH drops below 7.35 *acidosis* occurs, and when the pH rises above 7.45 *alkalosis* occurs. Both of these conditions can be fatal.

For a buffer to resist pH changes when H_3O^+ is added, a buffer must have a base in it that will react with and neutralize the added H_3O^+ ions. Similarly, to resist pH changes when OH^- is added, a buffer must also have an acid in it that will react to neutralize the added OH^- ions. Acids and bases automatically react with each other, so why don't buffer solutions 'neutralize' themselves? To prevent this neutralization, buffers are made from a weak acid (HA) and its conjugate weak base (A^-); when the two chemicals react, the weak acid loses H^+ to the weak base (becoming A^-) and when the weak base gains H^+, it becomes HA:

$$HA(aq) + A^-(aq) \rightleftharpoons A^-(aq) + HA(aq)$$

Example: Why can't buffer solutions be made from a strong acid and its conjugate (i.e., HNO_3 and NO_3^-) or from a strong base and its conjugate (i.e., NaOH and $NaOH_2^+ = Na^+ + H_2O$)?

Solution: The reason buffer solutions cannot be made from a strong acid and its conjugate because the anions of strong acids (like NO_3^-) are nonbases and will not react with added H_3O^+ ions. Similarly, buffer solutions cannot be made from strong base and its conjugate because the cations of strong bases (like Na^+) are nonacids and will not react with added OH^- ions.

Example: How does a buffer made from hydrofluoric acid (HF) and lithium fluoride (LiF) resist pH changes from added H_3O^+ ions and from added OH^- ions?

Solution: When H_3O^+ ions are added, they will react with the F^- of the LiF (the conjugate base of the buffer); when OH^- ions are added, they will react with HF (the conjugate acid of the buffer).

$$F^-(aq) + H_3O^+(aq) \rightleftharpoons HF(aq) + H_2O(\ell)$$

$$HF(aq) + OH^-(aq) \rightleftharpoons F^-(aq) + H_2O(\ell)$$

Buffers can be made by the direct addition of the conjugate acid and the conjugate base (neutral molecules or ionic salts) to water. However, they can also be made by adding a small amount of a strong acid (like HCl) to a sample of the conjugate base or by adding a small amount of a strong base (like NaOH) to a sample of the conjugate acid.

Although buffers resist pH changes when small amounts of H_3O^+ or OH^- are added, every buffer has its limit, called the **buffer capacity**. When large amounts of H_3O^+ or OH^- are added to a buffer solution, the solution initially resists changes in pH, but when the buffer solution fails, the pH drops dramatically (with added H_3O^+) or rises dramatically (with added OH^-). The buffer fails when added H_3O^+ completely reacts with the conjugate base or when added OH⁻ completely reacts with the conjugate acid. Once one of the chemicals in the buffer is gone, added H_3O^+ or OH^- ions are not neutralized and the pH changes.

Calculating pH Values and pH Changes in Buffer Solutions (17.1)

The pH of a buffer solution can be calculated using the acid ionization constant expression (K_a) or the **Henderson-Hasselbalch equation**. This equation is derived by taking the negative logarithm of the K_a expression (just like taking the negative logarithm of K_w gives the formula pH + pOH = 14.00).

$$pH = pK_a + \log\frac{[A^-]}{[HA]}$$

Example: What is the pH of a buffer containing 0.82 M HF and 0.54 M KF? $K_a(HF) = 6.8 \times 10^{-4}$

Solution: If you are going to use the K_a expression, solve for $[H_3O^+]$, then take the negative logarithm to get pH. If you are going to use the Henderson-Hasselbalch equation, calculate $pK_a = -\log K_a$ then plug in the [HF] and [F⁻] values to get pH. Both methods should give the same answer.

$$HF(aq) + H_2O(\ell) \rightleftharpoons H_3O^+(aq) + F^-(aq)$$

$$K_a = \frac{[H_3O^+][F^-]}{[HF]}, \ [H_3O^+] = K_a \times \frac{[HF]}{[F^-]} = 6.8 \times 10^{-4} \ M \times \frac{0.82 \ M}{0.54 \ M} = 1.0 \times 10^{-3} \ M$$

$$pH = -\log[H_3O^+] = -\log[1.0 \times 10^{-3}] = 2.99$$

$$pK_a = -\log K_a = -\log(6.8 \times 10^{-4}) = 3.17$$

$$pH = pK_a + \log\frac{[F^-]}{[HF]} = 3.17 + \log\frac{[0.54 \ M]}{[0.82 \ M]} = 3.17 + (-0.18) = 2.99$$

You can also use the K_a expression or the Henderson-Hasselbalch equation to calculate how the addition of small amounts of H_3O^+ or OH^- changes the pH of a buffer solution.

Example: Calculate the pH of the buffer solution in the previous example after 0.03 M KOH is added.

Solution: When KOH is added to the buffer, the OH⁻ ions produced will react with the conjugate acid of the buffer—$HF(aq) + OH^-(aq) \rightarrow F^-(aq) + H_2O(\ell)$. So, 0.03 M of the HF will react, and [HF] now equals 0.82 M – 0.03 M = 0.79 M. When 0.02 M of the HF reacts, it produces 0.03 M F⁻ ions, so [F⁻] now equals 0.54 M + 0.03 M = 0.57 M. Solving, for $[H_3O^+]$ and then pH shows that the pH has only increased slightly upon the addition of 0.03 M KOH.

$$K_a = \frac{[H_3O^+][F^-]}{[HF]}, \ [H_3O^+] = K_a \times \frac{[HF]}{[F^-]} = 6.8 \times 10^{-4} \ M \times \frac{0.79 \ M}{0.57 \ M} = 9.4 \times 10^{-4} \ M$$

$$pH = -\log[H_3O^+] = -\log[9.4 \times 10^{-4}] = 3.03$$

$$pH = pK_a + \log\frac{[F^-]}{[HF]} = 3.17 + \log\frac{[0.57 \ M]}{[0.79 \ M]} = 3.17 + (-0.14) = 3.03$$

Interpreting Acid-Base Titration Curves (17.2)

Acid-base titrations were introduced in Chapter 5 as a way to determine the concentration of an acid using a *standardized solution* of a base, or vice versa. When a pH meter is placed in a solution of the acid and the base solution is added using a buret, a titration curve can be made. An acid-base **titration curve** is a plot of the pH of the solution versus the volume of **titrant** (the solution in the buret). Although titration curves are usually created by placing the pH meter in the acid solution and the base solution in the buret, a titration curve can be made by reversing the solutions. Although titration curves have the same general shape, they are slightly different depending on whether the acid and base are strong or weak. Titration curves for a strong acid and strong base, a weak acid and strong base, and a strong acid and weak base appear below.

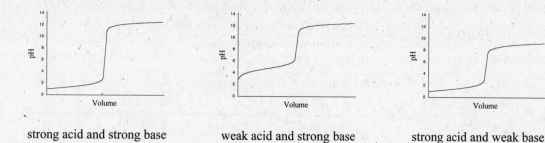

strong acid and strong base   weak acid and strong base   strong acid and weak base

One major difference between these plots is the pH of the **end point**, or *equivalence point* (the point on the curve where a stoichiometric equivalent of the acid and base have been added to the system, and where the pH jumps dramatically). The pH at the equivalence point is 7.0 for a strong acid/strong base curve, greater than 7.0 for a weak acid/strong base curve, and less than 7.0 for a strong acid/weak base curve.

The region of the titration curve before the equivalence point is determined by the type of acid. So, the first and third curves have much lower pH values before the end point (compared to the second graph) because they have strong acids that ionize 100% while the second curve has a weak acid that does not ionize completely. Similarly, the region of the titration curve after the equivalence point is determined by the type of base. So, the first and second curves have a much higher pH values after the end point because they have strong bases that ionize 100% while the third curve has a weak base that does not ionize completely.

For the strong acid or strong base regions of these curves, you can determine the pH of the solution by finding the concentration of unreacted $[H_3O^+]$ or $[OH^-]$ ions; in the weak acid or base regions of these curves, you must use the K_a or K_b expression.

Example: Calculate the pH when 20.0 mL (0.0200 L) of a 0.025 M solution of NaOH (a strong base) is added to 35.0 mL (0.0350 L) of a 0.036 M solution of HNO_3 (a strong acid).

Solution: Since this titration involves a strong acid and strong base, you need to calculate the amounts (mol) of H_3O^+ and OH^- ions present (because they are not the final answers, they have not been rounded).

$$n_{H_3O^+} = 0.0350 \text{ L solution} \times \frac{0.036 \text{ mol } HNO_3}{\text{L solution}} \times \frac{1 \text{ mol } H_3O^+}{\text{mol } HNO_3} = 1.26 \times 10^{-3} \text{ mol } H_3O^+$$

$$n_{OH^-} = 0.0200 \text{ L solution} \times \frac{0.025 \text{ mol NaOH}}{\text{L solution}} \times \frac{1 \text{ mol } OH^-}{\text{mol NaOH}} = 5.0 \times 10^{-4} \text{ mol } OH^-$$

Since there is more H_3O^+ ions present, this point appears in the strong acid region of the strong acid/strong base curve, and the pH will be defined by the concentration of unreacted $[H_3O^+]$ ions.

$$[H_3O^+] = \frac{(1.26 \times 10^{-3} \text{ mol } H_3O^+ - 5.0 \times 10^{-4} \text{ mol } H_3O^+)}{0.0200 \text{ L solution} + 0.0350 \text{ L solution}} = 0.014 \text{ M}$$

$$pH = -\log[H_3O^+] = -\log[0.014] = 1.86$$

Example: Calculate the pH when 20.0 mL of a 0.025 M solution of NaOH (a strong base) is added to 35.0 mL of a 0.036 M solution of HNO_2 (a weak acid). $K_a(HNO_2) = 4.5 \times 10^{-4}$

153

Solution: Since the volumes and concentrations are the same as the last example, $n_{HNO_2} = 1.26 \times 10^{-3}$ mol and $n_{OH^-} = 5.0 \times 10^{-4}$ mol. Since this falls in the weak acid region of the weak acid/strong base curve, the pH must be calculated using the K_a expression for HNO_2 after you have calculated $[HNO_2]$ and $[NO_2^-]$.

$$[HNO_2] = \frac{(1.26 \times 10^{-3} \text{ mol } HNO_2 - 5.0 \times 10^{-4} \text{ mol } HNO_2)}{0.0200 \text{ L solution} + 0.0350 \text{ L solution}} = 0.014 \text{ M}$$

$$[NO_2^-] = \frac{5.0 \times 10^{-4} \text{ mol } NO_2^-}{0.0200 \text{ L solution} + 0.0350 \text{ L solution}} = 0.0091 \text{ M}$$

$$K_a = \frac{[H_3O^+][NO_2^-]}{[HNO_2]}, \quad [H_3O^+] = K_a \times \frac{[HNO_2]}{[NO_2^-]} = 4.5 \times 10^{-4} M \times \frac{0.014 \text{ M}}{0.0091 \text{ M}} = 6.8 \times 10^{-4} \text{ M}$$

$$pH = -\log[H_3O^+] = -\log[6.8 \times 10^{-4}] = 3.16$$

Acid Rain (17.3)

Most samples of rain, snow, sleet, or fog are slightly acidic. This is due to the presence of carbon dioxide, which reacts with water to form carbonic acid (see the equation below). The equilibrium pH of carbon dioxide-saturated water is 5.6, so **acid rain** is usually defined as precipitation having a pH lower than 5.6.

$$CO_2(g) + 2 H_2O(\ell) \rightarrow H_2CO_3(aq) + H_2O(\ell) \rightleftharpoons H_3O^+(aq) + HCO_3^-(aq)$$

Acid rain is usually formed by the reaction of sulfur oxides (produced when sulfur-containing fuels are burned) or nitrogen oxides (produced in automobile engines) with water. Sulfur oxides react with water to produce sulfurous acid (H_2SO_3) and sulfuric acid (H_2SO_4); nitrogen oxides react with water to produce nitrous acid (HNO_2) and nitric acid (HNO_3). Acid rain causes serious environmental damage, leading to the death or damage of plant and animal life, as well as damage to buildings and marble statues.

Relating K_{sp} Expressions to Solubility Reactions (17.4)

In Chapter 5, you used solubility rules to determine whether an ionic solid will dissolve in water. Even insoluble salts listed in these rules dissolve to a small extent in water (and should be called *slightly* or *sparingly soluble*). These *solubility reactions* are written with the solid as the reactant and the aqueous ions as the products. The **solubility product constant expression** derived from this equation is labeled as K_{sp} (and is called the **solubility product constant**).

Example: Write the balanced equation and the equilibrium constant expression for the solubility of solid iron(III) carbonate, $Fe_2(CO_3)_3$, in water.

Solution: The reactant will be solid $Fe_2(CO_3)_3$, and the products will be the iron(III) and carbonate ions:

$$Fe_2(CO_3)_3(s) \rightleftharpoons 2 Fe^{3+}(aq) + 3 CO_3^{2-}(aq) \qquad K_{sp} = [Fe^{3+}]^2[CO_3^{2-}]^3$$

Note that the solid is left out of the K_{sp} expression, and its denominator is 1.

Calculating Solubility Product Constants (17.4)

The solubility of an ionic compound in water can be expressed in two different ways. The *mass solubility* is the mass of the compound that dissolves in water (with units of g/L); the *molar solubility* represents the concentration of the compound that dissolves in water (with units of mol/L). If you are given the mass or molar solubility of an ionic compound, you can calculate the value of K_{sp}.

Example: The mass solubility of aluminum hydroxide, $Al(OH)_3$, is 1.4×10^{-7} g/L. Write the balanced equation for this reaction and calculate the molar solubility and K_{sp} value for this reaction.

<u>Solution:</u> The balanced equation and the K_{sp} expression are:

$$Al(OH)_3(s) \rightleftharpoons Al^{3+}(aq) + 3\,OH^-(aq) \qquad K_{sp} = [Al^{3+}][OH^-]^3$$

The molar solubility of the compound is simply the mass solubility divided by its molar mass.

$$1.4 \times 10^{-7} \; \frac{g\,Al(OH)_3}{L\,solution} \times \frac{mol\,Al(OH)_3}{78.00\,g\,Al(OH)_3} = 1.8 \times 10^{-9}\,M$$

To calculate a value for K_{sp}, set up an *ICE* table.

	$Al(OH)_3(s)$ $\rightleftharpoons$	$Al^{3+}(aq)$	$+$	$3\,OH^-(aq)$
I		0.00 M		0.00 M
C		$+x$		$+3x$
E		x		$3x$

At equilibrium, you know that the amount of $Al(OH)_3$ solid that dissolved is $1.8 \times 10^{-9}\,M = x$. Plugging into the K_{sp} expression, you can find a numerical value for K_{sp}.

$$K_{sp} = [Al^{3+}][OH^-]^3 = [x][3x]^3 = 27x^4 = 27[1.8 \times 10^{-9}\,M]^4 = 2.8 \times 10^{-34}\,M^4 = 2.8 \times 10^{-34}$$

The Factors Affecting the Solubility of Ionic Compounds (17.5)

The textbook introduces four factors that affect the solubility of ionic compounds in water—the presence of H_3O^+ ions, the presence of common ions, the formation of complex ions, and amphoterism.

The addition of H_3O^+ ions increases the solubility of aqueous compounds if the anion in the solid is a weak base that reacts with H_3O^+ ions (in other words, all anions except those from strong acids like Cl^-, Br^-, I^-, NO_3^-, ClO_4^-, or SO_4^{2-}).

Example: Write balanced equations showing how $Mg(OH)_2$, $Mg(CH_3COO)_2$, $MgCO_3$, and $Mg_3(PO_4)_2$ solids will become more soluble upon the addition of H_3O^+ ions to the equilibrium system.

<u>Solution:</u> Since none of these anions come from strong acids, each of them will be protonated by H_3O^+.

$$Mg(OH)_2(s) + 2\,H_3O^+(aq) \rightarrow Mg^{2+}(aq) + 4\,H_2O(\ell)$$
$$Mg(CH_3COO)_2(s) + 2\,H_3O^+(aq) \rightarrow Mg^{2+}(aq) + 2\,H_2O(\ell) + 2\,CH_3COOH(aq)$$
$$MgCO_3(s) + 2\,H_3O^+(aq) \rightarrow Mg^{2+}(aq) + 3\,H_2O(\ell) + CO_2(g)$$
$$Mg_3(PO_4)_2(s) + 6\,H_3O^+(aq) \rightarrow 3\,Mg^{2+}(aq) + 6\,H_2O(\ell) + 2\,H_3PO_4(aq)$$

Le Chatelier's Principle and the Common Ion Effect (17.5)

The solubility of slightly soluble ionic compounds in water is dramatically decreased when the solid is dissolved in an aqueous solution containing the cation or anion from the solid (called *common ions*), or when a common ion is added to a saturated solution of the ionic solid. This **common ion effect** can be explained using Le Chatelier's principle. When you add a common ion to the saturated solution, the concentration of this ion increases. This causes the rate of the reverse reaction (precipitation of the solid) to increase until $Q = K_{sp}$ and the solution is again at equilibrium.

155

Calculating the Solubility of Ionic Compounds and the Common Ion Effect (17.5)

If you are given a value for the solubility product constant, you can calculate the mass and molar solubility of ionic compounds in pure water and in the presence of common ions.

Example: Determine the molar solubility of zinc hydroxide, $Zn(OH)_2$. $K_{sp} = 2.1 \times 10^{-16}$

Solution: The balanced equation and the K_{sp} expression are:

$$Zn(OH)_2(s) \rightleftharpoons Zn^{2+}(aq) + 2\,OH^-(aq) \qquad K_{sp} = [Zn^{2+}][OH^-]^2$$

	$Zn(OH)_2(s) \rightleftharpoons$	$Zn^{2+}(aq)$	$+$	$2\,OH^-(aq)$
I		0.00 M		0.00 M
C		$+x$		$+2x$
E		x		$2x$

Plug into the K_{sp} expression and solve for x, the molar solubility of the ionic solid.

$$K_{sp} = [Zn^{2+}][OH^-]^2 = [x][2x]^2 = 4x^3 = 2.1 \times 10^{-16}\ M^3$$

$$x^3 = \frac{2.1 \times 10^{-16}\ M^3}{4} = 5.3 \times 10^{-17}\ M^3, \quad x = \sqrt[3]{5.3 \times 10^{-17}\ M^3} = 3.7 \times 10^{-6}\ M$$

Example: Determine the molar solubility of iron(II) hydroxide, $Fe(OH)_2$ when the pH is 10.38.

Solution: In this solution, the common ion OH^- is present. First, you need to calculate $[OH^-]$ from pH.

$$pH + pOH = 14.00, \quad pOH = 14.00 - pH = 14.00 - 10.38 = 3.62$$

$$[OH^-] = 10^{-pOH}\ M = 10^{-3.62}\ M = 2.4 \times 10^{-4}\ M$$

	$Fe(OH)_2(s) \rightleftharpoons$	$Fe^{2+}(aq)$	$+$	$2\,OH^-(aq)$
I		0.00 M		2.4×10^{-4} M
C		$+x$		$+2x$
E		x		2.4×10^{-4} M $+ 2x$

$$K_{sp} = [Fe^{2+}][OH^-]^2 = [x][2.4 \times 10^{-4}\ M + 2x]^2 \approx [x][2.4 \times 10^{-4}\ M]^2$$

Because very little $Fe(OH)_2$ dissolves in water (i.e., x is very small), this calculation can be simplified by assuming that $2.4 \times 10^{-4}\ M + 2x \approx 2.4 \times 10^{-4}\ M$.

$$[x][2.4 \times 10^{-4}\ M]^2 = 8.0 \times 10^{-16}\ M^3, \quad x = \frac{8.0 \times 10^{-16}\ M^3}{[2.4 \times 10^{-4}\ M]^2} = 1.4 \times 10^{-8}\ M$$

$$\underline{\text{5\% rule:}} \quad \frac{2x}{2.4 \times 10^{-4}\ M} = \frac{2(1.4 \times 10^{-8}\ M)}{2.4 \times 10^{-4}\ M} \times 100\% = 0.012\% < 5\% \text{ (good)}$$

Complex Ion Formation and Dissolving Ionic Compounds (17.5)

Many metal cations form complex ions when they react with Lewis bases (like neutral NH_3 and H_2O molecules or negatively-charged Cl^-, CN^-, and OH^- ions). When these Lewis bases are added to saturated solutions of slightly soluble ionic compounds, these compound can dissolve. The balanced equation for the formation of these complex ions results in a K_f formula (the complex ion **formation constant** expression).

Example: Cyanide ions (CN^-) will cause the sparingly soluble ionic solid nickel(II) hydroxide to dissolve in water by making the complex ion $Ni(CN)_4^{2-}$. $K_{sp}(Ni(OH)_2) = 2.8 \times 10^{-16}$, $K_f(Ni(CN)_4^{2-}) = 1.7 \times 10^{30}$
(a) Write the balanced equations and the equilibrium constant expressions for the K_{sp} and K_f reactions.
(b) Use Le Chatelier's principle to explain how the formation of this complex ion will make $Ni(OH)_2(s)$ dissolve in water.
(c) Calculate K_c for forming the complex ion from solid nickel(II) hydroxide and aqueous cyanide ions.

Solution:
(a) The K_{sp} formula has the ionic solid as the reactant and the ions as products. The K_f formula has the complex ion as the product and the two ions in the complex ion as the reactants.

$$Ni(OH)_2(s) \rightleftharpoons Ni^{2+}(aq) + 2\,OH^-(aq) \qquad K_{sp} = [Ni^{2+}][OH^-]^2$$

$$Ni^{2+}(aq) + 4\,CN^-(aq) \rightleftharpoons Ni(CN)_4^{2-}(aq) \qquad K_f = \frac{[Ni(CN)_4^{2-}]}{[Ni^{2+}][CN^-]^4}$$

(b) When the complex ion is made from the Ni^{2+} and CN^- ions, the concentration of Ni^{2+} decreases. This decreases the reverse rate of the K_{sp} reaction (reducing the amount of $Ni(OH)_2(s)$ formed), and more of the solid will dissolve to counteract the loss of the Ni^{2+} ions.
(c) This reaction is simply the first two reactions added together. Since you are adding the two equations together, the K_c value for this reaction will simply be the product of the K_{sp} and K_f values. The large value of K_c shows that this reaction is product-favored and the solid will dissolve when cyanide ions are added.

$$Ni(OH)_2(s) + 4\,CN^-(aq) \rightleftharpoons Ni(CN)_4^{2-}(aq) + 2\,OH^-(aq) \qquad K_c = K_{sp} \times K_f$$

$$K_c = K_{sp} \times K_f = (2.8 \times 10^{-16})(1.7 \times 10^{30}) = 4.8 \times 10^{14}$$

Predicting Whether an Ionic Compound Will Precipitate (17.6)

You can predict whether a slightly soluble ionic compound will precipitate by determining a value of the reaction quotient (Q) for K_{sp} and comparing the Q and K_{sp} values—if $Q > K_{sp}$, a precipitate will form from the solution (which is supersaturated), but if $Q < K_{sp}$, no precipitate will form from the unsaturated solution.

Example: Will solid magnesium hydroxide, $Mg(OH)_2$, precipitate from a solution with a pH of 8.45 and a magnesium ion concentration of 0.050 M? $K_{sp} = 7.1 \times 10^{-12}$

Solution: First, solve for $[OH^-]$, then plug into the Q formula and compare this value to K_{sp}.

$$Mg(OH)_2(s) \rightleftharpoons Mg^{2+}(aq) + 2\,OH^-(aq) \qquad K_{sp} = [Mg^{2+}][OH^-]^2$$

$$pOH = 14.00 - pH = 14.00 - 8.45 = 5.55, \quad [OH^-] = 10^{-pOH}\,M = 10^{-5.55}\,M = 2.8 \times 10^{-6}\,M$$
$$Q = (\text{conc. } Mg^{2+})(\text{conc. } OH^-)^2 = (0.050\,M)(2.8 \times 10^{-6}\,M)^2 = 4.0 \times 10^{-13}\,M^3 < K_{sp}$$

Since $Q < K_{sp}$, solid magnesium hydroxide will not precipitate from this solution.

Selective Precipitation of Ionic Compounds (17.6)

Two different cations (or two anions) can be separated from each other by *selective precipitation*. In selective precipitation, an oppositely-charged ion is added to the solution that forms a much less soluble ionic compound with one of the ions to be separated than it does with the other ion. Selective precipitation can also be used when the added ion forms a slightly soluble ionic compound with one of the two ions to be separated and a very soluble ionic compound with the other ion.

Example: You have an aqueous solution that contains Ag^+ and Pb^{2+} ions (each at 0.010 M) and you decide to separate them by selective precipitation using Cl^- ions. $K_{sp}(AgCl) = 1.8 \times 10^{-10}$, $K_{sp}(PbCl_2) = 1.6 \times 10^{-5}$
(a) Write the balanced equations and the equilibrium constant expressions for these K_{sp} reactions.
(b) Calculate the Cl^- concentration needed to precipitate each solid.
(c) Explain how you can separate the Ag^+ and Pb^{2+} ions using Cl^- ions.

Solution:
(a) The balanced equations and the K_{sp} formulas are:

$$AgCl(s) \rightleftharpoons Ag^+(aq) + Cl^-(aq) \qquad\qquad K_{sp} = [Ag^+][Cl^-]$$

$$PbCl_2(s) \rightleftharpoons Pb^{2+}(aq) + 2\,Cl^-(aq) \qquad\qquad K_{sp} = [Pb^{2+}][Cl^-]^2$$

(b) For both solids, solve for $[Cl^-]$ in the K_{sp} expression.

$$K_{sp} = [Ag^+][Cl^-], \; [Cl^-] = \frac{K_{sp}}{[Ag^+]} = \frac{1.8 \times 10^{-10}\ M^2}{0.010\ M} = 1.8 \times 10^{-8}\ M$$

$$K_{sp} = [Pb^{2+}][Cl^-]^2, \; [Cl^-]^2 = \frac{K_{sp}}{[Pb^{2+}]}, \; [Cl^-] = \sqrt{\frac{K_{sp}}{[Pb^{2+}]}} = \sqrt{\frac{1.6 \times 10^{-5}\ M^3}{0.010\ M}} = 0.040\ M$$

(c) As long as the concentration of Cl^- remains below 0.040 M, $PbCl_2$ should not precipitate. So, the way to separate the Ag^+ and Pb^{2+} ions is to add Cl^- ions, making sure its concentration is more than 1.8×10^{-8} M but much less than 0.040 M. Then you can filter the precipitate from the solution; the Ag^+ ions should be in the precipitate as solid $AgCl$, and the Pb^{2+} should still be dissolved in the aqueous solution layer.

Chapter Review — Key Terms

The key terms that were introduced in this chapter are listed below, along with the section in which they were introduced. You should understand these terms and be able to apply them in appropriate situations.

acid rain (17.3)
buffer (17.1)
buffer capacity (17.1)
buffer solutions (17.1)
common ion effect (17.5)

end point (17.2)
formation constant (17.5)
Henderson-Hasselbalch equation (17.1)
ion product (17.6)
solubility product constant (17.4)

solubility product constant
 expression (17.4)
titrant (17.2)
titration curve (17.2)

Practice Test

After you have finished studying the chapter and the homework problems, the following questions can serve as a test to determine how well you have learned the chapter objectives.

1. Explain why acid-base buffer solutions are most effective when they are made with conjugate acid-conjugate base ratios between 0.10 and 10.0.

2. You have been asked to create 50.0 mL of a buffer solution with a pH of 9.25. You have 0.10 M solutions of HF ($K_a = 6.8 \times 10^{-4}$), HOCl ($K_a = 6.8 \times 10^{-8}$), and HCN ($K_a = 3.3 \times 10^{-10}$). You also have access to solid samples of LiF, NaOCl, and KCN.
   (a) Calculate the pK_a for each acid and determine which chemicals to use to make the buffer.
   (b) Calculate the ratio $[A^-]/[HA]$ for each buffer and explain why the other buffers would be ineffective.
   (c) Determine the volume of the acid solution and the mass of solid needed to make this buffer.

158

3. When 2.0 mL of a 0.10 M solution of HCl is added to the pH = 9.25 buffer made in the previous question, the pH decreases slightly.
   (a) Write the balanced equation for the reaction of this buffer with the added H_3O^+ ions.
   (b) Determine the pH change of the buffer solution after the HCl solution has been added.
   (c) Determine the pH change if 2.0 mL of a 0.10 M solution of HCl was added to 50.0 mL of an unbuffered solution with a pH of 9.25.

4. Does the titration curve to the right represent a strong acid/ strong base titration, a weak acid/strong base titration, or a strong acid/weak base titration? Why does this titration curve look different that the ones presented earlier in this book? (*Hint:* Look at the pH at the equivalence point for this titration curve).

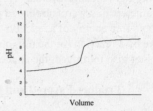

Volume

5. A 0.200 M solution of KOH (a strong base) is used to titrate 30.0 mL of a 0.200 M solution of HOBr (a weak acid, $K_a = 2.5 \times 10^{-9}$). Calculate the pH after the addition of the following volumes of KOH.
   (a) 0.0 mL of KOH   (c) 30.0 mL of KOH (the equivalence point)
   (b) 23.0 mL of KOH   (d) 42.0 mL of KOH

6. A saturated solution of barium phosphate, $Ba_3(PO_4)_2$ ($M = 601.8$ g/mol), contains 0.0059 g of dissolved solid in 15.0 L of solution.
   (a) Write the balanced equation and the K_{sp} expression for this solubility reaction.
   (b) Calculate the mass solubility and molar solubility for this reaction.
   (c) Determine the K_{sp} value for this reaction.

7. Determine the molar solubility of strontium fluoride, SrF_2, in the following solutions. $K_{sp} = 2.8 \times 10^{-9}$
   (a) pure water
   (b) 0.25 M strontium nitrate, $Sr(NO_3)_2$(aq)
   (c) 0.25 M sodium fluoride, NaF(aq)

8. Explain why the solubilities of strontium fluoride in question 7 are different for parts (b) and (c).

9. You have a solution with Ag^+, Au^{3+}, Cu^+, Hg^{2+}, and Pb^{2+} ions at a concentration of 3.0×10^{-6} M. When a drop of NaI(aq) is added to the solution, $[I^-] = 1.7 \times 10^{-6}$ M. K_{sp} values: $AgI = 8.3 \times 10^{-17}$, $AuI_3 = 1.0 \times 10^{-46}$, $CuI = 5.1 \times 10^{-12}$, $HgI_2 = 4.0 \times 10^{-29}$, $PbI_2 = 7.1 \times 10^{-9}$
   (a) Which of the ions listed above will precipitate as insoluble iodide salts?
   (b) If NaI(aq) is added dropwise, in what order should these solids precipitate?

10. In the last example of this chapter, you had an aqueous solution containing 0.010 M Ag^+ and Pb^{2+} ions and used Cl^- ions to selectively precipitate one solid. However, you could have used CrO_4^{2-} ions to separate these ions. $K_{sp}(Ag_2CrO_4) = 9.0 \times 10^{-12}$, $K_{sp}(PbCrO_4) = 1.8 \times 10^{-14}$
   (a) Write the balanced equations and the equilibrium constant expressions for these K_{sp} reactions.
   (b) Calculate the concentration of CrO_4^{2-} ions needed to precipitate each solid.
   (c) Explain how you can separate the Ag^+ and Pb^{2+} ions using CrO_4^{2-} ions.

11. When small amounts of aqueous NaOH are added to a solution containing Zn^{2+} ions, a white solid is formed. When more NaOH is added to this solution, the solid dissolves. Use the concepts in this chapter to explain these two events. Write balanced chemical equations and calculate K_c values for these two reactions. $K_{sp}(Zn(OH)_2) = 4.5 \times 10^{-17}$, $K_f(Zn(OH)_4^{2-}) = 4.5 \times 10^{17}$

159

Chapter 18
Thermodynamics: Directionality of Chemical Reactions

The goal of this chapter is to introduce Gibb's free energy as a way to predict whether a reaction is product-favored or reactant-favored. This new term is related to the entropy of the universe, and uses three parameters that were introduced previously—enthalpy, entropy, and temperature. This chapter also relates Gibb's free energy of a reaction to its equilibrium constant and the maximum amount of work that can be done by a reaction (or the minimum amount of work needed to force the reaction to occur). Finally, the chapter also introduces the concepts of thermodynamic and kinetic stability.

Product-Favored and Reactant-Favored Reactions (18.1)

Reactions that are predominantly shifted to the right and result in the production of the chemicals that appear on the right side of the equation are called product-favored reactions (or *spontaneous reactions*); reactions that are predominantly shifted to the left and result in the production of the chemicals that appear on the left side of the equation are referred to as reactant-favored reactions (or *nonspontaneous reactions*). Reactions that are product-favored in the forward direction are reactant-favored in the reverse direction and vice versa.

Probabilities and Energy Dispersal/Concentration (18.2)

Just as random motions of gas particles in a container result in the maximum dispersion of gas particles within the container, random motions of particles and random transfers of small amounts of energy from one particle to another results in the spontaneous dispersal of energy to all particles within the system. As a result, any reaction that disperses energy is considered to be spontaneous as written (product-favored) and any reaction that concentrates energy is considered to be nonspontaneous as written (reactant-favored). Exothermic reactions tend to be product-favored because the energy released from a few molecules before the reaction has occurred is now spread out over many molecules in the system, and endothermic reactions tend to be reactant-favored because they force energy to be concentrated among the product molecules. Similarly, reactions that result in the formation of more particles (especially gas molecules) are product-favored because the energy of the system is dispersed by the dispersal of the particles within the container.

Calculating Entropy Changes for Reversible Reactions (18.3)

The dispersal of energy within a system is measured by the thermodynamic parameter called *entropy* (S). The entropy change for a reversible reaction can be calculated as: $\Delta S = q_{rev}/T$, where q_{rev} is the heat change of the reversible reaction and T is the temperature in Kelvin units. A **reversible process** is a change in the system that can be reversed by making a small change in the parameters of the system (P, T, V, etc.).

Example: What is the entropy change that occurs when 1.00 g of liquid water boils into a gas at 100°C? $\Delta H^{\circ}_{vap} = 44.0$ kJ/mol

Solution: Since the water is boiling (an endothermic process), the q value will be positive. As a result, the entropy change will also be positive, which is consistent with the fact that the energy of the system will be more dispersed because the water molecules in the liquid state are allowed to escape the liquid and occupy a much larger volume as a gas, which also means an increased dispersal of energy.

$$q_{rev} = n\Delta H = \left(1.00 \text{ g H}_2\text{O} \times \frac{\text{mol H}_2\text{O}}{18.02 \text{ g H}_2\text{O}}\right)\left(44.0 \frac{\text{kJ}}{\text{mol H}_2\text{O}} \times \frac{10^3 \text{ J}}{1 \text{ kJ}}\right) = 2,440 \text{ J}$$

$$\Delta S = \frac{q_{rev}}{T} = \frac{2,440 \text{ J}}{(100 + 273.15 \text{ K})} = 6.54 \frac{\text{J}}{\text{K}}$$

Predicting the Sign of the Entropy Change for a Reaction (18.3)

The textbook introduces five general rules regarding the entropy of individual molecules or ions that can be used to predict whether the sign of the entropy change for a reaction is positive, negative, or close to zero. These rules are: (1) Entropies of gas molecules are much larger than liquid or solid particles, and liquid particles have higher entropies than solid particles; (2) As particles become more complex (more atoms within the molecule or more electrons within the atoms of the molecule), the entropies of the particles increase; (3) Entropies of ionic solids are higher when the ion-ion attractions are weaker; (4) When solids or liquids are dissolved in water to form aqueous solutions, the entropy of the solutions generally increase; and (5) When gases are dissolved in water to form aqueous solutions, the entropy of the solutions decrease.

Example: Predict whether the entropy change for these reactions are positive, negative, or close to zero.

(a) $H_2O(\ell) + HF(g) \rightarrow H_3O^+(aq) + F^-(aq)$

(b) $3 N_2H_4(g) \rightarrow N_2(g) + 4 NH_3(g)$

(c) $Zn(s) + Cu^{2+}(aq) \rightarrow Zn^{2+}(aq) + Cu(s)$

Solution:
(a) Since the reaction converts one gaseous and one liquid/aqueous particle into two liquid/aqueous particles, the energy is less dispersed after the reaction is over and $\Delta S°$ should be negative.
(b) In this reaction, three more complex gaseous molecules are converted into five simpler gaseous molecules. The number of gas particles is the stronger factor, so $\Delta S°$ should be positive for this reaction because the energy that was dispersed over three independent gas molecules is now dispersed over five.
(c) This reaction converts a solid and a liquid/aqueous ion into another solid and liquid/aqueous ion. Since neither molecule or ion is more complex than the other one, the dispersal of energy before and after the reaction should be just about the same, and $\Delta S°$ should be close to zero.

Calculating Entropy Changes for Reactions Using $S°$ Values (18.4)

The tabulated values for enthalpies are $\Delta H°_f$ because there is no way to determine the absolute amount of energy (enthalpy) that an atom or molecule contains. So, an arbitrary zero point (elements in their standard states) was defined. However, it is possible to estimate absolute values for the entropy (the dispersal of energy) of an atom or molecule, and absolute entropy values ($S°$) can be tabulated because the **third law of thermodynamics** states that the entropy values of all crystalline solids at zero Kelvin are zero. The $\Delta S°$ for any reaction can be calculated from tabulated $S°$ values using the following equation that looks just like the $\Delta H°/\Delta H°_f$ formula introduced in Chapter 6. The units of the $S°$ values are J/mol K and the units of n are mol, so the final answer has units of J/K.

$$\Delta S = \sum \{n_{product}S\ (product)\} - \sum \{n_{reactant}S\ (reactant)\}$$

Example: Calculate the entropy change for the following reaction. How does this value compare to your prediction above? $S°$ values (in J/mol K): $N_2H_4(g) = 238.6$, $N_2(g) = 191.6$, $NH_3(g) = 192.5$

$$3 N_2H_4(g) \rightarrow N_2(g) + 4 NH_3(g)$$

Solution: $\Delta S = \left\{ \left(1\ \text{mol}\ N_2 \times 191.6\ \dfrac{J}{\text{mol}\ N_2\ K}\right) + \left(4\ \text{mol}\ NH_3 \times 192.5\ \dfrac{J}{\text{mol}\ NH_3\ K}\right) \right\} -$

$\left\{ 3\ \text{mol}\ N_2H_4 \times 238.6\ \dfrac{J}{\text{mol}\ N_2H_4\ K} \right\} = \left\{ 191.6\ \dfrac{J}{K} + 770.0\ \dfrac{J}{K} \right\} - \left\{ 715.8\ \dfrac{J}{K} \right\} = 245.8\ \dfrac{J}{K}$

The sign of the $\Delta S°$ value is positive, just as you predicted in the previous example.

161

Using Enthalpy and Entropy Changes to Predict Spontaneity (18.5)

The **second law of thermodynamics** states that reactions are product-favored (spontaneous) when the entropy of the universe increases and reactant-favored (nonspontaneous) when the entropy of the universe decreases. The entropy of the universe is simply the sum of the entropy of the system ($\Delta S^\circ_{sys} = \Delta S^\circ_{rxn}$, calculated from tabulated S° values) plus the entropy of the surroundings (ΔS°_{surr}, calculated from the formula $\Delta S^\circ_{surr} = -\Delta H^\circ_{rxn}/T$).

Example: Calculate the entropy of the universe for this reaction. Is this reaction product-favored or reactant-favored at 25°C? ΔH°_f values (in kJ/mol): $N_2H_4(g) = 95.4$, $N_2(g) = 0.0$, $NH_3(g) = -46.1$

$$3 N_2H_4(g) \rightarrow N_2(g) + 4 NH_3(g)$$

Solution: The entropy of the system was calculated in the last example ($\Delta S^\circ_{sys} = \Delta S^\circ_{rxn} = 245.8$ J/K). You need to calculate ΔH° for this reaction from the ΔH°_f values and use this value to get ΔS°_{surr} and ΔS°_{univ}.

$$\Delta H = \left\{\left(1 \text{ mol } N_2 \times 0.0 \ \frac{kJ}{\text{mol } N_2}\right) + \left(4 \text{ mol } NH_3 \times -46.1 \ \frac{kJ}{\text{mol } NH_3}\right)\right\} - \left\{3 \text{ mol } N_2H_4 \times 95.4 \ \frac{kJ}{\text{mol } N_2H_4}\right\}$$

$$\Delta H = \{-184.4 \text{ kJ}\} - \{286.2 \text{ kJ}\} = -470.6 \text{ kJ} = -470,600 \text{ J}$$

$$\Delta S_{univ} = \Delta S_{sys} + \Delta S_{surr} = 245.8 \ \frac{J}{K} - \frac{-470,600 \text{ J}}{298.15 \text{ K}} = 245.8 \ \frac{J}{K} + 1580 \ \frac{J}{K} = 1820 \ \frac{J}{K}$$

Since the sign of ΔS°_{univ} is positive, this reaction should be product-favored (and this reaction would favor the formation N_2 and NH_3 from N_2H_4).

Calculating and Interpreting Gibbs Free Energy Values (18.6)

Chemists define a new thermodynamic parameter called **Gibbs free energy** ($\Delta G^\circ = -T\Delta S^\circ_{univ}$) to determine whether a reaction is product-favored or reactant-favored. Rearranging this mathematical relationship yields the following very important equation—$\Delta G^\circ = \Delta H^\circ - T\Delta S^\circ$. At low temperatures (when T is small), the Gibbs free energy (and therefore the spontaneity of a chemical reaction) is determined by the enthalpy (ΔH°) term and at high temperatures, the Gibbs free energy (reaction spontaneity) is determined by the entropy term ($T\Delta S^\circ$). When ΔG° is negative the reaction is product-favored, and when ΔG° is positive the reaction is reactant-favored. If the effect of the two terms in ΔG° counteract each other, you can calculate the temperature at which the reaction changes from being product-favored to reactant-favored (called the *equilibrium* or *transition temperature*) from the following equation—$T_{eq} = \Delta H^\circ/\Delta S^\circ$.

Example: Use this equation to answer the following questions. $3 N_2H_4(g) \rightarrow N_2(g) + 4 NH_3(g)$
(a) Calculate the Gibbs free energy (in J) for this reaction.
(b) Is this reaction product-favored or reactant-favored at low and high temperatures?
(c) What is the transition temperature for this reaction?

Solution:

(a) $\Delta G = \Delta H - T\Delta S = -470,600 \text{ J} - (298.15 \text{ K})\left(245.8 \ \frac{J}{K}\right) = -470,600 \text{ J} - 73,300 \text{ J} = -543,900 \text{ J}$

(b) Since ΔH° is negative (product-favored), this reaction is spontaneous at low temperatures; since ΔS° is positive (product-favored), this reaction is spontaneous at high temperatures as well.
(c) Since the ΔH° value and the ΔS° value both cause this reaction to be product-favored, there is no transition temperature. If you were to calculate a value for T_{eq}, it would be negative, which would be impossible for a temperature in units of Kelvin.

162

Calculating Gibbs Free Energy Changes for Reactions Using ΔG°_f Values (18.6)

You can also calculate ΔG° values at 25°C using tabulated Gibbs free energies of formation (ΔG°_f) values using a formula nearly identical to the $\Delta H^{\circ}/\Delta H^{\circ}_f$ formula introduced in Chapter 6. The units of the ΔG°_f values are kJ/mol (just like ΔH°_f) and the units of n are mol, so the final answer has units of kJ.

$$\Delta G = \sum \{n_{product}\Delta G_f(product)\} - \sum \{n_{reactant}\Delta G_f(reactant)\}$$

Calculating Standard Equilibrium Constants from Gibbs Free Energy Values (18.7)

Since the equilibrium constant and the Gibbs free energy value for a reaction can both be used to determine whether a reaction is product-favored or reactant-favored, these two values must be related to each other. This relationship is listed below, where K° is the **standard equilibrium constant**, which is the same as the equilibrium constant introduced in Chapter 14 except it has no units and it must contain concentrations for aqueous compounds and partial pressures for gases. When using this formula, the ΔG° value needs to be in units of kJ/mol or J/mol (i.e., it is a thermostoichiometric ratio).

$$\Delta G = -RT\ln K, \text{ or } K = e^{-\frac{\Delta G}{RT}}$$

<u>Example:</u> Calculate the equilibrium constant for this reaction at 25°C. $3 N_2H_4(g) \rightleftarrows N_2(g) + 4 NH_3(g)$

<u>Solution:</u> Use the ΔG° value calculated previously as a thermostoichiometric ratio ($-543,900$ J/mol N_2).

$$K = e^{-\frac{\Delta G}{RT}}, \quad -\frac{\Delta G}{RT} = -\frac{-543,900 \frac{J}{mol}}{\left(8.314 \frac{J}{mol\,K}\right)(273.15 + 25\,K)} = 219.42$$

$$K = e^{219.42} = 2.0 \times 10^{95} = \frac{P_{N_2}P^4_{NH_3}}{P^3_{N_2H_4}} = K_P$$

The ΔG° value (negative) and the K° value (10^{+x}) agree that this reaction is product-favored.

When the conditions of the system are not at standard states, the Gibbs free energy value at nonstandard conditions can be calculated. This value (ΔG) is related to ΔG° by the equation $\Delta G = \Delta G^{\circ} + RT\ln Q$, where Q is the reaction quotient of the reaction. ΔG and ΔG° must have units of kJ/mol or J/mol.

Coupling Product-Favored and Reactant-Favored Reactions (18.8)

Gibbs free energy for a product-favored reaction represents the maximum amount of work that can be done by this reaction on the surroundings. On the other hand, ΔG° for a reactant-favored reaction represents the minimum amount of work that must be done on the system to allow the reaction to occur. The only way for a reactant-favored (nonspontaneous) reaction to occur is for it to be coupled with and happen at the same time as a more product-favored reaction, so that the sum of ΔG° for both reactions is still negative.

Coupled Reactions in Biological Systems (18.9)

Many reactions in biological systems are **endergonic** (reactions with positive ΔG values) and are therefore nonspontaneous (reactant-favored) under normal conditions. In order for these reactions to occur under normal conditions, they must be coupled to more highly product-favored **exergonic** reactions (reactions with negative ΔG values). An example of an exergonic reaction used by biological systems to force other nonspontaneous reactions to occur is the conversion of ATP (adenosine triphosphate) to ADP (adenosine diphosphate) and a phosphate ion. ATP molecules (which are actually ions) are made in biological systems by the phosphorylation of ADP molecules (ions) and are created to serve as energy storage devices. Since this reaction is the reverse reaction of the exergonic reaction listed below, this reaction is endergonic and must be coupled to more product-favored reactions like the oxidation (combustion) of glucose, a sugar.

$$C_{10}H_{12}N_5O_{13}P_3^{\;4-}(aq) + H_2O(\ell) \rightarrow C_{10}H_{12}N_5O_{10}P_2^{\;3-}(aq) + H_2PO_4^{\;-}(aq) \qquad \Delta G = -30.5 \text{ kJ}$$
$$\text{ATP} \qquad\qquad\qquad\qquad\qquad \text{ADP}$$

Gibbs Free Energy and Energy Conservation (18.10)

When scientists discuss **energy conservation**, they are actually interested in the conservation of *useful* energy (i.e., Gibbs free energy). Energy conservation isn't really about the law of conservation of energy; instead, it is more about the efficient use of the free energy released by reactions that occur spontaneously. *Fuels* are chemicals that have the potential to release useful (free) energy, but many methods of releasing this energy (i.e., burning) to cause other reactions to occur have low efficiencies. In many cases, scientists working on energy conservation are interested in finding ways to use the free energy already being released from spontaneous reactions more efficiently to do more useful work.

Thermodynamic Stability and Kinetic Stability (18.11)

When chemicals do not react with other surrounding chemicals, they are said to be *stable*. There are two types of chemical stability. The first, *thermodynamic stability*, is measured by Gibbs free energy. If a reaction is thermodynamically stable, it has a positive ΔG value and will not spontaneously occur unless it is coupled to another more product-favored reaction. However, there are many chemical reactions that are thermodynamically unstable ($\Delta G < 0$) and should happen, but do not occur at any appreciable rate. This second kind of stability is called *kinetic stability*, and is measured by rate constants and activation energies. Reactions that are kinetically stable tend to have rather high activation energies, so very few molecules have enough energy to reach the transition state and react to form products (i.e., very small rate constants).

Chapter Review — Key Terms

The key terms that were introduced in this chapter are listed below, along with the section in which they were introduced. You should understand these terms and be able to apply them in appropriate situations.

endergonic (18.9)
energy conservation (18.10)
exergonic (18.9)
extent of reaction (18.7)

Gibbs free energy (18.6)
metabolism (18.9)
nutrients (18.9)
photosynthesis (18.9)

reversible process (18.3)
second law of thermodynamics (18.5)
standard equilibrium constant (18.7)
third law of thermodynamics (18.3)

Practice Test

After you have finished studying the chapter and the homework problems, the following questions can serve as a test to determine how well you have learned the chapter objectives.

1. What is the difference between an exothermic reaction and an exergonic reaction? What do these two terms tell you about the reaction?

2. Rank the following nanoscale pictures in order of increasing entropy values.

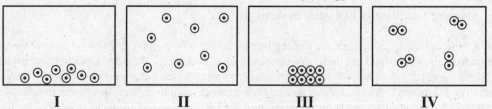

I   II   III   IV

3. For each pair, predict which molecule or ion has a greater absolute entropy. Explain your answer.
   (a) $CO_2(g)$ or $CO_2(s)$ (c) $NH_3(g)$ or $N_2H_4(g)$
   (b) $MgCl_2(aq)$ or $MgCl_2(s)$ (d) $HBr(aq)$ or $HBr(g)$

4. Under what conditions are the tabulated values of $\Delta H°_f$, $S°$, and $\Delta G°_f$ zero? Why are they zero?

5. Predict whether the entropy change for these reactions are positive, negative, or close to zero.
   (a) $SO_3(g) + 2\,NaOH(aq) \rightarrow Na_2SO_4(aq) + H_2O(\ell)$
   (b) $BrF_3(g) + 2\,H_2(g) \rightarrow HBr(g) + 3\,HF(g)$
   (c) $Fe(s) + NO_2(g) \rightarrow FeO(s) + NO(g)$
   (d) $HgCl_2(s) + 2\,Ag(s) \rightarrow 2\,AgCl(s) + Hg(\ell)$

6. Calculate the change in entropy ($\Delta S°$) for the reactions in question 5. Do these values match the predictions you made in the previous question? $S°$ values (in J/mol K): $Ag(s) = 43$, $AgCl(s) = 96$, $BrF_3(g) = 292.53$, $Fe(s) = 27.8$, $FeO(s) = 57.9$, $H_2(g) = 130.68$, $HBr(g) = 198.70$, $HF(g) = 173.78$, $H_2O(\ell) = 69.9$, $Hg(\ell) = 76$, $HgCl_2(s) = 146$, $NaOH(aq) = 48.1$, $Na_2SO_4(aq) = 138.1$, $NO(g) = 210.76$, $NO_2(g) = 240.06$, $SO_3(g) = 256.8$.

7. Calculate the Gibbs free energy for the following reactions at 25°C. Are these reactions product-favored or reactant-favored at this temperature? $\Delta G°_f$ values (in kJ/mol): $CS_2(\ell) = 65.27$, $CS_2(g) = 67.12$, $H_2O(\ell) = -237.13$, $H_2O(s) = -236.58$, $KOH(aq) = -440.5$, $KOH(s) = -379.1$.
   (a) $H_2O(\ell) \rightarrow H_2O(s)$ (c) $CS_2(\ell) \rightarrow CS_2(g)$
   (b) $KOH(s) \rightarrow KOH(aq)$

8. Calculate the Gibbs free energy for the following reactions at 250°C. Are these reactions product-favored or reactant-favored at this temperature? $\Delta H°_f$ values (in kJ/mol): $CS_2(\ell) = 89.70$, $CS_2(g) = 117.36$, $H_2O(\ell) = -285.83$, $H_2O(s) = -291.84$, $KOH(aq) = -482.4$, $KOH(s) = -424.8$. $S°$ values (in J/mol K): $CS_2(\ell) = 151.34$, $CS_2(g) = 237.84$, $H_2O(\ell) = 69.91$, $H_2O(s) = 47.91$, $KOH(aq) = 91.6$, $KOH(s) = 78.9$.
   (a) $H_2O(\ell) \rightarrow H_2O(s)$ (c) $CS_2(\ell) \rightarrow CS_2(g)$
   (b) $KOH(s) \rightarrow KOH(aq)$

9. Predict whether the reactions in question 8 are product-favored or reactant-favored at high and at low temperatures. Calculate the transition temperature for each equation as appropriate.

10. When copper(II) sulfide is burned in air, copper(II) sulfate can be formed according the following reaction that has $\Delta H° = -718$ kJ and $\Delta S° = -367$ J/K.

$$CS(s) + 2\,O_2(g) \rightleftharpoons CuSO_4(s)$$

   (a) Calculate the Gibb's free energy (in J) and the standard equilibrium constant for this reaction at 1000°C.
   (b) What is the Gibbs free energy (in J) for this reaction at 1000°C when $[O_2] = 2.125$ atm?

11. When propane (C_3H_8) burns in air, the following reaction occurs ($\Delta H° = -2044$ kJ, $\Delta S° = 101$ J/K):

$$C_3H_8(g) + 5\,O_2(g) \rightleftharpoons 3\,CO_2(g) + 4\,H_2O(g)$$

   (a) At what temperatures will this reaction be spontaneous (product-favored)?
   (b) Explain why liquid ethanol does not combust at room temperature without a spark.

Chapter 19
Electrochemistry and Its Applications

The goal of this chapter is to introduce the topic of **electrochemistry**, which is the relationship between electricity and chemical reactions. In voltaic (or galvanic) cells, product-favored reactions are used to produce electricity. Commercial batteries and fuel cells are examples of voltaic cells. In electrolytic (electrolysis) cells, electricity is used to force reactant-favored reactions to occur. This chapter starts with balancing oxidation-reduction reactions under acidic and basic conditions. It also relates the three variables that determine a reaction's spontaneity (Gibb's free energy, equilibrium constants, and cell potentials). Finally, the chapter describes the product-favored process of iron corrosion.

Identifying Oxidizing and Reducing Agents (19.1)

Oxidizing agents (chemicals that oxidize other chemicals and are reduced in the process) and *reducing agents* (chemicals that reduce other chemicals and are oxidized in the process) can be identified by looking for changes in oxidation numbers or by writing half-reactions. The rules for identifying oxidation numbers were introduced in Chapter 5. Oxidizing agents have an atom in them that undergoes an increase in its oxidation number; reducing agents have an atom in them that undergoes a decrease in its oxidation number.

Example: Identify the oxidizing agent and the reducing agent in the following reactions.
(a) $Al(s) + 3 Cu^+(aq) \rightarrow Al^{3+}(aq) + 3 Cu(s)$
(b) $CaS(s) + 2 O_2(g) \rightarrow CaSO_4(s)$

Solution:
(a) To identify oxidizing and reducing agents, first you need to determine the oxidation states for each chemical. The oxidation numbers for $Al(s)$ and $Cu(s)$ are 0, and the oxidation numbers for Cu^+ and Al^{3+} are +1 and +3 (respectively). Since the oxidation number for Al changes from 0 to +3 (increase), Al is oxidized and therefore $Al(s)$ is the reducing agent. Since the oxidation number for Cu changes from +1 to 0 (decrease), Cu is reduced and therefore $Cu^+(aq)$ is the oxidizing agent.
(b) Since CaS contains Ca^{2+} and S^{2-} ions, the oxidation number is +2 for Ca and –2 for S in this compound. The oxidation number for O in O_2 is 0. In $CaSO_4$, there are Ca^{2+} and SO_4^{2-} ions Ca has a +2 oxidation number, S has a +6 oxidation number, and O has a –2 oxidation number. Since the oxidation number for O changes from 0 to –2 (decrease), it is reduced and O_2 is the oxidizing agent; since the oxidation number for S changes from –2 to +6 (increase), it is oxidized and the S atom in CaS is the reducing agent.

Writing Half-Reactions for Redox Reactions (19.2)

Oxidation-reduction (*redox*) reactions can be separated into two **half-reactions**, one half-reaction showing the oxidation reaction and one showing the reduction reaction. Simple half-reactions that can be balanced by trial-and-error were introduced in Chapter 5 of this book. However, many half-reactions can be difficult to balance by trial-and-error. So, chemists have created a set of rules for balancing half-reactions under acidic and basic conditions.

The steps for balancing redox reactions under acidic conditions are: (1) Separate the reaction into two half-reactions; (2) Balance all atoms except O and H; (3) Balance the O atoms by adding $H_2O(\ell)$ molecules to one side of the equation; (4) Balance the H atoms by adding $H^+(aq)$ ions to one side of the equation; (5) Balance the charge by adding e^- to one side of the equation; (6) Add the two half-reactions together so that the electrons are completely canceled from the total equation, and cancel any chemicals that appear on both sides of the equation; (7) Add enough $H_2O(\ell)$ molecules to both sides of the equation to convert the $H^+(aq)$ ions to $H_3O^+(aq)$ ions; and (8) Cancel any chemicals that appear on both sides of the equation.

Example: Balance the following redox reaction, assuming acidic conditions.

$$Cr_2O_7^{2-}(aq) + VO^{2+}(aq) \rightarrow Cr^{3+}(aq) + VO_2^{+}(aq)$$

Solution: (1) One of the half-equations has the Cr-containing compounds and the other one has the V-containing compounds. (2) In the Cr reaction, place a 2 in front of the Cr^{3+} ions to balance Cr atoms; in the V reaction, the V atoms are already balanced. (3) In the Cr reaction, add 7 water molecules to the right side of the equation to balance the O atoms; in the V reaction, add one water molecule to the left side of the equation to balance the O atoms. (4) Add 14 H^+ ions to the left side of the Cr reaction and 2 H^+ ions to the right side of the V reaction to balance H atoms. (5) To balance charge, add 6 e^- to the left side of the Cr reaction and 1 e^- to the right side of the V reaction. (6) Multiply the V equation by 6 so that there will be the same number of e^- on both sides of the equation (so they cancel) and add the two half-reactions. Remove 12 H^+(aq) and 6 $H_2O(\ell)$ from both sides of the equation. (7) Add 2 H_2O to both sides of the equation, recognizing that 2 H^+ and 2 H_2O make 2 H_3O^+. (8) There are no chemicals that appear on both sides of the equation to be canceled.

$$Cr_2O_7^{2-}(aq) + 14\,H^+(aq) + 6\,e^- \rightarrow 2\,Cr^{3+}(aq) + 7\,H_2O(\ell)$$

$$VO^{2+}(aq) + H_2O(\ell) \rightarrow VO_2^{+}(aq) + 2\,H^+(aq) + e^-$$

$$Cr_2O_7^{2-}(aq) + 14\,H^+(aq) + 6\,VO^{2+}(aq) + 6\,H_2O(\ell) \rightarrow 2\,Cr^{3+}(aq) + 7\,H_2O(\ell) + 6\,VO_2^{+}(aq) + 12\,H^+(aq)$$

$$Cr_2O_7^{2-}(aq) + 2\,H^+(aq) + 6\,VO^{2+}(aq) \rightarrow 2\,Cr^{3+}(aq) + H_2O(\ell) + 6\,VO_2^{+}(aq)$$

$$Cr_2O_7^{2-}(aq) + 2\,H_3O^+(aq) + 6\,VO^{2+}(aq) \rightarrow 2\,Cr^{3+}(aq) + 3\,H_2O(\ell) + 6\,VO_2^{+}(aq)$$

When balancing redox reactions under basic conditions, use the first 6 steps from the acidic conditions and the following steps: (7') Add enough OH^-(aq) to both sides of the equation to convert the existing H^+ ions to water molecules; and (8') Cancel any chemicals that appear on both sides of the equation.

Example: Balance the following redox reaction, assuming basic conditions.

$$MnO_4^{-}(aq) + SO_2(aq) \rightarrow MnO_2(s) + SO_4^{2-}(aq)$$

Solution: (1) One of the half-equations has the Mn-containing compounds and the other one has the S-containing compounds. (2) In both reactions, the Mn and S atoms are already balanced. (3) In the Mn reaction, add 2 water molecules to the right side of the equation to balance the O atoms; in the S reaction, add 2 water molecules to the left side of the equation to balance the O atoms. (4) Add 4 H^+ ions to the left side of the Mn reaction and 4 H^+ ions to the right side of the S reaction to balance H atoms. (5) To balance charge, add 3 e^- to the left side of the Mn reaction and 2 e^- to the right side of the S reaction. (6) Multiply the Mn equation by 2 and the S equation by 3 so that there will be the same number of e^- on both sides of the equation. Cancel 8 H^+(aq) and 4 $H_2O(\ell)$ from both sides of the equation. (7') Add 4 OH^-(aq) to both sides of the equation, recognizing that 4 H^+ and 4 OH^- make 4 H_2O. (8') Remove 2 H_2O from both sides of the equation.

$$MnO_4^{-}(aq) + 4\,H^+(aq) + 3\,e^- \rightarrow MnO_2(s) + 2\,H_2O(\ell)$$

$$SO_2(aq) + 2\,H_2O(\ell) \rightarrow SO_4^{2-}(aq) + 4\,H^+(aq) + 2\,e^-$$

$$2\,MnO_4^{-}(aq) + 8\,H^+(aq) + 3\,SO_2(aq) + 6\,H_2O(\ell) \rightarrow 2\,MnO_2(s) + 4\,H_2O(\ell) + 3\,SO_4^{2-}(aq) + 12\,H^+(aq)$$

$$2\,MnO_4^{-}(aq) + 3\,SO_2(aq) + 2\,H_2O(\ell) \rightarrow 2\,MnO_2(s) + 3\,SO_4^{2-}(aq) + 4\,H^+(aq)$$

$$2\,MnO_4^{-}(aq) + 3\,SO_2(aq) + 2\,H_2O(\ell) + 4\,OH^-(aq) \rightarrow 2\,MnO_2(s) + 3\,SO_4^{2-}(aq) + 4\,H_2O(\ell)$$

$$2\,MnO_4^{-}(aq) + 3\,SO_2(aq) + 4\,OH^-(aq) \rightarrow 2\,MnO_2(s) + 3\,SO_4^{2-}(aq) + 2\,H_2O(\ell)$$

167

Identifying the Functions of the Parts of a Voltaic Cell (19.3)

When an oxidizing agent and a reducing agent react with each other, electrons are transferred and this movement of electrons represents an electrical current. To make use of this electrical current to do work (like lighting a light bulb or running a CD player), the oxidizing and reducing agents must be separated from each other so that the electrical current is forced to flow through an external circuit. **Voltaic cells** (also called *galvanic cells* or **batteries**, which are more generally called **electrochemical cells**) involve product-favored reactions that release free energy as electrical work. In these cells, there are two **half-cells**—one containing the reactants and products for the oxidation reaction and one containing the reactants and products for the reduction reaction.

Each half-cell contains an **electrode**, an electrically-conducting device (like a piece of metal or graphite) that allows electrons to enter or exit the half-cell through the conducting wire. Electrodes in oxidation half-cells are called **anodes** while electrodes in reduction half-cells are called **cathodes**. You can remember these definitions using the phrase 'an ox saw a red cat' (anode = oxidation; reduction = cathode). The two half-cells are linked by a **salt bridge**, a device that allows aqueous ions to flow between the half-cells to balance positive and negative charges.

Example: Describe the processes occurring at the anode and the cathode and describe the flow of charged particles (electrons and aqueous ions) occurring in a voltaic cell based on the following redox reaction:

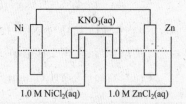

$$Ni^{2+}(aq) + Zn(s) \rightarrow Ni(s) + Zn^{2+}(aq).$$

Solution: The two half reactions are: $Ni^{2+}(aq) + 2\ e^- \rightarrow Ni(s)$ and $Zn(s) \rightarrow Zn^{2+}(aq) + 2\ e^-$. Since Ni^{2+} is reduced, the Ni electrode is the cathode, and the Zn electrode (where the oxidation occurs) is the anode.

At the zinc anode, Zn(s) atoms are releasing electrons that flow through the wire from left to right to the Ni cathode. When the Zn(s) atoms lose these electrons, they become $Zn^{2+}(aq)$ ions that enter the aqueous solution in the Zn half-cell. When the electrons reach the nickel cathode, they react at the cathode with the $Ni^{2+}(aq)$ ions in solution, producing Ni(s) atoms. So, as the reaction progresses, the Zn electrode will lose mass as Zn(s) atoms react and the Ni electrode will gain mass as Ni(s) atoms are produced.

As the reaction progresses, $Zn^{2+}(aq)$ ions are produced (leading to a build-up of positive ions in the right half-cell) and $Ni^{2+}(aq)$ ions are reacting (leading to a build-up of negative Cl^- ions in the left half-cell). The ions in the salt bridge move into the half-cells to balance these charges. In the right half-cell, NO_3^- ions enter the solution to balance the production of Zn^{2+} ions (two NO_3^- ions enter the solution every time one Zn^{2+} ion enters the solution). In the left half-cell, K^+ ions enter the solution to balance the removal of Ni^{2+} ions (two K^+ ions enter the solution every time one Ni^{2+} ion reacts and leaves the solution).

Using Standard Reduction Potentials to Predict Spontaneity (19.4 & 19.5)

When constructing a voltaic cell, you can determine the direction of electron flow (and which chemicals are oxidized or reduced) by performing the direct reaction—mixing the reactants and seeing what products are formed. Another way of determining the product-favored reaction is by measuring or calculating the **cell voltage**, also called the **electromotive force (emf)** or the *cell potential*, of the reaction. The cell potential represents the free energy (electrical work) transferred per charged particle (electron).

Cell potentials measured under **standard conditions** (25°C, 1 bar pressure for gases, and 1.0 M aqueous concentrations), are called **standard voltages** or *standard potentials*. The cell potential (also called the *potential difference*) for a reaction is the electron's potential at the end of the reaction (the reduction product at the cathode) minus its potential at the start of the reaction (the oxidation reactant at the anode).

$$E_{cell} = E_f - E_i = E_{red} - E_{ox} = E_{cathode} - E_{anode}$$

Potentials for the electrons in oxidation or reduction half-reactions are tabulated as **standard reduction potentials**. As with enthalpy values, it is impossible to measure absolute potentials for any half-reaction; only potential differences can be measured. So (just as with $\Delta H°_f$ values), chemists defined a zero point for the standard reduction potentials—the standard reduction potential for the **standard hydrogen electrode** was arbitrarily defined to be 0.00 V. The reactions associated with standard reduction potentials are written as equilibrium reactions because they can occur in either direction, depending on the reaction conditions.

$$2\ H_3O^+(aq)\ +\ 2\ e^-\ \rightleftarrows\ H_2(g)\ +\ 2\ H_2O(\ell) \qquad E° = 0.00\ V$$

A positive cell potential tells you that a redox reaction is product-favored; a negative cell potential means the reaction is reactant-favored. When comparing two or more reduction potentials, the reaction with the most positive/least negative standard reduction potential is most likely to be reduced, and the reaction with the most negative/least positive standard reduction potential is most likely to be oxidized.

Example: Predict whether the following (unbalanced) reactions are product-favored or reactant-favored and write the balanced redox equation for these reactions.

(a) $NO_3^-(aq)\ +\ Br^-(aq)\ \rightarrow\ NO(g)\ +\ BrO_3^-(aq)$

(b) $H_2SO_3(aq)\ +\ O_2(g)\ \rightarrow\ H_2O(\ell)\ +\ SO_4^{2-}(aq)$

$$BrO_3^-(aq)\ +\ 6\ H_3O^+(aq)\ +\ 6\ e^-\ \rightleftarrows\ Br^-(aq)\ +\ 9\ H_2O(\ell) \qquad E° = +1.44\ V$$

$$O_2(g)\ +\ 4\ H_3O^+(aq)\ +\ 4\ e^-\ \rightleftarrows\ 6\ H_2O(\ell) \qquad E° = +1.23\ V$$

$$NO_3^-(aq)\ +\ 4\ H_3O^+(aq)\ +\ 3\ e^-\ \rightleftarrows\ NO(g)\ +\ 6\ H_2O(\ell) \qquad E° = +0.96\ V$$

$$SO_4^{2-}(aq)\ +\ 4\ H_3O^+(aq)\ +\ 2\ e^-\ \rightleftarrows\ H_2SO_3(aq)\ +\ 5\ H_2O(\ell) \qquad E° = +0.17\ V$$

Solution: Each of the half-reactions above are written as reduction reactions, but redox reactions require one half-reaction to occur as a reduction and one to occur as an oxidation. This means that one of these half-reactions will be used as written and the other half-reaction must be written backwards. You can balance these half-reactions by using the rules previously introduced, but it is easier to use the balanced half-reactions given with the standard reduction potentials.

(a) The N half-reaction is used as written (NO_3^- as a reactant, NO as a product), and the Br half-reaction is written backwards (Br^- as a reactant, BrO_3^- as a product). So, NO_3^- is reduced and Br^- is oxidized. To get the balanced equation, multiply the N equation by 2 and cancel the e^- and any chemicals on both sides.

$$E_{cell} = E_{red} - E_{ox} = +0.96\ V - 1.44\ V = -0.48\ V\ (reactant\text{-}favored)$$

$$2\ NO_3^-(aq) + 8\ H_3O^+(aq) + Br^-(aq) + 9\ H_2O(\ell)\ \rightarrow\ 2\ NO(g) + 12\ H_2O(\ell) + BrO_3^-(aq) + 6\ H_3O^+(aq)$$

$$2\ NO_3^-(aq)\ +\ 2\ H_3O^+(aq)\ +\ Br^-(aq)\ \rightarrow\ 2\ NO(g)\ +\ 3\ H_2O(\ell)\ +\ BrO_3^-(aq)$$

(b) The O half-reaction is used as written (O_2 as a reactant, H_2O as a product), and the S half-reaction is written backwards (H_2SO_3 as a reactant, SO_4^{2-} as a product). So, O_2 is reduced and H_2SO_3 is oxidized. To get the balanced equation, multiply the S equation by 2 and cancel the e^- and any chemicals on both sides.

$$E_{cell} = E_{red} - E_{ox} = +1.23\ V - 0.17\ V = +1.06\ V\ (product\text{-}favored)$$

$$O_2(g)\ +\ 4\ H_3O^+(aq)\ +\ 2\ H_2SO_3(aq)\ +\ 10\ H_2O(\ell)\ \rightarrow\ 6\ H_2O(\ell)\ +\ 2\ SO_4^{2-}(aq)\ +\ 8\ H_3O^+(aq)$$

$$O_2(g)\ +\ 2\ H_2SO_3(aq)\ +\ 4\ H_2O(\ell)\ \rightarrow\ 2\ SO_4^{2-}(aq)\ +\ 4\ H_3O^+(aq)$$

Calculating ΔG° Values from Cell Potentials (19.6)

The cell potential represents the free energy transferred per charged particle (electron). The mathematical relationship for the cell voltage is listed below, where n is the number (mol) of electrons transferred and F is the **Faraday constant** (the charge of a mole of electrons = 96,500 C/mol e^-).

169

$$E_{\text{cell}} = \frac{\text{free energy of } e^- \text{ transfer}}{\text{total charge of } e^- \text{ transferred}} = \frac{\Delta G}{-nF}, \text{ or } \Delta G = -nFE_{\text{cell}}$$

$$\underline{\text{units:}} \quad J = \text{mol } e^- \times \frac{C}{\text{mol } e^-} \times V, \text{ or } J = C \times V$$

The electrical charge is measured in **coulombs** (C) and the cell potential are measured in **volts** (V); a coulomb-volt (C × V) is the same as a joule (J). The total charge of electrons transferred is the number of electrons times the charge per electron (nF); the negative sign comes from the fact that electrons are negatively-charged. Also note that when a reaction is product-favored, $\Delta G°$ will be negative and $E°$ will be positive (and vice versa for reactant-favored reactions).

Example: Calculate the free energy values (in kJ) for the following reactions.

(a) $2\ NO_3^-(aq) + 2\ H_3O^+(aq) + Br^-(aq) \rightarrow 2\ NO(g) + 3\ H_2O(\ell) + BrO_3^-(aq)$

(b) $O_2(g) + 2\ H_2SO_3(aq) + 4\ H_2O(\ell) \rightarrow 2\ SO_4^{2-}(aq) + 4\ H_3O^+(aq)$

Solution:
(a) This is the reaction from the previous example. In that example, the cell potential was calculated to be −0.48 V and the number of electrons transferred (that were cancelled out of the equation) is 6.

$$\Delta G = -nFE_{\text{cell}} = -(6\text{ mol } e^-)\left(\frac{96,500\text{ C}}{\text{mol } e^-}\right)(-0.48\text{ V}) \times \frac{1\text{ J}}{1\text{ C} \times \text{V}} \times \frac{1\text{ kJ}}{10^3\text{ J}} = 280\text{ kJ}$$

(b) In the previous example, the cell potential for this reaction was calculated to be 1.06 V and the number of electrons transferred (that were cancelled out of the equation) is 4.

$$\Delta G = -nFE_{\text{cell}} = -(4\text{ mol } e^-)\left(\frac{96,500\text{ C}}{\text{mol } e^-}\right)(+1.06\text{ V}) \times \frac{1\text{ J}}{1\text{ C} \times \text{V}} \times \frac{1\text{ kJ}}{10^3\text{ J}} = -409\text{ kJ}$$

Calculating Standard Equilibrium Constants from Cell Potentials (19.6)

Using the relationships $\Delta G° = -nFE°_{\text{cell}}$ and $\Delta G° = -RT \ln K°$, the relationship between $E°_{\text{cell}}$ and $K°$ is shown below. In these calculations, the n value (moles of e^- transferred) is used without units.

$$E_{\text{cell}} = \frac{RT}{nF} \ln K, \text{ or } K = e^{nFE_{\text{cell}}/RT}$$

Example: Calculate the standard equilibrium constant for the following reaction at 37°C.

$$Sn^{4+}(aq) + Sn(s) \rightleftharpoons 2\ Sn^{2+}(aq) \qquad\qquad E° = +0.29\text{ V}$$

Solution: In this reaction, $n = 2$ (Sn^{4+} gains 2 e^- to become Sn^{2+}, and $Sn(s)$ loses 2 e^- to become Sn^{2+}).

$$K = e^{nFE_{\text{cell}}/RT}, \quad \frac{nFE_{\text{cell}}}{RT} = \frac{(2)(96,500\text{ C/mol } e^-)(+0.29\text{ V})}{(8.314\text{ J/mol } e^-\text{ K})(273.15 + 37\text{ K})} \times \frac{1\text{ J}}{1\text{ C} \times \text{V}} = 21.71$$

$$K = e^{nFE_{\text{cell}}/RT} = e^{21.71} = 2.7 \times 10^9 = \frac{[Sn^{2+}]^2}{[Sn^{4+}]} = K_c$$

The Nernst Equation (19.7)

You can calculate the cell potential for redox reactions under nonstandard conditions using the **Nernst equation**. This equation is derived from the formulas $\Delta G° = -nFE°_{\text{cell}}$ and $\Delta G = \Delta G° + RT \ln Q$, where Q is the reaction quotient for the reaction and n is used without units.

$$E_{cell} = E_{cell} - \frac{RT}{nF} \ln Q$$

Using the Nernst Equation to Calculating Nonstandard Cell Potentials (19.7)

If you are given the concentrations of the reactants and products of a redox reaction, you can calculate the cell potential for the reaction under nonstandard conditions. The direction of the slight deviations of the nonstandard cell potential from $E°_{cell}$ can be explained using Le Chatelier's principle.

Example: Calculate the cell potential for this reaction at 37°C when $[Sn^{4+}] = 0.45$ M and $[Sn^{2+}] = 1.80$ M.

$$Sn^{4+}(aq) + Sn(s) \rightleftharpoons 2\,Sn^{2+}(aq) \qquad\qquad E° = +0.29\text{ V}$$

Solution: In this reaction, $n = 2$ (Sn^{4+} gains 2 e^- to become Sn^{2+}, and $Sn(s)$ loses 2 e^- to become Sn^{2+}).

$$E_{cell} = E_{cell} - \frac{RT}{nF} \ln Q = 0.29\text{ V} - \frac{(8.314\text{ J/mol }e^-\text{ K})(273.15 + 37\text{ K})}{(2)(96{,}500\text{ C/mol }e^-)} \times \frac{1\text{ C} \times \text{V}}{1\text{ J}} \ln \frac{[1.80]^2}{[0.45]}$$

$$E_{cell} = 0.29\text{ V} - 0.03\text{ V} = 0.26\text{ V}$$

Example: Use Le Chatelier's principle to explain the difference in E_{cell} and $E°_{cell}$ in the example above.

Solution: For the $E°_{cell}$ value, the concentrations of $[Sn^{4+}]$ and $[Sn^{2+}]$ are defined to be 1.00 M. Changing the $[Sn^{4+}]$ value from 1.00 M to 0.45 M shifts the reaction to the left (removing Sn^{4+} ions means the Sn^{2+} ions will react to make more Sn^{4+} ions and Sn atoms); changing the concentration of $[Sn^{2+}]$ from 1.00 M to 1.80 M also shifts the reaction to the left (adding more Sn^{2+} ions means the Sn^{2+} ions will react to make more Sn^{4+} ions and Sn atoms). Shifting the reaction to the left makes it slightly less product-favored or slightly more reactant-favored. Making the reaction less product-favored should make the nonstandard cell potential less positive than the standard cell potential, which it is. Note that the shift in the cell potential due to nonstandard concentrations is usually rather small.

Equilibrium Potentials Across Membranes in Neuron Cells (19.8)

Concentration cells are voltaic cells whose cell potential is based solely on differences in concentrations of aqueous compounds. The central nervous systems of biological creatures depend on electrical signals created as a result of different ion concentrations inside and outside the cell. The Nernst equation for these cells is shown below (assuming $T = 37°C$).

$$E_{cell} = \frac{RT}{nF} \ln \frac{(\text{conc. ion})_{outside}}{(\text{conc. ion})_{inside}} = 61.5\text{ mV} \log \frac{(\text{conc. ion})_{outside}}{(\text{conc. ion})_{inside}}$$

These **neuron** cells have very different ion concentrations inside and outside the cell for Ca^{2+}, Cl^-, Na^+, and K^+ ions. When neuron cells are stimulated, a brief electrical signal is created when ions flow into or out of the cell in an effort to equalize these concentrations. The difference in ion concentrations are created and maintained by active *ion pumps* within the cell.

Primary and Secondary Batteries (19.9)

Voltaic cells that provide electrical energy from product-favored reactions are referred to as *batteries*. **Primary batteries** are energy storage devices whose product-favored reaction cannot be easily reversed. This means that once primary batteries have been used, they cannot be recharged and should be discarded. **Secondary batteries**, on the other hand, are storage devices whose reaction can be reversed by recharging.

Dry cell batteries (under acidic conditions) were the first commercial batteries. Unfortunately, the reactions in these batteries still occurred even when the battery was not being used, and many of these batteries ruptured or leaked. The stability of these batteries was improved by making their solutions basic (*alkaline batteries*); alkaline batteries release energy according to the following equation.

$$Zn(s) + 2\,MnO_2(s) + H_2O(\ell) \rightarrow ZnO(s) + 2\,MnO(OH)(s) \qquad E° = +1.54\,V$$

Mercury batteries use the following reaction and tend to be used when small batteries are preferred.

$$Zn(s) + HgO(s) + H_2O(\ell) \rightarrow Hg(\ell) + Zn(OH)_2(s) \qquad E° = +1.35\,V$$

Lead-acid storage batteries use the reaction of lead metal and lead(IV) oxide in the presence of aqueous sulfuric acid (below). This battery is often used in automobiles and represents a secondary battery because it can be recharged and reused over and over. The major disadvantage of this battery is its large weight.

$$Pb(s) + PbO_2(s) + 2\,H_3O^+(aq) + 2\,HSO_4^-(aq) \rightarrow 2\,PbSO_4(s) + 4\,H_2O(\ell) \qquad E° = +2.04\,V$$

Fuel Cells (19.10)

A **fuel cell** is a voltaic cell that converts energy stored in a fuel (like H_2 or CH_4) into electrical energy. Unlike commercial batteries that are self-contained and have the reactants and products stored inside them, the reactants must be continually added and the products must be continually removed from a fuel cell. Fuel cells are more attractive than simple combustion engines because fuel cells are more efficient at converting the free energy released from the reaction into useful work.

Predicting the Products of Electrolysis Reactions (19.11)

Voltaic cells use product-favored reactions to produce electricity that can be used to do work. *Electrolytic cells*, on the other hand, use electricity from an external source to force reactant-favored reactions to occur. These chemical reactions are usually performed in a single container using inert electrodes that do not chemically react; however, the products of the **electrolysis** reaction must be separated.

Some properties are the same for both voltaic and electrolytic cells. These properties include the fact that oxidation is the loss of electrons/reduction is the gain of electrons, oxidation occurs at the anode/reduction occurs at the cathode, anions flow toward the anode/cations flow toward the cathode, and $E° = E°_{red} - E°_{ox}$. There are other properties that are opposite for voltaic and electrolytic cells (listed below).

property	voltaic cells	electrolytic cells
direction of spontaneity	product-favored	reactant-favored
sign of the cell potential	positive	negative
electricity requirements	generates electricity	requires electricity
container requirements	must separate reactants	must separate products

Example: Explain why the reactants of a voltaic cell must be separated (two-half cells) but reactants in electrolytic cells do not have to be separated (one container).

Solution: The reaction in voltaic cells is product-favored. So, if these reactants are mixed together in one container, they will react with each other directly and no electricity will flow through the external circuit. The reaction in electrolytic cells, on the other hand, is reactant-favored (nonspontaneous). So, the reactants do not need to be separated into different containers because they will not react with each other.

Electrolytic cells can occur in one of two forms—*molten cells* (where an ionic compound is heated until it melts into a liquid) or *aqueous cells* (where an ionic compound is dissolved in water). Predicting the products of electrolytic cells can be done using standard reduction potentials. First, you need to identify the chemicals that could be oxidized and reduced. If there is more than one chemical that could be oxidized or

172

reduced, use the relative values of their standard reduction potentials to predict which half-reaction will occur—reactions with more positive (less negative) reduction potentials are more likely to be reduced and reactions with more negative (less positive) reduction potentials are more likely to be oxidized.

Example: Predict the products of the electrolysis of molten lithium sulfate, $Li_2SO_4(\ell)$.

Solution: Molten lithium sulfate contains Li^+ and SO_4^{2-} ions. To determine which chemicals could be oxidized or reduced, look through the standard reduction potentials for reactions containing only Li^+ ion, SO_4^{2-} ions, or e^- on one side of the equation. The two reactions are (with reactants listed in boldface):

$$S_2O_8^{2-}(aq) + 2 e^- \rightleftharpoons \mathbf{2\ SO_4^{2-}\ (aq)} \qquad\qquad E° = +2.08 \text{ V}$$

$$\mathbf{Li^+(aq)} + e^- \rightleftharpoons Li(s) \qquad\qquad E° = -3.05 \text{ V}$$

In this molten salt, only Li^+ can be reduced (the only reactant appearing on the left side of the reduction equations) and only SO_4^{2-} can be oxidized (the only reactant appearing on the right side of the reduction equations). The net reaction and the cell potential (assuming the reduction potentials for the molten salt are the same as those for the aqueous ions at 25°C and standard conditions) appear below. The cell potential of electrolysis reactions are negative because these reactions are reactant-favored (nonspontaneous).

$$2 Li^+(\ell) + 2 SO_4^{2-}(\ell) \rightarrow 2 Li(s) + S_2O_8^{2-}(\ell)$$
$$E_{cell} = E_{red} - E_{ox} = -3.05 \text{ V} - 2.08 \text{ V} = -5.13 \text{ V}$$

Example: Predict the products of the electrolysis of aqueous lithium sulfate, $Li_2SO_4(aq)$.

Solution: Aqueous lithium sulfate contains Li^+ ions, SO_4^{2-} ions, and H_2O molecules. The reduction reactions containing only Li^+ ion, SO_4^{2-} ions, H_2O molecules, or e^- on one side of the equation are:

$$S_2O_8^{2-}(aq) + 2 e^- \rightleftharpoons \mathbf{2\ SO_4^{2-}\ (aq)} \qquad\qquad E° = +2.08 \text{ V}$$

$$O_2(g) + 4 H_3O^+(aq) + 4 e^- \rightleftharpoons \mathbf{6\ H_2O(\ell)} \qquad\qquad E° = +1.23 \text{ V}$$

$$\mathbf{2\ H_2O(\ell)} + 2 e^- \rightleftharpoons H_2(g) + 2 OH^-(aq) \qquad\qquad E° = -0.83 \text{ V}$$

$$\mathbf{Li^+(aq)} + e^- \rightleftharpoons Li(s) \qquad\qquad E° = -3.05 \text{ V}$$

In this aqueous solution, Li^+ and H_2O can be reduced (reactants on the left side of the equation). Since there is a choice of oxidizing agents, the reaction with the more positive (less negative) $E°$ value undergoes reduction (H_2O). In this aqueous solution, SO_4^{2-} and H_2O can be oxidized (reactants on the right side of the equation). Since there is a choice of reducing agents, the reaction with the more negative (less positive) $E°$ value undergoes oxidation (H_2O). Note that in the aqueous solution, the Li^+ and SO_4^{2-} ions are simply spectators. The net reaction and cell potential for this electrolysis reaction are:

$$10 H_2O(\ell) \rightarrow 2 H_2(g) + 4 OH^-(aq) + O_2(g) + 4 H_3O^+(aq)$$
$$E_{cell} = E_{red} - E_{ox} = -0.83 \text{ V} - 1.23 \text{ V} = -2.06 \text{ V}$$

Stoichiometry Calculations Involving Electrolysis Reactions (19.12)

You can perform stoichiometry calculations relating the amount of current (electron flow) transferred in an electrolytic cell and the amount of chemicals reacting or being produced as a result of the electrolysis reaction. Current flow (measured in **amperes**, A) is the amount of charged particles traveling past a point in the electrical circuit in a given timeframe. The relationship between current, charge, and time is:

$$\text{current flow} = \frac{\text{total charge of } e^- \text{ transferred}}{\text{time of } e^- \text{ transfer}}, \ i = \frac{q}{t}, \text{ or } q = it$$

units: $A = \dfrac{C}{s}$, or $C = A \times s$

Example: How much time (in s) is required to chrome-plate 283 g of chromium metal onto a car bumper if the current is 110 A and the applied voltage is 9.00 V. The reduction process for this reaction is:

$$CrO_4{}^{2-}(aq) + 8\ H_3O^+(aq) + 6\ e^- \rightarrow Cr(s) + 12\ H_2O(\ell)$$

Solution: Since you are interested in the time (t) for the reaction, you need to calculate the charge (q, in C) for the reaction transfer. Using the mass of Cr and the reduction reaction, you can determine the moles of e^- transferred, which can be converted to charge (in C) using the Faraday constant.

$$q = 283\ \text{g Cr} \times \frac{\text{mol Cr}}{52.00\ \text{g Cr}} \times \frac{6\ \text{mol } e^-}{1\ \text{mol Cr}} \times \frac{96,500\ \text{C}}{\text{mol } e^-} = 3.15 \times 10^6\ \text{C}$$

$$t = \frac{q}{i} = \frac{3.15 \times 10^6\ \text{C}}{110\ \text{A}} \times \frac{1\ \text{A} \times \text{s}}{1\ \text{C}} = 2.9 \times 10^4\ \text{s} = 8.0\ \text{h}$$

Electroplating Reactions (19.12)

Electroplating reactions involve using a piece of metal (or other conducting material) as the cathode in an electrolytic cell so that metal cations in the aqueous solution can be reduced into a thin layer of the new metal covering the old metal. Electroplating is often used to plate unreactive, and often more expensive, metals onto more reactive and less expensive metals.

Corrosion Processes and Their Prevention (19.13)

Corrosion is the oxidation of metals by oxygen in the atmosphere. Iron (steel) corrosion is probably the most important (and destructive) corrosion process. When iron corrodes, it reacts with oxygen in the air to make iron(III) oxide (*rust*). The iron corrosion reaction is particularly destructive because the rust formed by corrosion flakes off, revealing more iron metal that can corrode until the iron object completely rusts.

$$4\ Fe(s) + 6\ O_2(g) + 2x\ H_2O(\ell) \rightarrow 2\ Fe_2O_3 \cdot x\ H_2O(s)$$

Other metals (like aluminum or chromium) also corrode, but the metal oxide corrosion product forms a protective layer on the metal's surface, preventing further corrosion. ThomsonNOW contains a module (Chemistry Interactive) on the corrosion of aluminum.

Corrosion processes can be prevented by two different methods. The first method (**anodic inhibition**) involves protecting the metal and preventing it from reacting with oxygen in the air. This protection can involve painting the surface or by the formation of a protective oxide coating on the metal's surface.

The second method, called **cathodic protection**, involves attaching (or plating) a more reactive metal to this metal's surface, forcing the less reactive metal to act as a cathode (the site of reduction) instead of acting as the anode (and undergoing oxidation as part of the corrosion process). The more reactive metal (which is often zinc or magnesium) is called the *sacrificial anode* because it is sacrificed to oxidation so that the metal being protected doesn't oxidize.

Chapter Review — Key Terms

The key terms that were introduced in this chapter are listed below, along with the section in which they were introduced. You should understand these terms and be able to apply them in appropriate situations.

ampere (19.4)	anodic inhibition (19.13)	cathode (19.3)
anode (19.3)	battery (19.3)	cathodic protection (19.13)

174

cell voltage (19.4)   electromotive force (emf) (19.4)   salt bridge (19.3)
concentration cell (19.7)   Faraday constant (19.6)   secondary battery (19.9)
corrosion (19.13)   fuel cell (19.10)   standard conditions (19.4)
coulomb (19.4)   half-cell (19.3)   standard hydrogen electrode (19.4)
electrochemical cell (19.3)   half-reaction (19.2)   standard reduction potentials (19.5)
electrochemistry   Nernst equation (19.7)   standard voltages (19.4)
  (Introduction)   neurons (19.8)   volt (19.4)
electrode (19.3)   pH meter (19.7)   voltaic cell (19.3)
electrolysis (19.11)   primary battery (19.9)

Practice Test

After you have finished studying the chapter and the homework problems, the following questions can serve as a test to determine how well you have learned the chapter objectives.

1. When $Cl_2O(aq)$ and $N_2H_4(aq)$ are mixed, $Cl^-(aq)$ and $N_2(g)$ can be made.
   (a) Determine the oxidation numbers for these compounds or ions. Which chemical is oxidized and which chemical is reduced? What is the oxidizing and reducing agents in this reaction?
   (b) Balance this redox equation, assuming acidic conditions.
   (c) Balance this redox equation, assuming basic conditions.

2. A voltaic cell can be made from a silver and a platinum half-cell and a $NaNO_3(aq)$ salt bridge.

$$Pt^{2+}(aq) + 2\ e^- \rightleftharpoons Pt(s) \qquad\qquad E° = +1.20\ V$$

$$Ag^+(aq) + e^- \rightleftharpoons Ag(s) \qquad\qquad E° = +0.80\ V$$

   (a) Determine which chemical is oxidized and which is reduced.
   (b) Write the two half-reactions and the net reaction for this voltaic cell.
   (c) Calculate the cell potential for this reaction.
   (d) Which electrode is the anode and which is the cathode?
   (e) In which direction will the electrons flow in the external circuit?
   (f) In which direction do the $Ag^+(aq)$ and $Pt^{2+}(aq)$ ions move?
   (g) In which direction do the $Na^+(aq)$ and $NO_3^-(aq)$ ions from the salt bridge move?
   (h) What will happen to the mass of each electrode as the reaction progresses?

3. You can calculate the cell potential for reactions that are not actually oxidation-reduction reactions. Determine $E°_{cell}$ for these reactions. Are they product-favored or reactant-favored?
   (a) $PbCl_2(s) \rightleftharpoons Pb^{2+}(aq) + 2\ Cl^-(aq)$ (b) $H_3O^+(aq) + OH^-(aq) \rightleftharpoons 2\ H_2O(\ell)$

$$2\ H_3O^+(aq) + 2\ e^- \rightleftharpoons H_2(g) + 2\ H_2O(\ell) \qquad\qquad E° = +0.00\ V$$

$$Pb^{2+}(aq) + 2\ e^- \rightleftharpoons Pb(s) \qquad\qquad E° = -0.12\ V$$

$$PbCl_2(s) + 2\ e^- \rightleftharpoons Pb(s) + 2\ Cl^-(aq) \qquad\qquad E° = -0.26\ V$$

$$2\ H_2O(\ell) + 2\ e^- \rightleftharpoons H_2(g) + 2\ OH^-(aq) \qquad\qquad E° = -0.83\ V$$

4. Calculate the free energy change ($\Delta G°$) for the reactions listed in question 3. Are these reactions product-favored or reactant-favored? Is this consistent with the results in question 3?

5. Calculate the standard equilibrium constant ($K°$) for the reactions listed in question 3. How do these values match the tabulated values for these reactions?

6. Is the following reduction half-reaction product-favored or reactant-favored? Explain your answer.

$$Al^{3+}(aq) + 3 e^- \rightleftharpoons Al(s) \qquad\qquad E° = -1.68 \text{ V}$$

7. Standard reduction potentials assume all partial pressures are 1.0 atm and all aqueous concentrations are 1.0 M. However, under normal laboratory conditions ($T = 25°C$), pure water has a pH of 7.0 and the partial pressures of O_2 and H_2 are 0.21 atm and 5.0×10^{-7} atm (respectively).

$$O_2(g) + 4 H_3O^+(aq) + 4 e^- \rightleftharpoons 6 H_2O(\ell) \qquad\qquad E° = +1.23 \text{ V}$$

$$2 H_2O(\ell) + 2 e^- \rightleftharpoons H_2(g) + 2 OH^-(aq) \qquad\qquad E° = -0.83 \text{ V}$$

(a) Use the Nernst equation to calculate the standard reduction potentials for the half-reactions listed above under normal laboratory conditions.
(b) Calculate the cell potential for the electrolysis of water under these conditions.
(c) When pure water is placed in an electrolytic cell, no reaction occurs. Explain why no reaction occurs and explain what is needed for this reaction to occur.

8. Explain why lead-acid batteries will not function in very cold conditions (e.g., winters in Iowa).

9. Explain why the products produced at each electrode in electrolytic cells must be kept separate.

10. Use the following reduction half-reactions to predict the electrolysis products and calculate the cell potential for the following solutions.

$$O_2(g) + 4 H_3O^+(aq) + 4 e^- \rightleftharpoons 6 H_2O(\ell) \qquad\qquad E° = +1.23 \text{ V}$$

$$Br_2(\ell) + 2 e^- \rightleftharpoons 2 Br^-(aq) \qquad\qquad E° = +1.07 \text{ V}$$

$$2 H_2O(\ell) + 2 e^- \rightleftharpoons H_2(g) + 2 OH^-(aq) \qquad\qquad E° = -0.83 \text{ V}$$

$$Mg^{2+}(aq) + 2 e^- \rightleftharpoons Mg(s) \qquad\qquad E° = -2.36 \text{ V}$$

(a) Molten magnesium bromide, $MgBr_2(\ell)$.
(b) Aqueous magnesium bromide, $MgBr_2(aq)$.
(c) Which method would be better for producing Mg metal? Explain your answer.
(d) Which method would be better for producing liquid Br_2? Explain your answer.

11. Use the electrolysis method to show that iron(III) iodide, $FeI_3(aq)$, is unstable in water.

$$O_2(g) + 4 H_3O^+(aq) + 4 e^- \rightleftharpoons 6 H_2O(\ell) \qquad\qquad E° = +1.23 \text{ V}$$

$$Fe^{3+}(aq) + e^- \rightleftharpoons Fe^{2+}(aq) \qquad\qquad E° = +0.77 \text{ V}$$

$$I_2(s) + 2 e^- \rightleftharpoons 2 I^-(aq) \qquad\qquad E° = +0.54 \text{ V}$$

$$Fe^{3+}(aq) + 3 e^- \rightleftharpoons Fe(s) \qquad\qquad E° = -0.04 \text{ V}$$

$$2 H_2O(\ell) + 2 e^- \rightleftharpoons H_2(g) + 2 OH^-(aq) \qquad\qquad E° = -0.83 \text{ V}$$

12. Gold can be electroplated from $Au^{3+}(aq)$ ions, using a current of 5.00 A and a voltage of 7.50 V.
(a) How long (in h) will it take to plate out 875.0 g of Au under these conditions?
(b) How much gold (in g) can be plated out in 1.00 day (24.0 h)?

$$Au^{3+}(aq) + 3 e^- \rightarrow Au(s)$$

Chapter 20
Nuclear Chemistry

The goal of this chapter is to introduce nuclear chemistry, the study of chemical changes that occur within an atom's nucleus. This chapter describes how nuclear equations are balanced, and how unstable atoms are converted to more stable atoms. It also describes half-lives and how to determine the rates of nuclear decay reactions. This chapter also explains the concepts of nuclear fission and nuclear fusion reactions. Finally, the chapter introduces the units used to measure radiation dosage and how nuclear reactions can be used in food irradiation and medical imaging.

Radioactive Decay Processes (20.1)

Isotopes of atoms that spontaneously decompose, emitting particles and/or light with large amounts of energy, are said to be *radioactive*, and undergo the process of *radioactivity*. Although there are many radioactive isotopes, they all decay by the emission of the same particles—alpha particles, beta particles, gamma rays, or positron particles. **Alpha (α) particles** are $^4He^{2+}$ nuclei (particles containing $2\,p^+ + 2\,n^0$) that are emitted from an atom's nucleus. Similarly, **beta (β) particles** are electrons that are emitted from within an atom's nucleus. **Gamma (γ) rays** are simply a form of electromagnetic radiation (light) emitted from radioactive nuclei, often along with α or β particles.

Balanced Equations for Nuclear Reactions and Transmutations (20.2)

In normal (nonnuclear) chemical reactions, atoms are neither created nor destroyed (i.e., the reaction ends with the same number and types of atoms that were present before the reaction). However, when an atom undergoes a **nuclear reaction**, it is converted into another type of atom—this process is called a *nuclear transmutation*. When balancing nuclear reactions, there are two simple rules that must be followed: (1) The *law of conservation of matter* states that the mass of reactants and products should be the same (to the nearest amu), and (2) The *law of conservation of charge* states that the charge of the atoms' nuclei should be the same before and after the reaction. When balancing nuclear equations using nuclide symbols, the sum of the top numbers (A, the mass number) and the sum of the bottom numbers (Z, the nuclear charge number) on both sides of the equation must be equal.

Example: Determine the formula of the missing product in the following nuclear reactions.
(a) $^{26}Mg + {}^{2}H \rightarrow {}^{4}He +$ ____ , (b) $^{230}Th \rightarrow {}^{4}He +$ ____ , (c) $^{235}U + {}_{0}^{1}n^0 \rightarrow {}^{96}Y +$ ____ $+ 4\,{}_{0}^{1}n^0$

Solution:
(a) The sum of the top number (A) for the reactants is $26 + 2 = 28$, so the sum for the products must also be 28. That means that the unknown chemical has $A = 24$ (since $26 + 2 = 4 + 24$). The bottom numbers of nuclide symbols (Z) are often omitted but can be found from the periodic table. The sum of Z values for the reactants are $12 + 1 = 13$, so the sum for the products must also be 13. Since ^4He has a Z value of 2, the Z value for the unknown must be 11. When $Z = 11$, the atom is sodium; so, the unknown is ^{24}Na.
(b) The two mathematical formulas for these numbers are: $230 = 4 + A$ and $90 = 2 + Z$. So, the unknown atom has $A = 226$ and $Z = 88$. When $Z = 88$, the atom is radium and the unknown is ^{226}Ra.
(c) The two mathematical formulas are: $235 + 1 = 96 + A + 4 \times 1$, and $92 + 0 = 39 + Z + 4 \times 0$. So, the unknown atom has $A = 136$ and $Z = 56$. When $Z = 53$, the atom is iodine and the unknown is ^{136}I.

Predicting Modes of Nuclear Decay Reactions (20.2 & 20.3)

When a plot of stable and unstable (radioactive) isotopes are plotted on a graph with the $\#p^+$ on the *x*-axis and the $\#n^0$ on the *y*-axis, there is a very thin line of stable atoms (which is called the *band of stability*). This plot appears in Figure 20.2 of the textbook (p. 964). The important thing to know is that unstable (radioactive) isotopes will decay so that the products of the decay process are closer to the band of stability.

This leads to three general rules: (1) Radioactive isotopes with Z greater than 83 tend to decay by α-emission; (2) Radioactive isotopes above the band of stability (or isotopes with molar masses higher than the average mass from the periodic table) tend to decay by β-emission; and (3) Radioactive isotopes below the band of stability (or isotopes with molar masses lower than the average mass from the periodic table) tend to decay by positron (β^+) emission or by electron capture.

In order to see what happens inside the nucleus of an atom undergoing α-emission, look at the subatomic particles within the reactants and products of this reaction. As you can see, the ^{238}U nucleus (with $Z > 83$) emits a ^{4}He^{2+} ion (α-particle), producing ^{234}Th^{2-} ion. Once the reaction is over, 2 e$^-$ are transferred from Th to He, resulting in neutral atoms (in fact, most chemistry instructors and textbooks ignore these charges).

$^{238}_{92}$U	$\rightarrow$	$^4_2$He^{2+}	+	$^{234}_{90}$Th^{2-}
92 p$^+$		2 p$^+$		90 p$^+$
146 n^0		2 n^0		144 n^0
92 e$^-$		0 e$^-$		92 e$^-$

When an atom undergoes β-emission, an electron is emitted from the nucleus. But where did the electron come from (nuclei don't have electrons in them)? A ^{32}P atom (with $Z < 83$ and a molar mass higher than the value in the periodic table, 31) undergoes β-emission, producing an electron and an ^{32}S$^+$ ion. Later, the e$^-$ emitted from the nucleus reacts with ^{32}S$^+$, making a neutral ^{32}S atom. This reaction results in the loss of a neutron and the formation of a proton and electron ($^1_0$n^0 $\rightarrow$ $^1_1$p$^+$ + $^0_{-1}$e$^-$). So, when the neutron decays, the proton remains in the nucleus and the electron is emitted as a beta particle.

$^{32}_{15}$P	$\rightarrow$	$^0_{-1}$e$^-$	+	$^{32}_{16}$S$^+$
15 p$^+$		0 p$^+$		16 p$^+$
17 n^0		0 n^0		16 n^0
15 e$^-$		1 e$^-$		15 e$^-$

When an atom undergoes *positron-emission*, a **positron** (a positively-charged electron, e$^+$) is emitted. Positrons are also called *antielectrons,* because an electron and positron will annihilate each other, releasing two gamma rays ($^0_1$e$^+$ + $^0_{-1}$e$^-$ $\rightarrow$ 2 $^0_0\gamma$) when they collide. This reaction results in the loss of a proton and electron and the formation of a neutron ($^1_1$p$^+$ + $^0_{-1}$e$^-$ $\rightarrow$ $^1_0$n^0). So, when the proton decays, the neutron remains in the nucleus and the positron released from the nucleus annihilates itself and an electron.

$^{20}_{11}$Na $\rightarrow$ $^0_1$e$^+$ + $^{20}_{10}$Ne$^-$ = $^0_1$e$^+$ + $^{20}_{10}$Ne + $^0_{-1}$e$^-$ $\rightarrow$ $^{20}_{10}$Ne + 2 $^0_0\gamma$

$^{20}_{11}$Na	$\rightarrow$	2 $^0_0\gamma$	+	$^{20}_{10}$Ne
11 p$^+$		0 p$^+$		10 p$^+$
9 n^0		0 n^0		10 n^0
11 e$^-$		0 e$^-$		10 e$^-$

When an atom undergoes **electron capture**, an electron in a low-lying orbital (like the 1s orbital) is pulled into the nucleus of the radioactive isotope. But what does the nucleus do with this electron? When ^{37}Ar undergoes electron capture (the atom is written ^{37}Ar$^+$ + e$^-$ to show the 'captured' electron as a reactant), a neutral ^{37}Cl atom is made. This reaction (like the positron-emission) results in the loss of a proton and electron and the formation of a neutron ($^1_1$p$^+$ + $^0_{-1}$e$^-$ $\rightarrow$ $^1_0$n^0). So, when the electron is captured, it reacts with a proton in the nucleus, producing a neutron.

$^{37}_{18}$Ar$^+$	+	$^0_{-1}$e$^-$	$\rightarrow$	$^{37}_{17}$Cl
18 p$^+$		0 p$^+$		17 p$^+$
19 n^0		0 n^0		20 n^0
17 e$^-$		1 e$^-$		17 e$^-$

178

Both the positron-emission and the electron capture reactions will move unstable isotopes that are below the band of stability toward this region. Notice also that the starting atoms for the positron-emission (^{20}Na) and the electron capture (^{37}Ar) have molar masses less than their average molar masses in the periodic table (23 and 40, respectively) but the products (^{20}Ne and ^{37}Cl) have molar masses closer to their average molar masses in the periodic table (20 and 35.5). Since both of these reactions have the same effect (converting a proton and electron into a neutron), it is difficult to predict which one of these reactions will occur. In general, electron capture reactions become more likely than positron-emissions as Z increases.

Calculating Binding Energies (20.3)

Nuclear decay reactions occur because these particular atoms are unstable, and when they decay large amounts of energy are released. Einstein proposed the idea that this released energy (ΔE) comes from the annihilation of small amounts of matter (Δm) from the atom, using the following equation (where c is the speed of light, 3.00×10^8 m/s). When Δm is negative (mass loss) ΔE is negative (energy released), and when Δm is positive (mass gain) ΔE is positive (energy absorbed or added).

$$E = mc^2, \text{ or } \Delta E = \Delta mc^2$$

Example: What is the mass change (in g/mol) and energy change (in kJ/mol) for the following nuclear reaction? _m values (in g/mol)_: ^{1}H = 1.0078, ^{4}He = 4.0026, ^{7}Li = 7.0160

$$^7_3\text{Li} + ^1_1\text{H} \rightarrow 2\,^4_2\text{He}$$

Solution: First, use the masses to calculate Δm, then use Einstein's formula to calculate ΔE, remembering that 1 J = 1 kg m^2/s^2 or 1 J s^2 = 1 kg m^2.

$$\Delta m = m_f - m_i = 2\,m_{^4\text{He}} - (m_{^7\text{Li}} + m_{^1\text{H}}) = 2\left(4.0026\ \frac{\text{g}}{\text{mol}}\right) - \left(7.0160\ \frac{\text{g}}{\text{mol}} + 1.0078\ \frac{\text{g}}{\text{mol}}\right)$$

$$\Delta m = 8.0052\ \frac{\text{g}}{\text{mol}} - 8.0238\ \frac{\text{g}}{\text{mol}} = -0.0186\ \frac{\text{g}}{\text{mol}}$$

$$\Delta E = \Delta mc^2 = \left(-0.0186\ \frac{\text{g}}{\text{mol}} \times \frac{1\ \text{kg}}{10^3\ \text{g}}\right)\left(3.00 \times 10^8\ \frac{\text{m}}{\text{s}}\right)^2 \times \frac{1\ \text{J s}^2}{1\ \text{kg m}^2} \times \frac{1\ \text{kJ}}{10^3\ \text{J}} = -1.67 \times 10^9\ \frac{\text{kJ}}{\text{mol}}$$

Although the mass change is really small (about –0.02 g/mol), the energy change is incredibly large (about 2 billion kJ/mol). This huge energy release is why most nuclear reactions are considered dangerous.

Just as bond energies are defined as the energy needed to break the bond into its individual atoms, the nuclear **binding energy** is defined as the energy needed to break an atom into its subatomic components (noting that the mass of a p$^+$ and an e$^-$ is the same as the mass of an ^{1}H atom).

$$^A_Z\text{X} \rightarrow Z\,\text{p}^+ + (A - Z)\,\text{n}^0 + Z\,\text{e}^- = Z\,^1\text{H} + (A - Z)\,\text{n}^0 \qquad \Delta E = E_b$$

Example: Write the balanced equation for the binding energy of ^{79}Br and calculate its binding energy (in kJ/mol). _m values (in g/mol)_: e$^-$ = 0.0005, p$^+$ = 1.0073, n^0 = 1.0087, ^{79}Br = 78.9183

Solution: The balanced equation has ^{79}Br as the reactant and the subatomic particles as products.

$$\Delta m = \left(35 \times 1.0073\ \frac{\text{g}}{\text{mol}} + 44 \times 1.0087\ \frac{\text{g}}{\text{mol}} + 35 \times 0.0005\ \frac{\text{g}}{\text{mol}}\right) - \left(78.9183\ \frac{\text{g}}{\text{mol}}\right) = 0.7375\ \frac{\text{g}}{\text{mol}}$$

$$\Delta E = \Delta mc^2 = \left(0.7375\ \frac{\text{g}}{\text{mol}} \times \frac{1\ \text{kg}}{10^3\ \text{g}}\right)\left(3.00 \times 10^8\ \frac{\text{m}}{\text{s}}\right)^2 \times \frac{1\ \text{J s}^2}{1\ \text{kg m}^2} \times \frac{1\ \text{kJ}}{10^3\ \text{J}} = 6.64 \times 10^{10}\ \frac{\text{kJ}}{\text{mol}}$$

179

Calculating Nuclear Reaction Rates (20.4)

Nuclear decay reactions follow the first-order rate law $A = kN$, where A is the **activity** (rate) of nuclear decay (which equals –Rate), k is the decay constant (rate constant), and N is the number of radioactive atoms present (a measure of concentration). Therefore, these reactions follow the first-order rate law and you can use the first-order integrated rate law equation.

$$\text{Rate} = -A = -kN^1, \text{ or } \ln\frac{N}{N_0} = -kt$$

Example: Neutrons decay by β-emission. If you start with 1 million neutrons, after 1.00 h there are only 31,000 neutrons left.
(a) Write the balanced nuclear equation for this reaction.
(b) What is the value for the rate constant (in s^{-1}) of this reaction?
(c) What time (in s and in h) is needed for there to be only 1,000 neutrons left?

Solution:
(a) For this reaction, a neutron is the reactant, and one of the products is an electron (β-particle). The other product is a proton (p^+), which is the same as an $^1H^+$ ion.

$$^1_0n^0 \rightarrow {}^1_1H^+ + {}^0_{-1}e^-$$

(b) To solve for the rate constant, use the integrated rate law.

$$\ln\frac{N}{N_0} = -kt, \quad k = \frac{\ln\frac{N}{N_0}}{-t} = \frac{\ln\frac{31,000}{1,000,000}}{-1.00 \text{ h}} = 3.47 \frac{1}{h} \times \frac{1 \text{ h}}{60 \text{ min}} \times \frac{1 \text{ min}}{60 \text{ s}} = 9.65 \times 10^{-4} \text{ s}^{-1}$$

(c) To solve for time, use the integrated rate law and the rate constant from part (b).

$$\ln\frac{N}{N_0} = -kt, \quad t = \frac{\ln\frac{N}{N_0}}{-k} = \frac{\ln\frac{1,000}{1,000,000}}{-9.65 \times 10^{-4} \text{ s}^{-1}} = 7,160 \text{ s} \times \frac{1 \text{ min}}{60 \text{ s}} \times \frac{1 \text{ h}}{60 \text{ min}} = 1.99 \text{ h}$$

Reaction Rate Calculations Using Half-Lives (20.4)

For first-order reactions, the **half-life** ($t_{1/2}$) for the reaction is often tabulated in reference books because these values are constants. The half-life is the time needed for half of the atoms to undergo nuclear decay; it is also the time needed for half of the atoms to be remaining. The relationship between the half-life and the rate constant is $kt_{1/2} = \ln 2$.

Example: What is the half-life (in s and in h) for the β-decay of neutrons?

Solution: To determine the half-life, use the rate constant and formula $kt_{1/2} = \ln 2$.

$$kt_{1/2} = \ln 2, \quad t_{1/2} = \frac{\ln 2}{k} = \frac{0.693}{9.65 \times 10^{-4} \text{ s}^{-1}} = 718 \text{ s}$$

$$kt_{1/2} = \ln 2, \quad t_{1/2} = \frac{\ln 2}{k} = \frac{0.693}{3.47 \text{ h}^{-1}} = 0.200 \text{ h}$$

Nuclear Fission and Nuclear Fusion Reactions (20.6 & 20.7)

Figure 20.3 in the textbook (p. 967) shows that ^{56}Fe has the most **binding energy per nucleon** (a **nucleon** is simply a particle, proton or neutron, in an atom's nucleus.). This means that atoms with fewer nucleons in their nucleus than ^{56}Fe will become more stable if two atoms fuse together into a single larger atom (a process called **nuclear fusion**) and atoms with more nucleons in their nucleus than ^{56}Fe will become more stable if the atom breaks apart into two or more smaller atoms (a process called **nuclear fission**).

$$\underline{\text{nuclear fusion:}} \quad {}^{12}_{6}\text{C} + {}^{16}_{8}\text{O} \rightarrow {}^{28}_{14}\text{Si}$$

$$\underline{\text{nuclear fission:}} \quad {}^{172}_{78}\text{Pt} \rightarrow {}^{168}_{76}\text{Os} + {}^{4}_{2}\text{He}$$

The first nuclear weapons took advantage of the fact that when ^{235}U atoms are bombarded with neutrons, they produce two smaller atoms (i.e., a nuclear fission reaction) and two or three additional high-energy neutrons. These neutrons can react with other ^{235}U atoms, resulting in more fission reactions and even more high-energy neutrons. So, bombarding a ^{235}U atom with one neutron can cause a cascading reaction of several ^{235}U atoms (called a *chain reaction*). If there are very few ^{235}U atoms present, the high-energy neutrons will not have a good chance of hitting them, and a chain reaction cannot occur. The minimum amount of a nuclear fuel needed to support a chain reaction is called its **critical mass**. Many nuclear weapons have a total mass of nuclear fuel larger than the critical mass, but this fuel is broken into several pieces that are all below the critical mass. When the bomb is detonated these pieces are pushed together so that there is now a critical mass of the nuclear fuel and a nuclear explosion can occur.

Nuclear Power Plants (20.6)

Nuclear power plants use ^{235}U (in the form of pellets containing UO_2) as the nuclear fuel because it will decay (releasing huge amounts of energy) when bombarded with high-energy neutrons. These **nuclear reactors** often use ^{238}Pu and ^{9}Be as the neutron source and cadmium or hafnium *control rods* (objects that can absorb neutrons) as reaction *moderators* that can be added or removed from the reaction container. Nuclear reactors cannot undergo chain reactions that would lead to a nuclear explosion because the reactors do not have a critical mass of the nuclear fuel in them.

The heat released from the nuclear reaction is used to boil water and the steam created from this heating process is used to run electrical engines called *steam generators*. The coolant runs through self-contained (sealed) tubes inside the nuclear reactor, so that the coolant does not come in contact with the nuclear fuel or the by-products. Once the boiled water (steam) has created electricity in the steam generators, it must be cooled and condensed so that it can be reused to absorb energy from the nuclear reactor. Nuclear reactors use evaporative cooling to cool and condense the steam back into water. Many nuclear reactors use water supplies from nearby lakes or rivers to cool the coolant; in the past, this heated water was simply poured back into the lake or river. However, the hot water would raise the temperature of the lake or river, causing severe environmental threats to the creatures living in the lake or river (*thermal pollution*).

Measuring Background Radiation (20.8)

Figure 20.10 in the textbook (p. 982) shows the average **background radiation** exposure for inhabitants of the United States. About 82% of this radiation comes from natural sources; the other 18% comes from man-made sources. The largest source of background radiation is *radon exposure* (55%). Radon is a radioactive gas that is chemically inert (noble gas); one reason it is so dangerous is that it can be inhaled into the human body—if it reacts while in the lungs, it produces heavy metals (that are no longer gases) that become embedded in the lungs and undergo further nuclear decays. Other natural sources of background radiation are radiation from the sun (cosmic, 8%), radiation from naturally-occurring radioactive isotopes in the earth like ^{238}U and ^{232}Th (terrestrial, 8%), and radiation from naturally-occurring radioactive isotopes in the body like ^{40}K (internal, 11%). Most of the man-made radiation comes from medical procedures (11% from diagnostic X-rays and 4% from medical procedures using radioactive isotopes). Cigarette smoking exposes its users to far more radiation than these other man-made sources.

The radioactivity of a given sample is measured in units of curies or becquerels. A **becquerel** (Bq) equals one disintegration (nuclear transmutation) per second: 1 Bq = 1 s^{-1}. A **curie** (Ci) represents 3.7×10^{10} disintegrations per second, the radioactivity of 1.00 g of radium. Since each nuclear transmutation releases different amounts of energies, these measures of radioactivity don't show how much energy the object has absorbed from the radioactive isotopes. The **gray** (Gy) represents 1 J of energy absorbed per kilogram of material: 1 Gy = 1 J/kg; two older units still commonly used are the **rad** (which equals 0.01 Gy) and the **roentgen** (1 R = 9.33×10^{-9} Gy). The biological effects of radiation depend not only on the amount of energy absorbed by the object, but also on the type of radiation (α- or β-particles, etc.) and the type of body tissue absorbing the energy. The *effective dose* is measure in units of **sievert** (Sv) or **rem** (1 Sv = 100 rem).

Common Uses of Radioactive Isotopes (20.9)

Radioactive isotopes can be used for several beneficial purposes. One of these purposes is to identify the age of objects (*radiodating techniques*). Although *carbon-dating techniques* (using the radioisotope ^{14}C) can be used to identify the age of objects that used to be living (i.e., wooden objects, human or animal remains, etc.), other radioisotopes can be used to identify the age of rocks, or the earth itself.

Another use of radioisotopes is the irradiation of foods. Irradiated foods are exposed to gamma-emitters like ^{60}Co or ^{137}Cs. When bacteria, molds, and yeasts in the foods absorb energy from γ-rays, they are killed and this drastically prolongs the shelf life of these foods (γ-rays that are not absorbed simply pass through the food, so the foods are not made radioactive). While food irradiation is a widely accepted technique among scientists, some consumers are reluctant to eat irradiated food.

Radioisotopes are also used in medical procedures like medical imaging or cancer treatment. This branch of medicine is called **nuclear medicine**. X-rays use the fact that these high-energy electromagnetic waves (i.e., 'light') travel through different types of tissue (bone, muscle, organs, etc.) in different ways (rates) to produce anatomical images. Often, radioisotopes are injected into (or ingested by) a patient and tend to accumulate in specific tissues that depend on the identity of the radioisotope. Then, medical images can be made from the γ-rays (from γ-ray or positron emitters) or X-rays (from electron capture isotopes) that highlight the tissues containing the radioisotope.

Radioactivity tends to affect rapidly-growing cells the most. In normal humans, these cells include white blood cells, cells in the stomach lining, hair follicles, and fetal cells. Consequently, radiation poisoning is seen as effects on these cells—diminished immune system, nausea and vomiting, loss of hair, and even miscarriages. Cancer cells are also rapidly-growing cells, so they can be treated by radioactive isotopes. Because they are rapidly-growing, cancer cells are more likely to be damaged by absorbing high-energy radiation than normal (noncancerous) cells.

Chapter Review — Key Terms

The key terms that were introduced in this chapter are listed below, along with the section in which they were introduced. You should understand these terms and be able to apply them in appropriate situations.

activity (20.4)	curie (20.4)	nucleons (20.2)
alpha particles (20.1)	electron capture (20.2)	plasma (20.7)
alpha radiation (20.1)	gamma radiation (20.1)	positron (20.2)
background radiation (20.8)	gray (20.8)	rad (20.8)
becquerel (20.4)	half-life (20.4)	radioactive series (20.2)
beta particles (20.1)	nuclear fission (20.6)	rem (20.8)
beta radiation (20.1)	nuclear fusion (20.7)	roentgen (20.8)
binding energy (20.3)	nuclear medicine (20.9)	sievert (20.8)
binding energy per nucleon (20.3)	nuclear reactions (20.2)	tracers (20.9)
critical mass (20.6)	nuclear reactor (20.6)	

Practice Test

After you have finished studying the chapter and the homework problems, the following questions can serve as a test to determine how well you have learned the chapter objectives.

1. Determine the formula of the missing reactant or product in the following nuclear reactions.

    (a) ___ + ^{4}He → ^{12}C

    (c) ___ + ^{4}He → ^{13}C + ^{1}H

    (b) ^{12}C + ^{16}O → ___ + ^{4}He

    (d) ^{252}Cf → ^{106}Mo + ___ + 4$^1_0$n^0

2. Write a balanced nuclear equation for the following nuclear decay reactions.

    (a) ^{144}Nd, by α-emission

    (c) ^{14}O, by positron-emission

    (b) ^{31}Si, by β-emission

    (d) ^{195}Au, by electron capture

3. Predict how these unstable isotopes will decay and write a balanced equation for the decay reactions.

    (a) ^{214}At

    (d) ^{230}Pa

    (b) ^{59}Cu

    (e) ^{109}Pd

    (c) ^{20}F

    (f) ^{45}Ti

4. Determine the mass change (Δm) and energy change (ΔE) for the following nuclear reactions. m values (in g/mol): e$^+$ = e$^-$ = 0.0005, ^{4}He = 4.0026, ^{12}C = 12.0000, ^{12}N = 12.0186, ^{16}O = 15.9949, ^{28}Si = 27.9769

    (a) $^0_1$e$^+$ + $^0_{-1}$e$^-$ → 2 $^0_0$γ

    (c) $^{16}_8$O → $^{12}_6$C + $^4_2$He

    (b) $^{12}_6$C + $^{16}_8$O → $^{28}_{14}$Si

    (d) $^{12}_7$N → $^{12}_6$C$^-$ + $^0_1$e$^+$

5. Atomic nuclei contain positively-charged protons and neutral neutrons. Why don't the protons in these atoms repel each other and force the atom's nucleus to fall apart into protons and neutrons?

6. A sample of dilute phosphoric acid (H_3PO_4) was created using the radioisotope ^{32}P (a beta-emitter). This sample had an activity of 91,000 Ci when it was created; 21.0 d later, its activity was 33,000 Ci.

    (a) Write a balanced nuclear equation for this decay process.

    (b) Calculate the rate constant (in d^{-1}) and the half-life (in d) for this process.

    (c) What will be the activity of this sample 84.0 d after it was created?

    (d) At what time will the activity of this sample be 15,000 Ci?

7. In 1988, three independent chemists were allowed to carbon-date a small piece of the linen from the Shroud of Turin. The activity of ^{14}C (a beta-emitter with a half-life of 5730 yr) in living objects is 15.3 disintegrations per gram per minute (15.3 g^{-1} min^{-1}).

    (a) Write a balanced nuclear equation for this decay process.

    (b) Calculate the rate constant (in yr^{-1}) for this process.

    (c) If the shroud were created in 33 A.D., what would the activity of this sample be?

    (d) The activity measured for the Shroud of Turin was 14.15 g^{-1} min^{-1}. How old was the Shroud in 1988? When was it created?

    (e) Explain why radiocarbon dating cannot be used to date objects that were not living.

8. The textbook estimates that the average radon level in U.S. houses is 1.5 pCi/L = 1.5 × 10^{-12} Ci/L

    (a) ^{222}Rn is an α-emitter. Write a balanced nuclear equation for this decay process.

    (b) What is the number of ^{222}Rn atoms decaying per second in a house that is 2,000 ft^2 with 8.0 ft ceilings? 1 ft^3 = 28.3 L.

9. Smoking one pack of cigarettes per day exposes you to 900 mrem of background radiation per year.

    (a) How many chest X-rays would you have to have per year to get the same amount of background radiation as smoking a pack of cigarettes a day? 1 chest X-ray = 10 mrem

    (b) How many hours of television would you have to watch per year to get the same amount of background radiation as smoking a pack of cigarettes a day? 1 h TV watching = 0.15 mrem

    (c) What fraction of the year would be spent watching TV to get exposed to 900 mrem?

Chapter 21
The Chemistry of the Main Group Elements

The goal of this chapter is to introduce the chemistry of the main group (*s*- and *p*-block) elements. This chapter describes the formation of these elements in the stars and their terrestrial abundances. The chapter also describes how several of these main group elements are purified in their elemental forms. Finally, the chapter briefly describes the periodic properties of the eight groups in the main group elements.

The Formation of Main Group Elements in Stars (21.1)

The lighter elements (from He to Fe) are made in stars by a process called **nuclear burning**, the fusion of lighter elements to make heavier elements. The first process, **hydrogen burning**, represents the fusion of ^1_1H atoms into ^4He atoms: $4\,^1_1H \rightarrow\, ^4_2He^{2-} + 2\,^0_1e^+ + 2\,^0_0\gamma =\, ^4_2He + 6\,^0_0\gamma$. Once a critical amount of the hydrogen in the stars is converted to helium, the star collapses and the process of **helium burning** begins. In helium burning, larger atoms like ^8Be, ^{12}C, ^{16}O, ^{20}Ne, ^{24}Mg, ^{28}Si, ^{56}Fe, and ^{58}Ni are made from the fusion of ^4He atoms in the stars.

The Principal Elements in the Earth's Crust (21.2)

The textbook introduced the composition of the earth's gas phase (*air*, which is 78% N_2, 21% O_2, and 1% Ar) in Chapter 10 and the composition of the earth's liquid phase (*seawater*, which is mostly H_2O, and 0.55 M Cl^- ions and 0.47 M Na^+ ions) in Chapter 15. In this chapter, the textbook describes the composition of the earth's solid phase (the *earth's crust*). The earth's crust is mostly oxygen (49.5%) and silicon (25.7%), with smaller amounts of aluminum (7.4%), iron (4.7%), calcium (3.4%), sodium (2.6%), potassium (2.4%), and magnesium (1.9%).

The General Structure of Silicates (21.2)

Silicates are a class of **minerals** that contain silicon and oxygen atoms. The silicon atoms are surrounded by 4 O atoms in a tetrahedral arrangement (AX_4E_0); the neutral oxygen atoms are surrounded by two Si atoms with a bent shape (AX_2E_2) and the negative oxygen atoms are surrounded by one Si atom (AX_1E_3). In addition to the Si and O atoms, silicates generally have metal cations to balance the overall charge.

Zero-dimensional silicates have isolated SiO_4^{4-} ions in them and do not have any bridging O atoms linking the Si atoms together. Examples of 0-D silicates are Mg_2SiO_4 and Zn_2SiO_4. *One-dimensional silicates* have Si and O atoms covalently bonded together in one of the three dimensions. These solids have chains or fibers of Si and O atoms (with the general formulas SiO_3^{2-} or $Si_4O_{11}^{6-}$) that are loosely attached to each other by intermolecular forces. Examples of 1-D silicates are Na_2SiO_3 and $Mg_4Fe^{II}_3(Si_4O_{11})_2(OH)_2$. *Two-dimensional silicates* have Si and O atoms covalently bonded together in two of the three dimensions. These solids have sheets of Si and O atoms (with the general formula $Si_2O_5^{2-}$) that are loosely attached to each other by intermolecular forces. Examples of 2-D silicates are $Na_2Si_2O_5$ and $LiAl(Si_2O_5)_2$. *Three-dimensional silicates* (also called *silica*) have Si and O atoms bonded to each other in a three-dimensional structure (with all O atoms bridged between two Si atoms), and have the formula SiO_2.

Example: Show that $Mg_4Fe^{II}_3(Si_4O_{11})_2(OH)_2$ contains $Si_4O_{11}^{6-}$ ions. What are the oxidation numbers of the Si and O atoms in this ion?

Solution: This compound contains Mg^{2+} ions, Fe^{2+} ions, and OH^- ions. The four Mg^{2+} and the three Fe^{2+} ions have a total charge of +14. The two OH^- ions have a total charge of –2, so the remaining two (Si_4O_{11}) ions must have a total charge of –12, so each (Si_4O_{11}) ion has a charge of –6: $Si_4O_{11}^{6-}$. The rules for assigning oxidation numbers say that $4 \times O.N._{Si} + 11 \times O.N._O = -6$. Since $O.N._O = -2$ (Rule 3d), $4 \times O.N._{Si} + 11 \times (-2) = -6$, and $4 \times O.N._{Si} = -6 + 22 = +16$ and $O.N._{Si} = +4$.

Methods for Extracting Elements from the Earth's Crust (21.2)

Some elements exist on the earth in their elemental form and do not need to undergo chemical reactions to be extracted as elements. These elements include nitrogen, oxygen, the noble gases (He, Ne, Ar, Kr, Xe), sulfur, and the less reactive (noble) metals including copper, silver, and gold. The most reactive elements (alkali metals, alkaline earth metals, and the more reactive halogens fluorine and chlorine) must be made by electrochemical redox equations, while less reactive elements (like phosphorus, bromine, and iodine) can be made by simple chemical redox reactions not requiring electrolytic cells.

The Major Components in the Atmosphere (21.3)

The earth's atmosphere consists of 78% nitrogen (N_2), 21% oxygen (O_2), and 1% argon (Ar), with trace amounts of carbon dioxide (CO_2), water vapor (H_2O), several noble gases (He, Ne, Kr, Xe), methane (CH_4), and hydrogen (H_2).

Applying Chemical Principles to the Extraction and Purification of Elements (21.3-21.5)

The extraction of elements that exist in their uncombined (elemental) forms on earth simply requires phase changes to purify these elements (e.g., condensing gaseous air into a liquid and distilling it, melting and resolidifying solids like sulfur or the noble metals, etc.). Ions of the very reactive metals (like the alkali metals the alkaline earth metals, and aluminum) do not react with reducing agents because they are much more stable in their ion (oxidized) forms. Similarly, ions of the very reactive nonmetals (halogens like fluorine and chlorine) do not react with oxidizing agents because they are much more stable in their ion (reduced) forms. Therefore the only way to reduce these very reactive metals and to oxidize these very reactive halogens is to perform electrolysis reactions. In these reactions, it is important to separate the electrolysis products from each other and to keep strong oxidizing and reducing agents (like oxygen) away from the newly produced elements. Less reactive metals (like titanium or iron) can be made from chemical reactions involving strong reducing agents (like sodium metal, carbon, or hydrogen) and less reactive nonmetals (like phosphorus, bromine, or iodine) can be made from chemical reactions involving strong oxidizing agents (like fluorine, chlorine, or oxygen) or reducing agents.

Obtaining Elements from the Liquefaction of Air (21.3)

Once carbon dioxide and water are removed from air, it is essentially a mixture of nitrogen ($bp = -198°C$), oxygen ($bp = -183°C$), and argon ($bp = -189°C$). If its temperature is decreased and its pressure is increased, this mixture will liquefy. Because these three liquids have different boiling points, they can be separated by fractional distillation (introduced in Chapter 12 for the separation of gasoline fractions). This distillation process results in ultrapure liquid nitrogen, liquid oxygen, and liquid argon (99.5% pure).

The Frasch Process for Extracting Elemental Sulfur (21.3)

Most elemental sulfur that is currently made comes from petroleum feedstocks—it is removed from these fuels to prevent the formation of acid rain when the fuels are burned. However, the **Frasch process** is a method still used to extract and purify elemental sulfur that is found in the earth's crust. In this process, superheated steam at 165°C is injected into the ground (where the solid sulfur exists) in order to melt the solid sulfur into a liquid ($mp = 119°C$). Then, compressed air is forced underground, which forces the liquid sulfur-water mixture up to the surface. This mixture is allowed to cool and the sulfur resolidifies.

Obtaining Sodium, Chlorine, Magnesium, and Aluminum by Electrolysis (21.4)

Sodium metal and chlorine gas are commercially prepared using a *Downs cell*. In this cell, molten sodium chloride is electrolyzed into sodium metal at the cathode and chlorine gas at the anode. A 1:3 mixture of NaCl and $CaCl_2$ is used as the electrolyte solvent because it has a much lower melting point (600°C) than pure NaCl (801°C). At this temperature, the sodium metal produced is a liquid ($mp = 98°C$).

$$2\,Na^+(\ell) + 2\,Cl^-(\ell) \rightarrow 2\,Na(\ell) + Cl_2(g)$$

Although the Downs cell is used to make some chlorine, most of the chlorine made commercially comes from the **chlor-alkali process**, the electrolysis of aqueous sodium chloride. The major difference between the Downs process and the chlor-alkali process is that sodium metal is produced at the cathode in the Downs cell and sodium hydroxide and hydrogen gas are produced at the cathode in the chlor-alkali process. Although the reduction potential for the oxidation of water (+1.23 V) is less than the reduction potential for the oxidation of Cl^- ions (+1.36 V) and you should expect water to be oxidized, the chloride ions are in fact oxidized due to a large overpotential for the oxidation of water (which makes the effective reduction potential for the oxidation of water larger than +1.36 V).

$$2\,NaCl(aq)\ +\ 2\,H_2O(\ell)\ \rightarrow\ 2\,NaOH(aq)\ +\ H_2(g)\ +\ Cl_2(g)$$

Magnesium metal (along with chlorine gas) is made by the electrolysis of molten magnesium chloride. Magnesium ions are precipitated from seawater, using calcium hydroxide. The solid magnesium hydroxide is reacted with hydrochloric acid to make aqueous $MgCl_2$, which is dried before the electrolysis process.

Aluminum metal is made from the *Hall-Heroult process*, which involves the electrolysis of aluminum oxide. Aluminum oxide has a very high melting point (2030°C) and it would take a huge amount of energy to melt Al_2O_3. Both Hall and Heroult used the ionic solid cryolite (Na_3AlF_6) as a solvent in this electrolysis because it dissolves Al_2O_3 and has a much lower melting point (so, the reaction can be done at 1000°C). The Al^{3+} ions are oxidized to liquid aluminum metal at the cathode, and the O^{2-} ions react with the graphite (C) anode to produce carbon dioxide. Because they react, the graphite electrodes must be replaced often.

$$4\,Al^{3+}(\ell)\ +\ 6\,O^{2-}(\ell)\ +\ 3\,C(s)\ \rightarrow\ 4\,Al(\ell)\ +\ 3\,CO_2(g)$$

Example: What is the purpose of the following objects in the Hall-Heroult electrolytic cell?
(a) Na_3AlF_6, (b) Al_2O_3, (c) the graphite electrode

Solution:
(a) Cryolite (Na_3AlF_6) is the electrolyte solvent. This ionic compound (containing Na^+ and AlF_6^{3-} ions) dissolves the solid Al_2O_3, producing Al^{3+} and O^{2-} ions that can be electrolyzed into Al and CO_2. The reason it is used as the solvent instead of molten Al_2O_3 is that it has a much lower melting point.
(b) Al_2O_3 is the source of the Al^{3+} ions (reactant) that are reduced to aluminum, and the source of O^{2-} ions that react with the carbon electrode to make CO_2.
(c) The carbon (graphite) electrode acts as the anode, allowing the oxidation reaction to occur. However, the electrode is also a reactant—it is oxidized (and reacts with the O^{2-} ions) to produce carbon dioxide.

Obtaining Phosphorus, Bromine, and Iodine by Redox Reactions (21.5)

Elemental phosphorus is made from calcium phosphate. A mixture of $Ca_3(PO_4)_2$, carbon, and silicon dioxide are heated to a temperature of 1400-1500°C to produce P_4 and CO gases and calcium silicate. The mixture of gases is bubbled through water to precipitate $P_4(s)$, allowing CO(g) to pass through it.

$$2\,Ca_3(PO_4)_2(\ell)\ +\ 10\,C(s)\ +\ 6\,SiO_2(\ell)\ \rightarrow\ 6\,CaSiO_3(\ell)\ +\ P_4(g)\ +\ 10\,CO(g)$$

Elemental bromine and iodine are oxidized from their halide ions by the reaction of seawater containing Br^- and I^- ions with elemental chlorine, a more reactive halogen (and a better oxidizing agent).

$$2\,Br^-(aq)\ +\ Cl_2(g)\ \rightarrow\ Br_2(\ell)\ +\ 2\,Cl^-(aq)$$
$$2\,I^-(aq)\ +\ Cl_2(g)\ \rightarrow\ I_2(s)\ +\ 2\,Cl^-(aq)$$

The Production of Sulfuric Acid (21.6)

Sulfuric acid is the most commonly used acid in industrial processes (most likely because it is the cheapest acid to produce). To produce sulfuric acid, elemental sulfur is burned in the presence of oxygen to produce sulfur dioxide. The sulfur dioxide produced is reacted with additional oxygen and a catalyst to produce sulfur trioxide. When dissolved in water, SO_3 produces sulfuric acid—$SO_3(g) + H_2O(\ell) \rightarrow H_2SO_4(aq)$. However, the dissolution of SO_3 in water is rather slow. A much quicker reaction is to dissolve gaseous sulfur trioxide in sulfuric acid. The intermediate product, $H_2S_2O_7$, is called pyrosulfuric acid. Once the SO_3 has dissolved and produced pyrosulfuric acid, the solution is diluted with water, which reacts with the pyrosulfuric acid to produce concentrated sulfuric acid.

$$SO_3(g) + H_2SO_4(\ell) \rightarrow H_2S_2O_7(\ell)$$
$$H_2S_2O_7(\ell) + H_2O(\ell) \rightarrow 2\ H_2SO_4(aq)$$

Periodic Trends Among the Main Group Elements (21.6)

Metallic properties among the main group elements increase as you look down a column and decrease as you look across a row. Therefore, the elements at the extreme left (the alkali metals, and to a lesser extent the alkaline earth metals and aluminum) are very reactive metals and are good reducing agents. When looking at the elements within a column (groups), the metal reactivity increases down the column so that the elements at the top of the column are more likely to be nonmetals, the middle elements tend to be metalloids, and the elements at the bottom are metals. Nonmetallic properties are opposite of metallic properties and therefore decrease as you look down a column and increase as you look across a row. Therefore, the elements at the extreme top left (the halogens and oxygen) are the most reactive nonmetals and are good oxidizing agents.

Chemists noticed several chemical similarities between atoms in period 2 and their bottom-right diagonal atoms in period 3 (e.g., Li and Mg; Be and Al; B and Si), and they called this the *diagonal relationship*. These atoms have similar chemical properties because these atoms have similar atomic and ionic radii.

Chemists also noticed that the metals at the bottom of the columns tend to form positive charges that are two less than the column number. For example, Al and Ga metals form only Al^{3+} and Ga^{3+} ions (which is consistent with the column number) and although In^{3+} and Tl^{3+} ions are known, the +1 ion becomes more stable for these heavier elements (Tl^+ is very common and Tl^{3+} is rather rare). Similarly, Sn and Pb atoms (column 4) form Sn^{2+} and Pb^{2+} ions (Pb^{2+} is more common than Pb^{4+}) and Bi (column 5) forms Bi^{3+} ions.

Chapter Review — Key Terms

The key terms that were introduced in this chapter are listed below, along with the section in which they were introduced. You should understand these terms and be able to apply them in appropriate situations.

catenation (21.6)
chlor-alkali process (21.4)
cryogen (21.6)
Frasch process (21.3)

helium burning (21.1)
hydrogen burning (21.1)
mineral (21.2)
nitrogen cycle (21.6)

nitrogen fixation (21.6)
nuclear burning (21.1)
ore (21.2)

Practice Test

After you have finished studying the chapter and the homework problems, the following questions can serve as a test to determine how well you have learned the chapter objectives.

1. Determine the formula of the missing product of the following element-forming nuclear reactions.

   (a) $^{13}C + {}^{1}H \rightarrow$ ____ $+ {}^{4}He$

   (b) $^{13}C + {}^{4}He \rightarrow$ ____ $+ {}^{1}_{0}n^0$

   (c) $2\ {}^{12}C \rightarrow$ ____ $+ {}^{4}He$

   (d) $2\ {}^{20}Ne \rightarrow$ ____ $+ {}^{16}O$

2. Silicon-28 is made by the following nuclear reaction: $^{12}C + {}^{16}O \rightarrow {}^{28}Si$.

   (a) Would you expect this reaction to be exothermic or endothermic? Explain your answer.

   (b) Explain why this reaction needs temperatures of about 10^8 K to occur.

3.   Determine the empirical formulas for the Si-O anions present in the following silicate compounds and classify these silicates as zero-, one-, two-, or three-dimensional silicates.
(a)   cristobalite, SiO_2 (d)   talc, $Mg_3Si_4O_{10}(OH)_2$
(b)   kaolinite, $Al_2Si_2O_5(OH)_4$ (e)   tremolite, $Ca_2Mg_5Si_8O_{22}(OH)_2$
(c)   spodumene, $LiAlSi_2O_6$ (f)   zircon, $ZiSiO_4$

4.   Describe the physical (state of matter) changes used to purify elemental nitrogen and sulfur. How are these physical changes accomplished?

5.   In a Downs cell, a molten mixture of sodium chloride and calcium chloride is electrolyzed.

$$Cl_2(g) + 2\,e^- \rightleftharpoons 2\,Cl^-(aq) \qquad\qquad E° = +1.36\ V$$

$$Na^+(aq) + e^- \rightleftharpoons Na(s) \qquad\qquad E° = -2.71\ V$$

$$Ca^{2+}(aq) + 2\,e^- \rightleftharpoons Ca(s) \qquad\qquad E° = -2.87\ V$$

(a)   What ions are present in this mixture?
(b)   Use the following reduction potentials to predict the electrolysis products.
(c)   Why is sodium metal formed instead of calcium metal?

6.   Before the Hall-Heroult process, aluminum metal was made by the reaction of solid aluminum chloride and sodium metal. Write a balanced equation for this reaction and identify the oxidizing agent and reducing agent in this reaction.

7.   Aluminum recycling is quite profitable because the formation of aluminum from aluminum oxide using the Hall-Heroult process is very expensive. List three expenses associated with this process.

8.   What is the purpose of these chemicals in the conversion of calcium phosphate into phosphorus?
(a)   calcium phosphate, $Ca_3(PO_4)_2$ (c)   silicon dioxide, SiO_2
(b)   coke (graphite), C (d)   water, H_2O

9.   In describing the diagonal relationship, the textbook points out that lithium metal and magnesium metal will react with nitrogen gas to make ionic nitride compounds.
(a)   What are the formulas of these nitride compounds? Write the balanced equations for the formation of these compounds from their elements. What kind of reactions are these?
(b)   These nitride compounds react with water to produce ammonia. Write the balanced equations for these reactions. What kind of reactions are these?

Chapter 22
Chemistry of Selected Transition Elements and Coordination Compounds

The goal of this chapter is to introduce the chemistry of the transition (d-block) elements. This chapter describes the general properties of the transition elements, and discusses the specific properties, extraction methods, and uses of specific transition elements including iron, copper, gold, silver, and chromium. This chapter also introduces coordination complexes, explains how to name these compounds, and describes examples of isomerism in these complexes. Finally, this chapter describes the crystal-field theory and uses it to explain the optical and magnetic properties of these complexes.

General Properties of Transition Elements (22.1)

The *transition elements* are all metals (also called *transition metals*), and have several properties indicative of metals—high electrical conductivity, high malleability and ductility (the ability to be deformed into different shapes without breaking), and are paramagnetic (i.e., they have unpaired electrons). These metals also have properties that are slightly different than those of the main group metals—transition metals tend to have higher melting and boiling points, higher electronegativities, and higher densities. The transition elements also tend to form brightly-colored compounds and complex ions, and also tend to have more than one stable oxidation number.

Writing Electron Configurations for Transition Element Atoms and Ions (22.1)

The electron configurations of transition element atoms and ions were discussed in Chapter 7 of the textbook. Transition element atoms have their outer (valence) electrons in the s and d orbitals. When the transition element atoms form cations by losing electrons, they lose electrons from the s orbital before losing them from the d orbitals. The textbook also explained the anomalous electron configurations of Cr ($[Ar] 4s^1 3d^5$), Cu ($[Ar] 4s^1 3d^{10}$), Mo ($[Kr] 5s^1 4d^5$), Ag ($[Kr] 5s^1 4d^{10}$), and Au ($[Xe] 6s^1 4f^{14} 5d^{10}$) in Chapter 7 by recognizing that empty, half-filled, or completely filled orbital sets are more stable than other electron configurations.

Example: Write the electron configurations (orbital box diagram) for the manganese atom ($Z = 25$) and the Mn^{2+}, Mn^{4+}, and Mn^{7+} ions.

Solution: Since manganese has 25 protons (Z), the Mn atom also has 25 electrons. The first 18 e^- appear in the noble gas configuration [Ar] and the last 7 e^- are the valence electrons. The first 2 e^- are added to the $4s$ orbital, and the last 5 e^- are added to the $3d$ orbitals. To form cations from Mn, remove the first 2 e^- from the $4s$ orbital, then remove additional e^- from the $3d$ orbitals. Since Mn^{2+} has lost 2 e^-, these electrons will be removed from the $4s$ orbital—$[Ar] 3d^5$. For Mn^{4+}, remove 2 e^- from the $4s$ orbital and 2 e^- from the $3d$ orbital—$[Ar] 3d^3$. For Mn^{7+}, remove 2 e^- from the $4s$ orbital and 5 e^- from the $3d$ orbital—$[Ar]$.

Multiple Oxidation Numbers of Transition Elements (22.1)

Most of the transition elements have multiple oxidation numbers. The exceptions are the very early (left-most) and the very late (right-most) transition elements, which form only one ion charge (Sc^{3+}, Y^{3+}, La^{3+}, Zn^{2+}, and Cd^{2+}). Most of the transition elements form stable +2 and +3 ions, and many of the early transition elements form ions (or oxidation numbers) that match their column number and have noble gas configurations (e.g., Ti(IV), V(V), Cr(VI), and Mn(VII) oxidation numbers).

Trends in the Atomic Radii of the Transition Elements (22.1)

The general trends for the atomic radii of atoms in the periodic table are that atomic radii get smaller as you look across a row and get bigger as you look down a column. However, these trends are not as clear-cut for the transition elements. The transition elements in the middle of the *d*-block are smaller than the early transition elements because the added protons in the nucleus pull the outer (valence) electrons in closer to the nucleus. However, the late transition elements are larger than those atoms in the middle of the *d*-block because the added electrons are forced to pair up with other electrons in the *d* orbitals. These electrons repel each other, moving farther apart from each other and leading to a larger atomic radius. The elements in period 4 (the 3*d* elements) are smaller than the elements in period 5 (the 4*d* elements) because the outer electrons are placed in larger orbitals. You might expect that the elements in period 6 (the 5*d* elements) would be larger than the elements in period 5, but they are actually about the same size. This is due to the fact that these atoms have added 14 e^- into the 4*f* orbitals; this has also resulted in 14 more protons in the nucleus that pull the outer electrons in closer (so that their sizes match the period 5 atoms). The fact that these atoms have smaller sizes than would be expected if the 4*f* orbitals were not filled is referred to as the **lanthanide contraction**, named after the lanthanides (the 14 elements with valence e^- in 4*f* orbitals).

Processing Iron Ore to Make Iron Metal and Steel (22.2)

Iron does not exist in its metallic form in nature because it is a reactive metal. Iron metal reacts with oxygen in the atmosphere to produce hydrated Fe_2O_3 (*rust*). Chemists use two iron oxide ores to produce iron metal—Fe_2O_3 (hematite) and Fe_3O_4 (magnetite). The process of extracting a metal from its ore using high-temperature methods is called **pyrometallurgy**. In this process, the iron oxides are mixed with coke (carbon), calcium carbonate, and air (oxygen). The oxygen in the air reacts with the carbon to make carbon monoxide, which reduces the iron oxide to iron metal and carbon dioxide.

$$Fe_2O_3(s) + 3\,CO(g) \rightarrow 2\,Fe(\ell) + 3\,CO_2(g)$$

Calcium carbonate decomposes into calcium oxide and carbon dioxide and the CaO reacts with any silicon impurities (SiO_2) to form $CaSiO_3$ (slag), which is a liquid and can be drained out of the reaction container. The resulting iron metal (which has a carbon content of up to 4.5%) is called *pig* or *cast iron*. This form of iron is rather brittle due to the excess carbon (which produces the brittle compound Fe_3C, cementite). Pig iron can be converted into **steel** (iron with 0.5-1.3% C content) by removing phosphorus, sulfur, and silicon impurities and reducing the carbon content to below 1.3% (the excess C in pig iron is removed by forcing oxygen through the mixture, which oxidizes the solid C into gaseous carbon dioxide). Steel can be *alloyed* with other metals to fine-tune the properties of the steel. Table 22.3 in the textbook (p. 1044) shows examples of these alloys and their properties.

Extracting and Purifying Copper Metal (22.3)

Copper metal is a much less reactive (*inert*, or *noble*) metal than iron. As a result, copper can exist in its elemental form, but it normally appears in sulfur- and iron-containing ores like chalcocite (Cu_2S) and chalcopyrite ($CuFeS_2$). Copper is separated from the iron ore by *roasting*, a pyrometallurgic process that involves adding air (oxygen) to the mixture at high temperatures. The iron is converted to Fe_3O_4 and the copper is converted to Cu_2S. Once silicon dioxide is added, Fe_3O_4 is converted to a liquid iron-silicate slag that can be separated from liquid Cu_2S. The Cu_2S liquid is partially oxidized by oxygen gas into sulfur dioxide and Cu_2O, which reacts with more of the Cu_2S to produce liquid copper and more sulfur dioxide.

$$3\,Cu_2S(\ell) + 3\,O_2(g) \rightarrow Cu_2S(\ell) + 2\,Cu_2O(\ell) + 2\,SO_2(g) \rightarrow 6\,Cu(\ell) + 3\,SO_2(g)$$

This impure form of copper (called *blister copper*, which is 96-99.5% pure) can be purified by a process called *electrorefining*. In electrorefining, the blister copper is used as the anode (which is oxidized and dissolves), and ultrapure copper is produced at the cathode. The cell potential of this concentration cell is carefully set so that copper and all metals that are more reactive than copper (including Ni, Fe, Pb, and Zn) are oxidized at the anode. Metals that are less reactive than copper are never oxidized and end up as solids at the bottom of the container, called *anode sludge*. The anode sludge contains precious metals like silver, gold, and platinum, which are collected and sold (which is usually enough to pay for the electrorefining process). The potential of the cathode is set so that copper and all metals that are more reactive than copper (which are in the anode sludge and not the aqueous solution) can be reduced to their metal form at the cathode. As a result, only copper metal is reduced and this pure form is more than 99.9% pure copper.

The Chemistry of Gold and Silver (22.4)

Silver and gold are even less reactive metals than copper, and are commonly found in nature in their metallic, uncombined forms (although silver is also found in nature as argentite, Ag_2S). Since these are unreactive metals, they will not dissolve in acid solutions unless it is an oxidizing acid, like nitric acid or *aqua regia* (a mixture of HCl and HNO_3). Silver and gold are purified from their ores using cyanide ion extraction, in which the ore is mixed with NaCN and air (oxygen). This leads to the formation of $Ag(CN)_2^-$ and $Au(CN)_2^-$ ions that dissolve in water and can then be oxidized by zinc metal.

$$Ag_2S(s) + 4\,CN^-(aq) \rightarrow 2\,Ag(CN)_2^-(aq) + S^{2-}(aq)$$

$$4\,Au(s) + 8\,CN^-(aq) + O_2(g) + 2\,H_2O(\ell) \rightarrow 4\,Au(CN)_2^-(aq) + 4\,OH^-(aq)$$

$$2\,M(CN)_2^-(aq) + Zn(s) \rightarrow 2\,M(s) + Zn(CN)_4^{2-}(aq),\ M = Au\ or\ Ag$$

The Chemistry of Chromium (22.5)

As a transition element in the middle of the *d*-block, chromium is a reactive metal that does not exist in nature as its metallic form. However, once chromium metal is formed, it is resistant to oxidation—not because it is unreactive, but because it forms an oxide coating with air that protects the metal from further oxidation. Chromium has several stable oxidation numbers, including +2, +3, and +6. Chromium exists in nature as chromite, $FeCr_2O_4$, which can be oxidized to sodium chromate. The chromate ion (CrO_4^{2-}) is stable in basic solutions, but is converted into the dichromate ion ($Cr_2O_7^{2-}$) in acidic solutions.

$$4\,FeCr_2O_4(s) + 8\,CaO(s) + 8\,Na_2CO_3(s) + 7\,O_2(g) \rightarrow 8\,Na_2CrO_4(aq) + 2\,Fe_2O_3(s) + 8\,CaCO_3(s)$$

Example: When the chromate ion acts as an oxidizing agent in basic solutions, it is reduced to $Cr_2O_3(s)$. Write the half-reaction for this reaction.

Solution: Place a 2 in front of CrO_4^{2-} to balance Cr atoms. Then, add 5 H_2O to the right side to balance O atoms and 10 H^+ ions to the left to balance H atoms. Finally, add 6 e^- on the left to balance charge. Since this reaction is basic, add 10 OH^- to both sides (10 H^+ + 10 OH^- = 10 H_2O) and cancel 5 H_2O on both sides.

191

$$2\,CrO_4^{2-}\,(aq)\,+\,10\,H^+(aq)\,+\,6\,e^-\,\rightarrow\,Cr_2O_3(s)\,+\,5\,H_2O(\ell)$$

$$2\,CrO_4^{2-}\,(aq)\,+\,5\,H_2O(\ell)\,+\,6\,e^-\,\rightarrow\,Cr_2O_3(s)\,+\,10\,OH^-(aq)$$

Coordinate Covalent Bonding in Complex Ions and Compounds (22.6)

Like most metals, transition elements have empty orbitals that can accept electrons from other ions or molecules that have lone pairs to form *coordinate covalent bonds* (introduced in Chapter 16). The ions or molecules that serve as Lewis bases to the metal ions are called **ligands**, and the chemical species produced from the metal ion and the ligands is called a **coordination compound**. The number of coordinate covalent bonds formed by the metal ion and the ligands is called the **coordination number**.

Most ligands bind to the metal ion by one pair of electrons, forming one coordinate covalent bond. These ions are called **monodentate** ligands. Ligands that have more than one atom that can bind to the metal ions are called **polydentate** ligands or **chelating ligands** (ligands that bind to the metal ion using two different Lewis base atoms are called **bidentate** ligands).

Names and Formulas of Coordination Ions and Compounds (22.6)

The formula of coordination ions (also called *complex ions*) and coordination compounds (neutral molecules) are usually written in square brackets (not to be confused with concentration values), with the ion charge written outside the brackets. For complex ions, the ionic solids also have cations or anions that are present to balance the overall charge. For example, $K_2[PtCl_6]$ contains 2 K^+ ions and the complex ion $[PtCl_6]^{2-}$, which has a coordination number of 6 for the Pt(IV) metal ion.

Example: When a gold ore is mixed with sodium cyanide (NaCN) and an oxidizing agent, one of the products formed is $Na[Au(CN)_4]$.
(a) What is the formula of the complex ion?
(b) What are the oxidation number (charge) and coordination number of gold in this ion?

Solution:
(a) Since sodium ions have the formula Na^+, the complex ion must be $[Au(CN)_4]^-$.
(b) In the complex ion $[Au(CN)_4]^-$, the ligand is the CN^- ion. Since there are four CN^- ions (with a total charge of -4), the Au ion must have a $+3$ charge—Au(III) or Au^{3+}. The coordination number for Au is 4.

Complex ions are named using a systematic set of rules (nomenclature). The names of the ligands are listed first (in alphabetic order) using the Greek prefixes introduced in Chapter 3 to show how many of each ligand is present. Special names are often used for some of the more common neutral ligands, including 'ammine' for ammonia (NH_3), 'aqua' for water (H_2O), and 'carbonyl' for carbon monoxide (CO). The names for the anionic ligands are changed so that they end in '-o' (e.g., when serving as ligands, the chloride ion Cl^- is called 'chloro', the carbonate ion CO_3^{2-} is called 'carbonato', etc.). After the ligands are listed, the metal ion's name is written with its charge (oxidation number) in parentheses (except for atoms that form only one ion like Ag^+, Cd^{2+}, Zn^{2+}, or Sc^{3+}). If the complex ion is an anion, the name of the anion is changed so that it ends in '-ate'; if the atom has a Latin name, this Latin name with the '-ate' ending is used for anionic complexes (e.g., 'argentate' for Ag, 'aurate' for Au, 'cuprate' for Cu, 'ferrate' for Fe, 'hydrargyrate' for Hg, 'plumbate' for Pb, 'stannate' for Sn, and 'wolfrate' for W). The names for the complex ions and compounds are written as one word. For complex ions, the ion and its counter ions are named as separate ions, with the cation first and the anion last.

Example: Write the names for the complex compounds $Li_3[Co(CN)_5]$ and $[Ru(NH_3)_5(H_2O)]Br_2$.

Solution: The CN⁻ (cyanide) ligands are named 'cyano'. The $[Co(CN)_5]^{3-}$ ion is 'pentacyanocobaltate(II)' (the charge for Co must be 2+ so that when combined with the 5– of the 5 cyanide ions, the overall charge is 3–). The compound $Li_3[Co(CN)_5]$ is named lithium pentacyanocobaltate(II). The names of the ligands in $[Ru(NH_3)_5(H_2O)]Br_2$ are 'ammine' (NH_3) and 'aqua' (H_2O). The $[Ru(NH_3)_5(H_2O)]^{2+}$ ion has a Ru^{2+} ion, so it is named 'pentammineaquaruthenium(II)' and the $[Ru(NH_3)_5(H_2O)]Br_2$ compound is named pentammineaquaruthenium(II) bromide.

Example: Write the formula for the complex compounds tetraamminediaquanickel(II) nitrate and iron(III) hexacyanoferrate(II).

Solution: In the first compound, the ligands are NH_3 (ammine = ammonia) and H_2O (aqua = water), and there are four NH_3 and two H_2O ligands. The complex ion is $Ni^{2+} + 4\ NH_3 + 2\ H_2O = [Ni(NH_3)_4(H_2O)_2]^{2+}$, so you need two NO_3^- ions to balance charge, so its formula is $[Ni(NH_3)_4(H_2O)_2](NO_3)_2$. In the second compound, the ligands are CN⁻ (cyano = cyanide) and there are six CN⁻ ligands. The complex ion is $Fe^{2+} + 6\ CN^- = [Fe(CN)_6]^{4-}$, to balance charge with the Fe^{3+} counter ion, its formula must be $Fe_4[Fe(CN)_6]_3$.

Isomerism in Coordination Compounds and Ions (22.6)

Transition elements tend to make coordination compounds involving coordination numbers of 2 (linear), 4 (tetrahedral and square planar), and 6 (octahedral). *Linear complexes* do not form isomers, and *tetrahedral complexes* form optical isomers (enantiomers) only when there are four different ligands on the metal (just like carbon atoms with four different substituents are chiral). *Square planar complexes* (where the four ligands appear in a single plane with bond angles of 90°) with formulas MX_2Y_2 or MX_2YZ form *cis* and *trans* isomers just like alkenes do. *Octahedral complexes* with formulas like MX_4Y_2 or MX_4YZ can also form *cis* and *trans* isomers (the Y_2 or YZ ligands are 90° apart in the *cis* isomer and 180° apart in the *trans* isomer). Octahedral complexes with formulas like MX_3Y_3 also form *fac* and *mer* isomers. In the *fac* (facial) isomer, the X_3 ligands (and the Y_3 ligands) are all 90° apart, but in the *mer* (meridinal) isomer, two pairs of the X_3 (and Y_3) ligands are 90° apart and one pair is 180° apart.

Example: Draw the *cis* and *trans* $[Co(NH_3)_4Cl_2]$ isomers and the *fac* and *mer* $[Cr(H_2O)_3Br_3]$ isomers.

Solution: The two Cl⁻ ligands are 90° apart in the *cis* isomer and they are 180° apart for the *trans* isomer. The *fac* isomer has all 3 Br⁻ ligands (and all 3 H_2O ligands) 90° apart (all Br⁻ ions are across from H_2O molecules), and the *mer* isomer has two pairs of Br⁻ ions 90° apart and one pair 180° apart (two Br⁻ ions are across from each other, and the other Br⁻ ion is across from an H_2O molecule).

cis isomer   *trans* isomer   *fac* isomer   *mer* isomer

In addition to the isomers listed above, there are some ligands that have more than two atoms with lone pairs that can bind to the metal. For example, the SCN⁻ ion can bind to the metal ion using the S atom (thiocyanato ligand) or using the N atom (isothiocyanato). Other ligands that can bind the metal using two different atoms are the CN⁻ ion ('cyano' when bonded by the C atom, 'isocyano' when bonded by the N atom) and the NO_2^- ion ('nitro' when bonded by the N atom, 'nitrito' when bonded by one of the O atoms).

193

Examples and Uses of Coordination Compounds (22.6)

When ionic compounds containing transition element ions are dissolved in water, aqua complexes are usually formed. Most of these ions (like Mn^{2+}, Fe^{3+}, Cr^{3+}, Co^{3+}, or Ni^{2+} ions) form octahedral complexes, but late transition metals like Zn^{2+} or Cu^{2+} form tetrahedral complexes. Transition element complexes formed using a *porphyrin* ligand, a tetradentate cyclic ligand containing four N atoms (which appears in Figure 22.15 on p. 1064 of the textbook), are important in biological systems. For example, Fe^{3+} porphyrin complexes are present in blood (hemoglobin) and muscles (myoglobin), vitamin B-12 contains a Co^{3+} porphyrin complex, and chlorophyll (the green pigments responsible for photosynthesis) contains an Mg^{2+} porphyrin complex. Coordination compounds are also used in chemotherapy; the most common of these compounds is cisplatin (*cis*-[Pt(NH$_3$)$_2$Cl$_2$]).

Interpreting Colors and Magnetic Properties of Coordination Compounds (22.7)

Crystal field theory is used by chemists to explain the magnetic and optical (colors) properties of transition element coordination compounds. Crystal field theory assumes that the when the ligands are brought close to the metal ion, the electrons from the ligands and the electrons on the metal ion in the d orbitals will repel each other. These repulsions cause the five d orbitals that were at equal energy to have different energies depending on whether they are very close or farther away from the lone-pair electrons from the ligands.

In octahedral complexes, for example, three d orbitals (collectively called the t_{2g} orbitals) are at lower energies and two d orbitals (collectively called the e_g orbitals) are at higher energies. In tetrahedral complexes, the three t_2 orbitals are at higher energies and the two e orbitals are at lower energies. The difference in energy between these two sets of orbitals is called the **crystal-field splitting energy** and is labeled as Δ_o for octahedral complexes and Δ_t for tetrahedral complexes. The t_{2g}/t_2 orbitals are the d_{xy}, d_{xz}, and d_{yz} orbitals; the e_g/e orbitals are the $d_{x^2-y^2}$ and d_{z^2} orbitals.

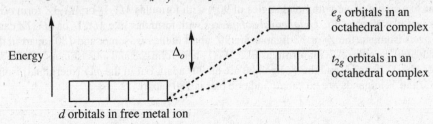

Energy

Δ_o

e_g orbitals in an octahedral complex

t_{2g} orbitals in an octahedral complex

d orbitals in free metal ion

<u>Example</u>: Explain why both the t_{2g} and the e_g orbitals in an octahedral complex are at higher energy than the d orbitals in the free metal ion. Why are the e_g orbitals higher in energy than the t_{2g} orbitals?

<u>Solution</u>: When the ligands are brought close to the free metal ion, the electrons in the d orbitals will repel the ligands' lone-pair electrons. This repulsion causes the energy of all of the d orbitals to be higher than in the free metal ion. The ligands' lone-pair electrons must be repelling the electrons in the e_g orbitals to a greater extent than the electrons in the t_{2g} orbitals, causing them to be at a higher energies.

The first three electrons in an octahedral complex will go into each of the t_{2g} orbitals (Hund's rule). For the fourth electron, you have two choices: (1) The electron can go into a t_{2g} orbital, pairing up with one of the other electrons in the t_{2g} orbitals, or (2) The electron can go into an e_g orbital with spin parallel to the t_{2g} electrons. Each option requires energy (option 1 requires energy to overcome the repulsions of putting two electrons in the same orbital; option 2 requires energy to put the electron in a higher-energy orbit).

If the Δ_o value is very small (smaller than the e^- repulsions), the electrons will enter the e_g orbitals before being paired in the t_{2g} orbitals (**high-spin complexes**); if the Δ_o value is very large (larger than the e^-/e^- repulsions), the electrons will pair with the other e^- in the t_{2g} orbitals before entering the e_g orbitals (**low-spin complexes**). The **spectrochemical series** shows how the identity of the ligands affects whether a complex is high-spin or low-spin. Ligands like CN^- and NO_2^- (*strong-field ligands*) that bond to metals using low *EN* atoms (C or N) tend to form low-spin complexes while ligands like H_2O, F^-, and Cl^- (*weak-field ligands*) that bond to metals using high *EN* atoms (like O, F, or Cl) tend to form high-spin complexes.

Transition element coordination compounds or ions are often brightly colored because these complexes absorb light in the visible region due to *d-to-d transitions* (electrons moving from lower-energy *d* orbitals to higher-energy *d* orbitals). The color of the coordination compounds is the complementary color of the light absorbed by the complex. For example, if a complex absorbs red light, when white light shines on the complex, it absorbs red light and transmits all colors except red (blue and green) to its surroundings (like your eyes) and it will appear blue-green. Figure 22.22 in the textbook (p. 1069) uses a color wheel to show complementary colors. For the same transition element ion, complexes with strong-field ligands will absorb higher-energy (shorter wavelength) light and complexes with weak-field ligands will absorb lower-energy (longer wavelength) light.

Chapter Review — Key Terms

The key terms that were introduced in this chapter are listed below, along with the section in which they were introduced. You should understand these terms and be able to apply them in appropriate situations.

bidentate (22.6)	hexadentate (22.6)	polydentate (22.6)
chelating ligands (22.6)	high-spin complex (22.7)	pyrometallurgy (22.2)
coordination compound (22.6)	lanthanide contraction (22.1)	spectrochemical series (22.7)
coordination number (22.6)	ligands (22.6)	steel (22.2)
crystal-field splitting energy (22.7)	low-spin complex (22.7)	
d-to-*d* transitions (22.7)	monodentate (22.6)	

Practice Test

After you have finished studying the chapter and the homework problems, the following questions can serve as a test to determine how well you have learned the chapter objectives.

1. Lead(II) chromate ($PbCrO_4$) is a very intense yellow solid used as a paint pigment.
    (a) What is the oxidation number of chromium in this compound?
    (b) Which metal is responsible for the yellow color?

2. Write the electron configuration (orbital box diagram) for the ions listed below.
    (a) Cu^{2+} (c) Ru^{3+}
    (b) Co^{2+} (d) Ti^{4+}

3. Determine the oxidation numbers of the transition elements in the following compounds.
    (a) $BaRuO_3$ (c) $LiNb_3O_8$
    (b) Cs_2CuCl_4 (d) $Pt(NO_3)_2(OH)_2$

4. The following compounds contain two different transition elements. Predict the possible charges for each transition element in these compounds.
    (a) $CuFeS_2$ (b) $MnCr_2O_4$

5. Magnetite, $Fe_3O_4(s)$, can be converted to iron metal using solid carbon (coke) and oxygen gas.
    (a) Write balanced equations to show this conversion.
    (b) What are the oxidation numbers of the iron atoms in Fe_3O_4?

6. Given the standard reduction potentials, determine whether the following metal atoms would be oxidized into the aqueous solution or would remain as solids in the anode sludge during the copper electrorefining process ($E°$ for Cu^{2+}(aq)/Cu(s) is +0.34 V).
    (a) Fe^{2+}(aq)/Fe(s), $E° = -0.44$ V   (c) Pt^{2+}(aq)/Pt(s), $E° = +1.20$ V
    (b) Ni^{2+}(aq)/Ni(s), $E° = -0.25$ V   (d) Rh^{3+}(aq)/Rh(s), $E° = +0.80$ V

195

7. Write the names for the following complexes that appear in the purification of silver and gold.
   (a) $[Ag(CN)_2]^-$ (c) $[Zn(CN)_4]^{2-}$
   (b) $Na[Au(CN)_2]$

8. The dichromate ion is a very good oxidizing agent in acidic conditions. Balance the following oxidation-reduction reaction involving this ion.

$$Cr_2O_7^{2-}(aq) + H_2SO_3(aq) \rightarrow Cr^{3+}(aq) + SO_4^{2-}(aq)$$

9. The dichromate ion, $Cr_2O_7^{2-}(aq)$, is unstable in basic solutions and is converted to the chromate ion, $CrO_4^{2-}(aq)$. Write a balanced equation for this reaction.

10. Write the names for the following complex compounds or ions.
    (a) $[Co(CO_3)_3]^{3-}$ (d) $[Pt(en)(H_2O)_2]Cl_2$
    (b) $H_2[WOCl_5]$ (e) $[V(H_2O)_6](NO_3)_2$
    (c) $K[Hg(CN)_3]$ (f) $[Zn(NH_3)_4][ZnCl_4]$

11. Write the formulas for the following complex compounds or ions.
    (a) cobalt(II) ethylenediaminetetraacetatocobalt(II)
    (b) hexacarbonyl molybdenum(0)
    (c) potassium tetraoxomanganate(IV)
    (d) triaquatrifluoroiron(III)

12. Identify the isomer shown below (*cis*, *trans*, *fac*, *mer*, or none of these).

    (a)

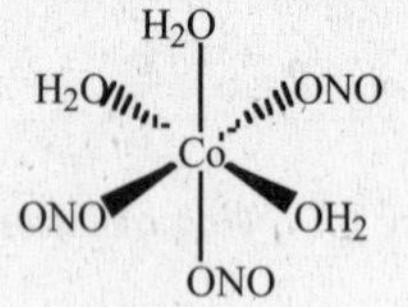

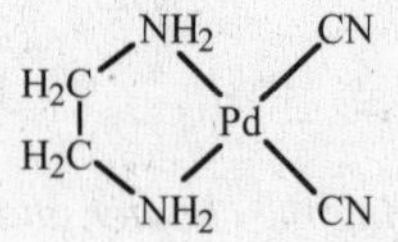

    (b)

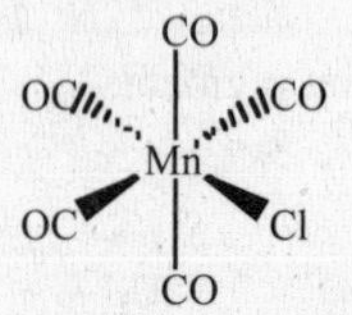

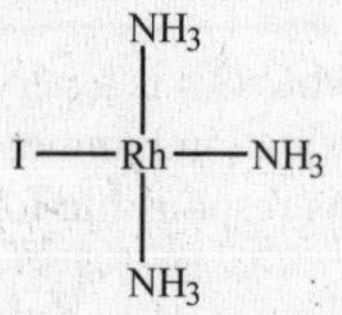

    (c)

    (d)

    (e)

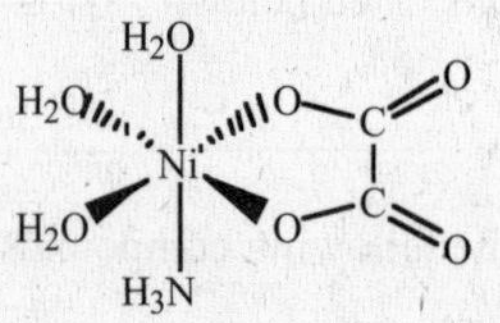

    (f)

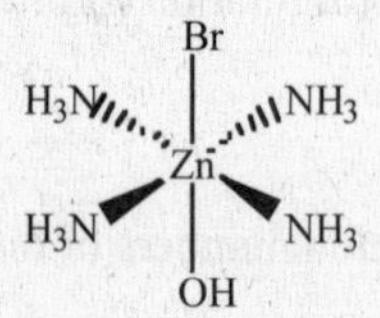

13. Determine the number of unpaired electrons in the following octahedral or tetrahedral complexes.
    (a) high-spin $[MnBr_4]^{2-}$ (c) low-spin $[Cr(CO)_6]$
    (b) low-spin $[Fe(NH_3)_4]^{2+}$ (d) high-spin $[Co(H_2O)_6]^{2+}$

14. There are many cobalt(III) ammine/chloro coordination complexes. The complex $[Co(NH_3)_6]Cl_3$ is yellow, the complex $[Co(NH_3)_5Cl]Cl_2$ is purple, and the complex $[Co(NH_3)_4Cl_2]Cl$ is green.
    (a) What are the names of these three ions?
    (a) What color of light do these ions absorb?
    (b) Use the spectrochemical series in the textbook (p. 1067) to explain the colors of these three complexes.

196

ANSWERS — PRACTICE TEST

Chapter 1

1. (a) The first thing the scientist should do is collect more data to see if there was a mistake made during the first data collection process. Another way to confirm the data would be to ask another scientist to try the same experiment and see if that scientist gets the same results.

   (b) If the data and theory are found to be incompatible, then the theory must be changed. The data represent information about the world, and theories are simply our attempt to explain the world. When theories and data conflict, the data prevail (we have to change our explanation of the world because we can't actually change the world itself).

2. (a) Quantitative (mentions numerical values of hydrogen and oxygen atoms present).
   (b) Qualitative (non-numerical description of temperature).
   (c) Quantitative (mentions numerical value of mass).

3. (a) Chemical, the plants chemically convert carbon dioxide and water into oxygen and plant sugar.
   (b) Physical, the liquid gasoline evaporates to the gas phase, leaving the pavement dry.
   (c) Physical, the salt dissolves in water (goes from solid to aqueous phase), and the water melts (goes from solid to liquid phase).

4. (a) 240°F is higher. Water boils at 100°C, which corresponds to 212°F. So, 240°F is warmer than 80°C.
   (b) 37°C is higher. 37°C is body temperature (98.6°F), which is warmer than 73°F.
   (c) 20°C is higher. 20°C is warmer than the melting point of ice (0°C) while 20°F is cooler than the melting point of ice (32°F).
   (d) They are the same. –40 is the temperature where the Celsius and Fahrenheit scales are equal.

5. This is actually a trick question. Since the pound is a unit of weight (mass), a pound of any two things weigh the same. Many students mistakenly believe that because lead is 'heavy' (more dense) and feathers are 'light' (less dense), that a sample of lead always weighs more than a sample of feathers. The real difference between a pound of lead and a pound of feathers is their volumes—a pound of lead will occupy a rather small volume but a pound of feathers will occupy a much larger volume. Since $V = m/d$, given the same mass (one pound), the volume of a very dense substance will be smaller and the volume of a less dense substance will be larger.

6. To compare masses of the two liquids, you must calculate the masses of each liquid from the liquids' volumes and their densities:

$$\text{ethanol:} \quad 313 \text{ mL} \times \frac{0.789 \text{ g}}{1 \text{ mL}} = 247 \text{ g}$$

$$\text{benzene:} \quad 286 \text{ mL} \times \frac{0.880 \text{ g}}{1 \text{ mL}} = 252 \text{ g}$$

So, the sample of benzene weighs more than the sample of ethanol.

7. (a) Cheerios is a homogeneous mixture. It is made of oats, sugar, etc., but appears uniform to the eye.
   (b) Froot Loops is a heterogeneous mixture of different-colored circles. It is heterogeneous because you can see these differences.
   (c) Although milk appears to be a homogeneous mixture when you look at it, it is actually a heterogeneous mixture of water, fats, and lactose (a sugar in milk).
   (d) Sugar (sucrose) is a chemical compound made of carbon, hydrogen, and oxygen ($C_{12}H_{22}O_{11}$).

8. (a) Von Baeyer proposed a theory that benzene turns blue with the indophenine test. In fact, this was used as a test to confirm that a sample contained benzene. Meyer collected additional data to confirm Von Baeyer's theory using this test, and his data and the theory were in conflict. Meyer collected additional data to make sure he hadn't made a mistake. When he performed the indophenine test on the purest sample of benzene he could buy (from coal tar), the sample only turned faint blue.

(b)   Meyer figured out that it was not actually the benzene that was causing the blue color. It was an impurity from the coal tar causing the color (the chemical causing the blue color was thiophene, a compound that has several chemical and physical properties similar to benzene). The fact that the color did not always occur when testing benzene and appeared much weaker in a purified sample of benzene all pointed to the presence of an impurity that was actually causing the color.

9.   (a)   Molecules in a gas sample are not touching each other and they move independently of each other. Because individual gas molecules are so small (nanoscale), they cannot be seen by the human eye (macroscale). Molecules in a liquid do touch each other, and while individual liquid molecules are too small for humans to see, the conglomeration of liquid molecules is big enough to be seen.

(b)   Molecules in liquids and solids are very close to each other (touching); therefore, they cannot be pushed closer together. Molecules in gases, however, are very far apart from each other and can be pushed closer together (eliminating some of the empty space between the gas molecules).

(c)   Within the liquid, the molecules have the freedom to move about and the food coloring molecules and the water molecules will slowly mix together. Solid particles, on the other hand, cannot move unless all of the particles move in the same direction at the same time. Since it is extremely unlikely that all of the random motions will occur in the same direction at the same time, the solid objects do not move.

10.   (a)   Solid. The particles do not occupy the volume of the entire container, and these particles show an organized pattern (lined up in rows and columns).

(b)   Heterogeneous mixture. This is a mixture because there are two substances present (square and diamond-diamond); it is a heterogeneous mixture because there are some regions (upper middle) that contain only diamonds and other regions (bottom left and bottom right) that contain only squares.

(c)   Both. This mixture contains atoms of square and diatomic molecules of diamond-diamond.

(d)   Elements. The atoms of square are elements (all atoms are elements); the molecules of diamond-diamond are also elements because they contain only one type of atom.

11.   Elements are composed of a single type of atom. Under normal chemical conditions, there is no way to break this atom into smaller (or different) parts, and any rearrangements of the particles within the element can only lead to the same element. You can't rearrange atoms into something new if there is only one type of atom.

12.   Based on the law of constant composition, any two compounds with the same mass ratios could be the same compound; any two compounds with different mass ratios must be different compounds.

$$\underline{\text{Compound A}}: \frac{mass\ O}{mass\ N} = \frac{42.0 \text{ g O}}{18.4 \text{ g N}} = 2.28, \quad \underline{\text{Compound B}}: \frac{mass\ O}{mass\ N} = \frac{40.8 \text{ g O}}{23.8 \text{ g N}} = 1.71$$

$$\underline{\text{Compound C}}: \frac{mass\ O}{mass\ N} = \frac{97.3 \text{ g O}}{42.6 \text{ g N}} = 2.28, \quad \underline{\text{Compound D}}: \frac{mass\ O}{mass\ N} = \frac{60.0 \text{ g O}}{52.5 \text{ g N}} = 1.14$$

So, Compounds A and C could be the same, but Compound B and D are two other (different) compounds.

13.   (a)   Nonmetals. Although metals are mostly solids at room temperature (only mercury is liquid), so are many nonmetals and metalloids. However, only nonmetals exist as gases at room temperature.

(b)   Metals. Metals are known for their high electrical conductivities.

(c)   Metals or metalloids. Most metals are shiny and lustrous, but some metalloids also have this property.

14.   (a)   Since silicon appears in between the metals and the nonmetals (on the metal-nonmetal separation), it is a metalloid (semimetal).

(b)   Silver is a metal because it appears to the left of the metal-nonmetal separation.

(c)   Since sodium appears to the left of the metal-nonmetal separation, it is a metal.

(d)   Sulfur is a nonmetal because it appears to the upper right of the metal-nonmetal separation.

15.   (a)   Nanoscale. This statement describes the relationship between the two atoms, which is nanoscale.

(b)   Macroscale. This statement describes observations based on the five senses (colorless, odorless) that are made without any assistance from microscopes.

(c)   Symbolic. The chemical formula for water uses several symbols: H for hydrogen, O for oxygen, and the '2' to show that there are two H atoms for every O atom in the molecule.

ANSWERS — PRACTICE TEST

Chapter 2

1.   (a)   Both protons and neutrons are found in the nucleus.
      (b)   Electrons have a negative charge.
      (c)   Protons have a positive charge.
      (d)   Both protons and neutrons weigh about 1 amu.
      (e)   Neutrons have no charge.
      (f)   Electrons are found outside the atom's nucleus.

2.   (a) $486.1 \text{ nm} \times \dfrac{10^{-9} \text{ m}}{1 \text{ nm}} = 4.861 \times 10^{-7} \text{ m}$ (4 s.f. $\times \infty$ s.f. = 4 s.f.).

      (b) $0.11 \text{ mi} \times \dfrac{5280 \text{ ft}}{1 \text{ mi}} \times \dfrac{12 \text{ in}}{1 \text{ ft}} = 7.0 \times 10^{3} \text{ in}$ (2 s.f. $\times \infty$ s.f. $\times \infty$ s.f. = 2 s.f.).
           The only way to write 7000 in and show 2 s.f. is to write it is scientific notion to show that the zero in the hundreds place is significant but the zeroes in the tens and units place are not.

      (c) $1.5 \text{ L} \times \dfrac{1.056710 \text{ qt}}{1 \text{ L}} \times \dfrac{1 \text{ gal}}{4 \text{ qt}} = 0.40 \text{ gal}$ (2 s.f. $\times$ 7 s.f. $\quad \infty$ s.f. = 2 s.f.).

      (d) $50 \text{ oz} \times \dfrac{1 \text{ lb}}{16 \text{ oz}} \times \dfrac{453.59237 \text{ g}}{1 \text{ lb}} \times \dfrac{1 \text{ kg}}{10^{3} \text{ g}} = 1 \text{ kg}$ (1 s.f. $\quad \infty$ s.f. $\times$ 8 s.f. $\quad \infty$ s.f. = 1 s.f.).

3.   (a)   Although pi (π) is known to several digits, the student only used 3 digits, so his/her answer can only be reported to 3 significant figures (187 m^2).

      (b)   Student 2 assumed that by dividing 15.426 m by 2, the total number of significant figures in that answer would be 1; however, we are infinitely confident that the radius of a circle is one-half the diameter of the circle. So, the students should have listed his/her answer to 5 significant figures.

      (c) $A = \pi r^2 = (3.14159)\left(\dfrac{15.426 \text{ m}}{2}\right)^2 = 186.89 \text{ m}^2$.

4.   (a) 8.25×10^{-3} (c) 7.10×10^{1} (you have to add the zero to get 3 s.f.).
      (b) 8.50×10^{5} (d)   1.35 (it can also be written 1.35×10^{0}).

5.   Remember that $Z = \#p^+$ (so column 2 = column 5), Z determines the atom identity (X in the nuclide symbol), that $A = \#p^+ + \#n^0$, and $\#n^0 = A - Z$.

nuclide symbol	$\#p^+ (= Z)$	$\#n^0 (= A - Z)$	$A (= \#p^+ + \#n^0)$	$Z (= \#p^+)$
^{88}Sr	**38**	**50**	**88**	**38**
^{193}Ir	77	116	**193**	77
^{92}Zr	**40**	**52**	92	40

6.   To determine the percents, you need to know the total mass of the compound. Since the total mass is not your final answer, don't round this number (if it were your final answer, it should be reported as 466 g).

$$m_{compound} = m_{Fe} + m_{Br} = 88.1 \text{ g} + 378 \text{ g} = 466.1 \text{ g total (unrounded)}$$

$$\%\text{Fe} = \frac{m_{Fe}}{m_{total}} \times 100\% = \frac{88.1 \text{ g}}{466.1 \text{ g}} \times 100\% = 18.9\% \text{ Fe}$$

$$\%\text{Br} = \frac{m_{Br}}{m_{total}} \times 100\% = \frac{378 \text{ g}}{466.1 \text{ g}} \times 100\% = 81.1\% \text{ Br}$$

Since both masses were listed to 3 significant figures, the percents are also listed to 3 s.f. (100% has ∞ s.f.).

7. (a) Since $Z = 29$ for copper, ^{63}Cu has $63 - 29 = 34$ neutrons and ^{65}Cu has $65 - 29 = 36$ neutrons.

(b) To determine the fractional abundances of the two isotopes, you must assume one has a value of x (^{63}Cu), and the other (^{65}Cu) has a value of $(1 - x)$, then solve for x and $(1 - x)$.

$$63.55 \text{ amu} = (x)(62.9296 \text{ amu}) + (1 - x)(64.9278 \text{ amu})$$

$$63.55 \text{ amu} = (62.9296)x + 64.9278 \text{ amu} - (64.9278 \text{ amu})x$$

$$63.55 \text{ amu} - 64.9278 \text{ amu} = (62.9296 - 64.9278 \text{ amu})x$$

$$-1.3778 \text{ amu} = (-1.9982 \text{ amu})x$$

$$x = \frac{-1.3778 \text{ amu}}{-1.9982 \text{ amu}} = 0.690$$

So, the fractional abundance of ^{63}Cu is 0.690 and its percent abundance is 69.0%. The fractional abundance for ^{65}Cu is $1 - 0.690 = 0.310$, and its percent abundance would be 31.0%. In this example, 1 (just like 100% in problem 5 above) has ∞ s.f. Because 63.55 amu has error in the hundredths digit, the value -1.3778 amu has only 3 s.f. (-1.38 amu) and x has only 3 s.f.

8. (a) 1.2×10^{24} molec $C_4H_4S \times \dfrac{9 \text{ atoms}}{1 \text{ molec } C_4H_4S} = 1.1 \times 10^{25}$ atoms.

(b) 8.24 mol $H_2S \times \dfrac{6.022 \times 10^{23} \text{ molec } H_2S}{1 \text{ mol } H_2S} \times \dfrac{3 \text{ atoms}}{1 \text{ molec } H_2S} = 1.49 \times 10^{25}$ atoms.

(c) 1000 g Sl $\times \dfrac{\text{mol S}}{32.07 \text{ g S}} \times \dfrac{6.022 \times 10^{23} \text{ atom S}}{1 \text{ mol S}} = 2 \times 10^{25}$ atom S.

So, 1000 g of S has the most atoms in it.

9. (a) 2.325×10^{23} atom ^{12}C $\times \dfrac{1 \text{ mol } ^{12}\text{C}}{6.022 \times 10^{23} \text{ atom } ^{12}\text{C}} \times \dfrac{12 \text{ g } ^{12}\text{C}}{\text{mol } ^{12}\text{C}} = 4.633$ g ^{12}C.

(b) 2×10^{22} atom Cs $\times \dfrac{1 \text{ mol Cs}}{6.022 \times 10^{23} \text{ atom Cs}} \times \dfrac{132.91 \text{ g Cs}}{\text{mol Cs}} = 4$ g Cs.

(c) 1.20 mol He $\times \dfrac{4.003 \text{ g He}}{\text{mol He}} = 4.80$ g He.

(d) 0.025 mol Hg $\times \dfrac{200.6 \text{ g Hg}}{\text{mol Hg}} = 5.0$ g Hg.

So, 0.025 mol of Hg atoms weighs the most.

10. (a) Scandium (Sc)   (f) Samarium (Sm)
    (b) Iodine (I)   (g) Phosphorus (P)
    (c) Radon (Rn)   (h) Selenium (Se)
    (d) Mendelevium (Md)   (i) Radium (Ra)
    (e) Lithium (Li)   (j) Manganese (Mn)

ANSWERS — PRACTICE TEST

Chapter 3

1. The atoms in the molecular formula are usually listed with carbon first, hydrogen second, then all other elements in alphabetic order. Also, the lines for double and triple bonds are drawn in the condensed formula.

molecular formula	condensed formula	structural formula
$C_2H_8N_2$	$NH_2CH_2CH_2NH_2$	
C_5H_9N	$(CH_3)_2CHCH_2C{\equiv}N$	
Cl_2O_7	$O_3ClOClO_3$	
$C_5H_6F_4$	$(CH_3)_2C{=}CFCF_3$	

2. (a) In order to determine if these compounds could be isomers of each other, you need to write the molecular formulas for these compounds. Structure I is $C_2H_4Cl_2$, structure II is $C_2H_4Cl_2$, structure III is $C_2H_3Cl_3$, and structure IV is $C_2H_4Cl_2$. So, structures I, II, and IV could be isomers of each other, and III cannot be an isomer of structures I, II, or IV.

    (b) Looking at structures I, II, and IV, you can see that structures I and II are the same compound (both have a C atom with 3 H atoms attached to it and the other C atom has 2 Cl atoms and 1 H atom). Structure IV, however, is different (it has 2 H atoms and 1 Cl atom attached to both C atoms).

    (c) Structure IV is a structural isomer to both structure I and structure II (which are the same molecule) because it has different types of atom connections, which are described in answer (b) above.

3. (a) Water could be called dihydrogen monoxide (or dihydrogen oxide).
    (b) Ammonia could be called nitrogen trihydride.
    (c) Hydrazine could be called dinitrogen tetrahydride.
    (d) Nitric oxide could be called nitrogen monoxide (or nitrogen oxide).
    (e) Nitrous oxide could be called dinitrogen monoxide (or dinitrogen oxide).
    (f) Phosphine could be called phosphorus trihydride (similar to ammonia).

4. (a) Since carbon is a nonmetal, its negative charge would be $8 - 4 = 4$; and its monatomic ion is C^{4-}.
    (b) Fluorine is a nonmetal so its negative charge would be $8 - 7 = 1$; the fluoride ion is F^-.
    (c) Iron is a transition element, so predicting charges is difficult; the chart in the book shows that iron makes a +2 or +3 cation, so two answers are correct: Fe^{2+} and Fe^{3+}.
    (d) Lithium is a main group metal, so its charge equals its column number, and its ion is Li^+.
    (e) Selenium is a nonmetal so its negative charge would be $8 - 6 = 2$; the selenide ion is Se^{2-}.

(f)   Strontium is a main group metal, so its charge equals its column number, and its ion is Sr^{2+}.

5.   (a)   Bromine is a nonmetal in column 7A, so its ion is Br^- ($8 - 7 = 1$);; calcium is a metal in column 2A, so it is Ca^{2+}; the formula for calcium bromide is $CaBr_2$ (remember that the cation is always written first).

     (b)   Carbon is a nonmetal in column 4A (C^{4-}), and potassium is a metal in column 1A (K^+). The formula for potassium carbide is K_4C (4 K^+ ions for every 1 C^{4-} ion).

     (c)   Chromium is a transition element that can have two charges (Cr^{2+} or Cr^{3+}); the hydroxide ion is OH^-. There are two formulas: Chromium(II) hydroxide, $Cr(OH)_2$, and chromium(III) hydroxide, $Cr(OH)_3$.

     (d)   Magnesium is a metal in column 2A, so its ion is Mg^{2+}; since oxygen is a nonmetal in column 6A, its ion is O^{2-}, and the formula for magnesium oxide is MgO (don't write it as Mg_2O_2 since this is not the simplest ratio of ions needed to get a neutral compound).

     (e)   Zinc is a transition element that has a charge of +2 (Zn^{2+}), and the cyanide ion is CN^-. The neutral compound zinc cyanide is $Zn(CN)_2$.

     (f)   Sulfur is a nonmetal in column 6A, so its ion is S^{2-}, and the ammonium ion is NH_4^+. The formula for ammonium sulfide is $(NH_4)_2S$.

6.   Explanations for each entry are listed below. Note that only ionic and ionic hydrate compounds have ions.

molecular formula	compound type	ions present	chemical name
B_2O_3	molecular	none	diboron trioxide
C_2H_6	organic	none	propane
$CoCl_2\cdot6H_2O$	ionic hydrate	Co^{2+} and Cl^-	cobalt(II) chloride hexahydrate
$FeCO_3$	ionic	Fe^{2+} and CO_3^{2-}	iron(II) carbonate
IF_7	molecular	none	iodine heptafluoride
K_2Se	ionic	K^+ and Se^{2-}	potassium selenide
S_5N_6	molecular	none	pentasulfur hexanitride

     (a)   Boron is a metalloid and oxygen is a nonmetal, so this compound is molecular. There are no ions in a molecular compound. It is named using the Greek prefixes ('*di-*' for the 2 B atoms, and '*tri-*' for 3 O atoms), and the ending of oxygen is replaced with '*-ide*' to give oxide: diboron trioxide.

     (b)   Compounds made of carbon and hydrogen are organic compounds, which are molecular and have no ions. This alkane is called propane, using the prefix '*prop-*' which means there are 3 carbon atoms in the compound.

     (c)   The metal (Co), nonmetal (Cl), and "$6H_2O$' tell you it is an ionic hydrate. The ions in this compound are the chloride ion (Cl^-) and the cobalt(II) ion (Co^{2+}). Although cobalt can form +2 or +3 ions, the +2 ion must be present in this compound to balance the charge of 2 Cl^- ions. For this compound the cation and anion are named first, followed by the Greek prefix for 6 ('*hexa-*') and the word 'hydrate': cobalt(II) chloride hexahydrate.

     (d)   The presence of a metal (Fe) and nonmetals (C and O) tells you it is ionic. The ions in this compound are the carbonate ion (CO_3^{2-}) and the iron(II) ion (Fe^{2+}). Although iron can form +2 or +3 ions, the +2 ion must be present in this compound to balance the charge of CO_3^{2-} ion and the name of this compound is iron(II) carbonate.

     (e)   Since iodine and fluorine are both nonmetals, this compound is molecular (no ions). The Greek prefix '*mono-*' is omitted for iodine and the Greek prefix for fluorine (changed to fluoride) is '*hepta-*' (7).

     (f)   The presence of a metal (K) and nonmetal (Se) tells you it is ionic. The ions in this compound are the potassium ion (K^+) and the selenide ion (Se^{2-}), and its name is potassium selenide (the Greek prefixes are used only on molecular compounds and not ionic compounds such as this one).

     (g)   Since sulfur and nitrogen are both nonmetals, this compound is molecular (no ions). The Greek prefix for sulfur is '*penta-*' (5) and the Greek prefix for nitrogen (changed to nitride) is '*hexa-*' (6), so the name is pentasulfur hexanitride.

7. Explanations for each entry are listed below. Note that only ionic and ionic hydrate compounds have ions.

molecular formula	compound type	ions present	chemical name
$Al_2(SO_4)_3$	ionic	Al^{3+} and SO_4^{2-}	aluminum sulfate
C_4H_{10}	organic	none	butane
ClF_3	molecular	none	chlorine trifluoride
$Fe(NO_3)_3 \cdot 9H_2O$	ionic hydrate	Fe^{3+} and NO_3^-	iron(III) nitrate nonahydrate
$Mg(OH)_2$	ionic	Mg^{2+} and OH^-	magnesium hydroxide
C_9H_{20}	organic	none	nonane
Na_3N	ionic	Na^+ and N^{3-}	sodium nitride

(a) Since there are no Greek prefixes, this is probably an ionic compound (it also contains a metal and nonmetal). Aluminum is in column 3A so its ion is Al^{3+}, and the sulfate ion is SO_4^{2-}. The formula of aluminum sulfate must be neutral, so 2 Al^{3+} ions are needed for every 3 SO_4^{2-} ions: $Al_2(SO_4)_3$.

(b) The '-ane' ending tells you this is an alkane (an organic compound). The 'but-' prefix tells you there are 4 C atoms; the formula of C_nH_{2n+2} (using $n = 4$), tells you the formula is C_4H_{10}.

(c) The Greek prefix in front of fluoride tells you this is a molecular compound. 'Chlorine' means 1 Cl atom (if a prefix is missing, it is assumed to be 'mono-'), and 'trifluoride' means 3 F atoms, so the formula is ClF_3.

(d) The lack of Greek prefixes in the first part of the name and the 'nonahydrate' tells you this is an ionic hydrate. The ions present are the iron(III) ion (Fe^{3+}) and the nitrate ion (NO_3^-). The balanced ionic compound is $Fe(NO_3)_3$. The Greek prefix 'nona-' tells you there are 9 waters of hydration in this compound.

(e) Since there are no Greek prefixes, this is probably an ionic compound. Magnesium is in column 2A (metal) so its ion is Mg^{2+}; and the hydroxide ion is OH^- (nonmetals). The formula of magnesium hydroxide must be neutral, so 1 magnesium ion is needed for every 2 hydroxide ions: $Mg(OH)_2$.

(f) The '-ane' ending tells you this is an alkane (an organic compound). The 'non-' prefix tells you there are 9 C atoms; the formula of C_nH_{2n+2} (using $n = 9$), tells you the formula is C_9H_{20}.

(g) The lack of Greek prefixes tells you this is an ionic compound. Sodium is in column 1A (metal), so its ion is Na^+. The '-ide' ending tells you that the anion is monatomic. Nitride comes from nitrogen, and the monatomic anion of this group 5A nonmetal would be N^{3-} ($8 - 5 = 3$). The neutral compound would have 3 sodium ions for every 1 nitride ion, and its formula is Na_3N.

8. (a) This compound appears to be a molecular compound—it has a low melting point, it is a liquid at room temperature, and it does not conduct electricity in its liquid (molten) state.

(b) This is surprising since it is made of a metal (titanium, Ti) and a nonmetal (chlorine, Cl); most compounds made of metals and nonmetals are ionic compounds.

(c) Since it is a molecular compound, it should be named titanium tetrachloride.

9. (a) To determine the molar mass of calcium phosphate, find the mass of 1 mol $Ca_3(PO_4)_2$, recognizing that calcium phosphate has 3 Ca atoms, 2 P atoms, and 8 O atoms.

$$1 \text{ mol } Ca_3(PO_4)_2 = 3 \text{ mol Ca} \times \frac{40.08 \text{ g Ca}}{\text{mol Ca}} + 2 \text{ mol P} \times \frac{30.97 \text{ g P}}{\text{mol P}} + 8 \text{ mol O} \times \frac{16.00 \text{ g O}}{\text{mol O}}$$

$$1 \text{ mol } Ca_3(PO_4)_2 = 310.18 \text{ g } Ca_3(PO_4)_2$$

So, calcium phosphate has a molar mass of 310.18 g/mol.

(b) $1.25 \text{ mol } Ca_3(PO_4)_2 \times \dfrac{310.18 \text{ g } Ca_3(PO_4)_2}{1 \text{ mol } Ca_3(PO_4)_2} = 388 \text{ g } Ca_3(PO_4)_2$.

(c) To find the number of moles of P atoms, first determine the number of moles of $Ca_3(PO_4)_2$ present in 100.0 g. The number of P atoms will be two times this number since $Ca_3(PO_4)_2$ has 2 P atoms in it.

$$100.0 \text{ g Ca}_3(\text{PO}_4)_2 \times \frac{\text{mol Ca}_3(\text{PO}_4)_2}{310.18 \text{ g Ca}_3(\text{PO}_4)_2} \times \frac{2 \text{ mol P atoms}}{1 \text{ mol Ca}_3(\text{PO}_4)_2} = 0.6448 \text{ mol P atoms}$$

(d) $\% \text{ P} = \dfrac{2 \text{ mol P} \times \dfrac{30.97 \text{ g P}}{\text{mol P}}}{310.18 \text{ g Ca}_3(\text{PO}_4)_2} \times 100\% = 19.97\% \text{ P}$

10.   (a) $1 \text{ mol H}_2\text{S} = 2 \text{ mol H} \times \dfrac{1.008 \text{ g H}}{\text{mol H}} + 1 \text{ mol S} \times \dfrac{32.07 \text{ g S}}{\text{mol S}} = 34.09 \text{ g H}_2\text{S}$

           $\% \text{ S} = \dfrac{1 \text{ mol S} \times \dfrac{32.07 \text{ g S}}{\text{mol S}}}{34.09 \text{ g H}_2\text{S}} \times 100\% = 94.07\% \text{ S}.$

   (b) $1 \text{ mol FeS}_2 = 1 \text{ mol Fe} \times \dfrac{55.85 \text{ g Fe}}{\text{mol Fe}} + 2 \text{ mol S} \times \dfrac{32.07 \text{ g S}}{\text{mol S}} = 119.99 \text{ g FeS}_2$

           $\% \text{ S} = \dfrac{2 \text{ mol S} \times \dfrac{32.07 \text{ g S}}{\text{mol S}}}{119.99 \text{ g FeS}_2} \times 100\% = 53.45\% \text{ S}.$

   (c) $1 \text{ mol SO}_2 = 1 \text{ mol S} \times \dfrac{32.07 \text{ g S}}{\text{mol S}} + 2 \text{ mol O} \times \dfrac{16.00 \text{ g O}}{\text{mol O}} = 64.07 \text{ g SO}_2$

           $\% \text{ S} = \dfrac{1 \text{ mol S} \times \dfrac{32.07 \text{ g S}}{\text{mol S}}}{64.07 \text{ g SO}_2} \times 100\% = 50.05\% \text{ S}.$

   (d) $1 \text{ mol H}_2\text{SO}_4 = 2 \text{ mol H} \times \dfrac{1.008 \text{ g H}}{\text{mol H}} + 1 \text{ mol S} \times \dfrac{32.07 \text{ g S}}{\text{mol S}} + 4 \text{ mol O} \times \dfrac{16.00 \text{ g O}}{\text{mol O}} = 98.09 \text{ g H}_2\text{SO}_4$

           $\% \text{ S} = \dfrac{1 \text{ mol S} \times \dfrac{32.07 \text{ g S}}{\text{mol S}}}{98.09 \text{ g H}_2\text{SO}_4} \times 100\% = 32.69\% \text{ S}.$

11.   (a)   If you assume exactly 100 g of the compound, then you have 33.3 g Fe, 28.6 g C, and 38.1 g O.

$$33.3 \text{ g Fe} \times \frac{\text{mol Fe}}{55.85 \text{ g Fe}} = 0.59624 \text{ mol Fe (unrounded)}$$

$$28.6 \text{ g C} \times \frac{\text{mol C}}{12.01 \text{ g C}} = 2.38135 \text{ mol C (unrounded)}$$

$$38.1 \text{ g O} \times \frac{\text{mol O}}{16.00 \text{ g O}} = 2.38125 \text{ mol O (unrounded)}$$

$$\text{Fe}_{0.59624}\text{C}_{2.38135}\text{O}_{2.38125} = \text{Fe}_{\frac{0.59624}{0.59624}}\text{C}_{\frac{2.38135}{0.59624}}\text{O}_{\frac{2.38125}{0.59624}} = \text{Fe}_{1.000}\text{C}_{3.994}\text{O}_{3.994} = \text{FeC}_4\text{O}_4$$

   (b)   To find the molecular formula, calculate the molar mass of the empirical formula and divide the actual molar mass by this value. The resulting whole number is then multiplied by the empirical formula to get the molecular formula.

$$1 \text{ mol FeC}_4\text{O}_4 = 1 \text{ mol Fe} \times \frac{55.85 \text{ g Fe}}{\text{mol Fe}} + 4 \text{ mol C} \times \frac{12.01 \text{ g C}}{\text{mol C}} + 4 \text{ mol O} \times \frac{16.00 \text{ g O}}{\text{mol O}} = 167.89 \text{ g FeC}_4\text{O}_4$$

$$n = \frac{M \text{ molecular}}{M \text{ empirical}} = \frac{500 \text{ g/mol}}{167.89 \text{ g/mol}} = 2.978 = 3$$

So, the molecular formula is 3 times the empirical formula, or $\text{Fe}_3\text{C}_{12}\text{O}_{12}$.

ANSWERS — PRACTICE TEST

Chapter 4

1. (a) $S_8(s) \rightarrow 4\,S_2(g)$.

 (b) This is a decomposition reaction because one reactant molecule produces four product molecules.

 (c) The subscripts tell you how many S atoms are attached together (in the reactant, there are 8 S atoms attached as one molecule; in the product, there are 2 S atoms attached as molecules). The coefficient tells you how many of these atoms are present (there is 1 molecule present in the reactant, there are 4 molecules present in the products).

 (d)

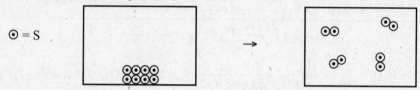

2. (a) $K_2CO_3(s) + 2\,HBr(aq) \rightarrow 2\,KBr(aq) + H_2CO_3(aq)$, $H_2CO_3(aq) \rightarrow H_2O(\ell) + CO_2(g)$.
 The first reaction is an exchange reaction because the potassium ion exchanges the carbonate ion for the bromide ion and the hydrogen ion exchanges the bromide ion for the carbonate ion. The second reaction is a decomposition reaction because two products are made from one reactant.

 (b) $CuO(s) + H_2SO_4(aq) \rightarrow CuSO_4(aq) + H_2O(\ell)$, $2\,CuSO_4(aq) + Ti(s) \rightarrow 2\,Cu(s) + Ti(SO_4)_2(aq)$.
 The first reaction is an exchange reaction because Cu^{2+} trades O^{2-} for SO_4^{2-}, and H^+ trades SO_4^{2-} for O^{2-}. The second reaction is a displacement reaction because elemental Ti reacts to form elemental Cu.

3. On the reactant side there are 4 atoms of C in the solid state (organized pattern) and 6 molecules of F_2 in the gas phase (occupying the entire container's volume). The product side has 3 molecules of CF_4 (1 C atom bonded to 4 F atoms) and 1 atom of C left over. Removing the C atom that is a spectator atom, and dividing the equation by 3 gives the proper answer. This is a combination reaction because there is only one product.

$$4\,C(s) + 6\,F_2(g) \rightarrow 3\,CF_4(g) + C(s)$$
$$3\,C(s) + 6\,F_2(g) \rightarrow 3\,CF_4(g)$$
$$C(s) + 2\,F_2(g) \rightarrow CF_4(g)$$

4. (a) $CoCl_2 \cdot 6H_2O(s) \rightarrow CoCl_2(s) + 6\,H_2O(g)$.

 (b) This is a decomposition reaction because one reactant produces two different products.

 (c) $5.000 \text{ g hyd. } CoCl_2 \times \dfrac{\text{mol hyd. } CoCl_2}{237.93 \text{ g hyd. } CoCl_2} \times \dfrac{1 \text{ mol } CoCl_2}{1 \text{ mol hyd. } CoCl_2} \times \dfrac{129.83 \text{ g } CoCl_2}{\text{mol } CoCl_2} = 2.728 \text{ g } CoCl_2$.

5. (a) This reaction is a displacement reaction because elemental Zn reacts to form elemental H_2.

 (b) $6.00 \text{ g } ZnSO_4 \times \dfrac{\text{mol } ZnSO_4}{161.45 \text{ g } ZnSO_4} \times \dfrac{1 \text{ mol } Zn}{1 \text{ mol } ZnSO_4} \times \dfrac{65.38 \text{ g } Zn}{\text{mol } Zn} = 2.43 \text{ g } Zn$.
 $2.52 \text{ g total} - 2.43 \text{ g } Zn = 0.09 \text{ g } Cu$.

 (c) $\% \text{ Zn} = \dfrac{2.43 \text{ g } Zn}{2.52 \text{ g total}} \times 100\% = 96.4\% \text{ Zn}$, $\% \text{ Cu} = 100\% \text{ total} - 96.4\% \text{ Zn} = 3.6\% \text{ Cu}$.

6. (a) $3\,Ca(CN)_2(aq) + 2\,Na_3PO_4(aq) \rightarrow \_\,Ca_3(PO_4)_2(s) + 6\,NaCN(aq)$.

 (b) This is an exchange reaction because the calcium ion exchanges the cyanide ion for the phosphate ion and the sodium ion exchanges the phosphate ion for the cyanide ion.

 (c) $16.4 \text{ g } Ca(CN)_2 \times \dfrac{\text{mol } Ca(CN)_2}{92.12 \text{ g } Ca(CN)_2} \times \dfrac{1 \text{ mol } Ca_3(PO_4)_2}{3 \text{ mol } Ca(CN)_2} \times \dfrac{310.18 \text{ g } Ca_3(PO_4)_2}{\text{mol } Ca_3(PO_4)_2} = 18.4 \text{ g } Ca_3(PO_4)_2$

$$25.3 \text{ g Na}_3\text{PO}_4 \times \frac{\text{mol Na}_3\text{PO}_4}{163.94 \text{ g Na}_3\text{PO}_4} \times \frac{1 \text{ mol Ca}_3(\text{PO}_4)_2}{2 \text{ mol Na}_3\text{PO}_4} \times \frac{310.18 \text{ g Ca}_3(\text{PO}_4)_2}{\text{mol Ca}_3(\text{PO}_4)_2} = 23.9 \text{ g Ca}_3(\text{PO}_4)_2$$

So, $Ca(CN)_2$ is the limiting reactant, and 18.4 g $Ca_3(PO_4)_2$ can be made.

(d) $\% \text{ yield} = \dfrac{\text{actual yield}}{\text{theoretical yield}} \times 100\% = \dfrac{8.25 \text{ g Ca}_3(\text{PO}_4)_2}{18.4 \text{ g Ca}_3(\text{PO}_4)_2} \times 100\% = 44.8\%.$

7. (a) Because CO_2 and H_2O are produced, you know the compound has C and H atoms in it, and it may have O atoms. First you need to determine the moles and masses of C and H in the compound to see if there are any O atoms in the compound.

$$99.9 \text{ g CO}_2 \times \frac{\text{mol CO}_2}{44.01 \text{ g CO}_2} \times \frac{1 \text{ mol C}}{1 \text{ mol CO}_2} = 2.2699 \text{ mol C (unrounded)}$$

$$2.2699 \text{ mol C} \times \frac{12.01 \text{ g C}}{\text{mol C}} = 27.3 \text{ g C}$$

$$40.9 \text{ g H}_2\text{O} \times \frac{\text{mol H}_2\text{O}}{18.016 \text{ g H}_2\text{O}} \times \frac{2 \text{ mol H}}{1 \text{ mol H}_2\text{O}} = 4.5404 \text{ mol H (unrounded)}$$

$$4.5404 \text{ mol H} \times \frac{1.008 \text{ g H}}{\text{mol H}} = 4.6 \text{ g H}$$

Since the total mass of C and H isn't 50.0 g, there must be some oxygen in this compound. The missing mass must be oxygen, and you need to determine how many moles of O are present.

$$m_\text{O} = 50.0 \text{ g total} - 27.3 \text{ g C} - 4.6 \text{ g H} = 18.1 \text{ g O}$$

$$18.1 \text{ g O} \times \frac{\text{mol O}}{16.00 \text{ g O}} = 1.1313 \text{ mol O (unrounded)}$$

$$C_{\frac{2.2699}{1.1313}} H_{\frac{4.5404}{1.1313}} O_{\frac{1.1313}{1.1313}} = C_{2.006} H_{4.013} O_{1.000} = C_2H_4O$$

(b) The empirical mass of C_2H_4O is 44.052 g/mol. Since the molecule's molar mass is almost exactly twice this value, the molecular formula must be $C_4H_8O_2$.

(c) $C_4H_8O_2(\ell) + 5 O_2(g) \rightarrow 4 CO_2(g) + 4 H_2O(g)$

8. (a) This compound contains 3.000 g Fe and 4.279 g total – 3.000 g Fe = 1.279 g O.

$$\% \text{ Fe} = \frac{3.000 \text{ g Fe}}{4.279 \text{ g total}} \times 100\% = 70.1\% \text{ Fe}$$

$$\% \text{ O} = \frac{1.279 \text{ g O}}{4.279 \text{ g total}} \times 100\% = 29.9\% \text{ O}$$

(b) You can use the percent composition by mass values you just calculated to determine the empirical formula, or you can simply use the masses already given to you in the problem.

$$3.000 \text{ g Fe} \times \frac{\text{mol Fe}}{55.85 \text{ g Fe}} = 0.053715 \text{ mol Fe (unrounded)}$$

$$1.279 \text{ g O} \times \frac{\text{mol O}}{16.00 \text{ g O}} = 0.079938 \text{ mol O (unrounded)}$$

$$Fe_{0.053763} O_{0.079938} = Fe_{\frac{0.053715}{0.053715}} O_{\frac{0.079938}{0.053715}} = Fe_{1.00} O_{1.488} = Fe_{2.00} O_{2.976} = Fe_2O_3$$

When you divide by the smaller value, you get subscripts of 1.000 and 1.488 (which is not a whole number); however, if you multiply by 2, then they become 2.000 and 2.976, or 2 and 3.

(c) $4 Fe(s) + 3 O_2(g) \rightarrow 2 Fe_2O_3(s).$

(d) This reaction can be classified as a combination reaction because two reactants make one product, and it can also be classified as a combustion reaction because O_2 is a reactant and the product is an oxide.

ANSWERS — PRACTICE TEST

Chapter 5

1. (a) This equation is wrong because the two Cl^- ions will not stay attached to each other; in fact, these two ions will repel each other because they have the same charge. The equation should show that the Cl^- ions move independently of each other: $ZnCl_2(s) \rightarrow Zn^{2+}(aq) + 2\,Cl^-(aq)$.

   (b) This equation is wrong because silver carbonate is insoluble in water. Most carbonate salts are insoluble and Ag^+ is not an exception: $Ag_2CO_3(s) \rightarrow Ag_2CO_3(s)$, or $Ag_2CO_3(s) \rightarrow$ N.R.

   (c) This equation is wrong because the two cations (Na^+ and H^+) were written together and the two anions (Cl^- and NO_3^-) were written together. These ions would repel each other (also, a combination of Na^+ and H^+ would be "NaH^{2+}", not neutral, and Cl^- and NO_3^- would make "$ClNO_3^{2-}$"). The correct equation is: $NaNO_3(aq) + HCl(aq) \rightarrow NaCl(aq) + HNO_3(aq)$.

   (d) This equation is wrong because the formulas for the products are wrong. The cations are K^+ and H^+, and the anions are PO_4^{3-} and SO_4^{2-}. So, the product of K^+ and SO_4^{2-} would be K_2SO_4 (and not "K_3SO_4"); similarly the product of H^+ and PO_4^{3-} would be H_3PO_4 (not "H_2PO_4"). To balance the equation, change the coefficients: $2\,K_3PO_4(aq) + 3\,H_2SO_4(aq) \rightarrow 3\,K_2SO_4(aq) + 2\,H_3PO_4(aq)$.

2. (a) The ions present are Co^{3+}, OH^-, H^+, and Br^-. When the ions exchange, Co^{3+} and Br^- will combine to make $CoBr_3$, and the H^+ and OH^- combine to make liquid H_2O. Because H^+ and OH^- are reacting, this is an acid-base reaction. From the solubility rules, HBr and $CoBr_3$ are soluble (aqueous), but $Co(OH)_3$ is not (solid): $Co(OH)_3(s) + 3\,HBr(aq) \rightarrow CoBr_3(aq) + 3\,H_2O(\ell)$.

   (b) The ions present are H^+, NO_3^-, Al^{3+}, and S^{2-}. When the ions exchange, H^+ and S^{2-} react to form H_2S, one of the gases listed in the chart appearing in this chapter, making this a gas-forming reaction. The other product is $Al(NO_3)_3$, made from the Al^{3+} and NO_3^- ions. From the solubility rules, HNO_3 and $Al(NO_3)_3$ are soluble, but Al_2S_3 is not: $6\,HNO_3(aq) + Al_2S_3(s) \rightarrow 2\,Al(NO_3)_3(aq) + 3\,H_2S(g)$.

   (c) The ions present are Ba^{2+}, OH^-, Li^+, and CH_3COO^-. When the ions exchange, Ba^{2+} and CH_3COO^- form $Ba(CH_3COO)_2$, and the Li^+ and OH^- combine to make LiOH. From the solubility rules, all four of these compounds are soluble, so this reaction is not a precipitation, acid-base, or gas-forming reaction (none of these): $Ba(OH)_2(aq) + 2\,LiCH_3COO(aq) \rightarrow Ba(CH_3COO)_2(aq) + 2\,LiOH(aq)$.

   (d) The ions present are Ag^+, NO_3^-, K^+, and I^-. When the ions exchange, Ag^+ and I^- form AgI, and the K^+ and NO_3^- form KNO_3. From the solubility rules, $AgNO_3$, KI, and KNO_3 are soluble but AgI is not, so this is a precipitation reaction: $AgNO_3(aq) + KI(aq) \rightarrow AgI(s) + KNO_3(aq)$.

   (e) The ions present are H^+, SO_4^{2-}, Li^+, and OH^-. When the ions exchange, H^+ and OH^- ions react to make liquid water (making this an acid-base reaction), and the Li^+ and SO_4^{2-} ions will combine to make Li_2SO_4. From the solubility rules, H_2SO_4, LiOH, and Li_2SO_4 are all soluble (aqueous) and water is a liquid: $H_2SO_4(aq) + 2\,LiOH(aq) \rightarrow Li_2SO_4(aq) + 2\,H_2O(\ell)$.

   (f) The ions present are Fe^{2+}, NO_3^-, K^+, and PO_4^{3-}. When the ions exchange, Fe^{2+} and PO_4^{3-} form $Fe_3(PO_4)_2$, and the K^+ and NO_3^- combine to make KNO_3. From the solubility rules, $Fe(NO_3)_2$, K_3PO_4, and KNO_3 are soluble (aqueous) but $Fe_3(PO_4)_2$ is not (solid), so this is a precipitation reaction: $3\,Fe(NO_3)_2(aq) + 2\,K_3PO_4(aq) \rightarrow Fe_3(PO_4)_2(s) + 6\,KNO_3(aq)$.

   (g) The ions present are NH_4^+, SO_4^{2-}, K^+, and OH^-. When the ions exchange, NH_4^+ and OH^- will react to form liquid water and ammonia (NH_3) gas (listed in the chart appearing in this chapter), making this one of the gas-forming reactions. The other product is K_2SO_4 is made from the K^+ and SO_4^{2-} ions. From the solubility rules, you can see that the three compounds ($NH_4)_2SO_4$, KOH, and K_2SO_4 are soluble (aqueous): $(NH_4)_2SO_4(aq) + 2\,KOH(aq) \rightarrow K_2SO_4(aq) + 2\,H_2O(\ell) + 2\,NH_3(g)$.

3. (a) The students did not need to mix the same solutions together (i.e., Unknown A with Unknown A, etc.) because these solutions will not react with each other (since they are the same compound). It is also unnecessary (but not wrong) to mix Unknown A with Unknown B after already testing Unknown B with Unknown A. The order of mixing will not affect whether a reaction occurs or not, and the students could double-check their work by mixing the solutions in different orders.

(b)   The chemicals present are KCH_3COO (potassium acetate), Na_2CO_3 (sodium carbonate), H_2SO_4 (sulfuric acid), and $ZnCl_2$ (zinc chloride). The six reactions are simple exchange reactions. The starting compounds are all soluble in water. Solubilities are determined from the solubility rules, remembering that H^+ and CO_3^{2-} ions react to produce $H_2CO_3(aq)$, which forms $H_2O(\ell)$ and $CO_2(g)$.

$$2\,KCH_3COO(aq) + Na_2CO_3(aq) \rightarrow K_2CO_3(aq) + 2\,NaCH_3COO(aq)$$
$$2\,KCH_3COO(aq) + H_2SO_4(aq) \rightarrow K_2SO_4(aq) + 2\,CH_3COOH(aq)$$
$$2\,KCH_3COO(aq) + ZnCl_2(aq) \rightarrow 2\,KCl(aq) + Zn(CH_3COO)_2(aq)$$
$$Na_2CO_3(aq) + H_2SO_4(aq) \rightarrow Na_2SO_4(aq) + H_2O(\ell) + CO_2(g)$$
$$Na_2CO_3(aq) + ZnCl_2(aq) \rightarrow 2\,NaCl(aq) + ZnCO_3(s)$$
$$H_2SO_4(aq) + ZnCl_2(aq) \rightarrow 2\,HCl(aq) + ZnSO_4(aq)$$

(c)   The fourth equation has bubbles being formed and the fifth equation has a precipitate formed; the other four are "no reactions" (N.R.). Looking at each chemical, you can see that KCH_3COO has 4 N.R., and that matches the reactivity pattern of Unknown D. Na_2CO_3 has 1 ppt, 1 bubble, and 1 N.R., which matches Unknown A. H_2SO_4 has 1 bubble and 2 N.R., and this matches Unknown B. Finally, ZnCl2 has 1 ppt, and 2 N.R., and that matches the reactivity pattern of Unknown C.

4.   (a)   For the ionic equation, the aqueous solutions are written as individual ions; for the net ionic equation, the spectator ions (those that are identical on both sides of the equation) are canceled, leaving the net ionic equation (which is written in boldface).

$$\mathbf{2\,Fe^{3+}(aq)} + 6\,Br^-(aq) + 6\,Na^+(aq) + \mathbf{3\,S^{2-}(aq)} \rightarrow \mathbf{Fe_2S_3(s)} + 6\,Na^+(aq) + 6\,Br^-(aq)$$
$$2\,Fe^{3+}(aq) + 3\,S^{2-}(aq) \rightarrow Fe_2S_3(s)$$

Remember to multiply both ions by the coefficient (i.e., 6 NaBr means 6 Na^+ and 6 Br^- ions) and to convert subscripts in formulas to coefficients (i.e., 6 Br^- ions from 2 $FeBr_3$ and not "2 Br_3^{3-}" ions).

(b)   Since all of these substances are aqueous solutions, the net ionic equation is nothing (N.R.).

$$2\,Ag^+(aq) + SO_4^{2-}(aq) + Cu^{2+}(aq) + 2\,ClO_4^-(aq) \rightarrow 2\,Ag^+(aq) + 2\,ClO_4^-(aq) + Cu^{2+}(aq) + SO_4^{2-}(aq)$$
$$\underline{\text{net ionic equation: N.R.}}$$

(c)   The aqueous chloride salts are written as ions for the ionic equation. Note that only the chloride ions cancel in the net ionic equation. Do not cancel $Mn^{2+}(aq)$ and $Mn(s)$ or $Al(s)$ and $Al^{3+}(aq)$ because these are different chemical species (also note that 6 Cl^- is not the same as "3 Cl_2^{2-}" or "2 Cl_3^{3-}").

$$\mathbf{2\,Al(s)} + \mathbf{3\,Mn^{2+}(aq)} + 6\,Cl^-(aq) \rightarrow \mathbf{2\,Al^{3+}(aq)} + 6\,Cl^-(aq) + \mathbf{3\,Mn(s)}$$
$$2\,Al(s) + 3\,Mn^{2+}(aq) \rightarrow 2\,Al^{3+}(aq) + 3\,Mn(s)$$

(d)   The solids and gases are not written as ions, and are left alone. Only the NO_3^- ions are spectators.

$$\mathbf{CaC_2(s)} + \mathbf{2\,H^+(aq)} + 2\,NO_3^-(aq) \rightarrow \mathbf{Ca^{2+}(aq)} + 2\,NO_3^-(aq) + \mathbf{C_2H_2(g)}$$
$$CaC_2(s) + 2\,H^+(aq) \rightarrow Ca^{2+}(aq) + C_2H_2(g)$$

(e)   In this equation there are no spectator ions, so the ionic and net ionic equations are the same.

$$\mathbf{2\,Ag^+(aq)} + \mathbf{SO_4^{2-}(aq)} + \mathbf{Ba^{2+}(aq)} + \mathbf{2\,OH^-(aq)} \rightarrow \mathbf{2\,AgOH(s)} + \mathbf{BaSO_4(s)}$$
$$2\,Ag^+(aq) + SO_4^{2-}(aq) + Ba^{2+}(aq) + 2\,OH^-(aq) \rightarrow 2\,AgOH(s) + BaSO_4(s)$$

5.   The two reactant for these reactions will be the acid of the anion, and the hydroxide salt of the cation. The products will be the compound of interest and water (use the solubility rules to predict states of matter).

(a) $3\,HClO_3(aq) + Cr(OH)_3(s) \rightarrow Cr(ClO_3)_3(aq) + 3\,H_2O(\ell)$

(b) $2\,HNO_3(aq) + Mg(OH)_2(aq) \rightarrow Mg(NO_3)_2(aq) + 2\,H_2O(\ell)$

(c) $H_3PO_4(aq) + 3\,NaOH(aq) \rightarrow Na_3PO_4(aq) + 3\,H_2O(\ell)$

(d) $H_2SO_4(aq) + Sr(OH)_2(aq) \rightarrow SrSO_4(s) + 2\,H_2O(\ell)$

6. (a) The one half-reaction will contain chlorine, and the other will contain iodine. Balance the charge by adding electrons (e^-) to either side of the equation. Electrons will always appear on opposite sides of the two half-reactions (when e^- are reactants, the reactant in the half-reaction is reduced; when e^- are products, the reactant in the half-reaction is oxidized).

$$Cl_2(aq) + 2\,e^- \rightarrow 2\,Cl^-(aq)$$
$$2\,I^-(aq) \rightarrow I_2(aq) + 2\,e^-$$

So, $Cl_2(aq)$ was reduced to $Cl^-(aq)$, and the $I^-(aq)$ ions were oxidized to $I_2(aq)$.

(b) The one half-reaction will contain iron, and the other will contain oxygen. Remember that $Fe_2O_3(s)$ contains Fe^{3+} and "O^{2-} ions in it.

$$4\,Fe(s) \rightarrow 4\,Fe^{3+}(s) + 12\,e^-$$
$$3\,O_2(g) + 12\,e^- \rightarrow 6\,O^{2-}(s)$$

So, $Fe(s)$ was oxidized to Fe^{3+} ions, and $O_2(aq)$ was reduced to O^{2-} ions.

7. (a) **CH_4:** Rule 4 says $O.N._C + 4 \times O.N._H = 0$; Rule 3c says $O.N._H = +1$, so $O.N._C = -4$. **O_2:** Rule 4 says $2 \times O.N._O = 0$, so $O.N._O = 0$. **CO_2:** Rule 4 says $O.N._C + 2 \times O.N._O = 0$, and Rule 3d says $O.N._O = -2$, so $O.N._C = +4$. **H_2O:** Rule 4 says $2 \times O.N._H + O.N._O = 0$; Rule 3c says $O.N._H = +1$, so $O.N._O = -2$. Since the oxidation number of carbon increases from -4 to $+4$, the C atom is oxidized; since the oxidation number of oxygen decreases from 0 to -2, the O atoms are reduced.

(b) **CaO:** Rule 4 says $O.N._{Ca} + O.N._O = 0$; Rule 3d says $O.N._O = -2$, so $O.N._{Ca} = +2$. **CF_4:** Rule 4 says $O.N._C + 4 \times O.N._F = 0$; Rule 3a says $O.N._F = -1$, so $O.N._C = +4$. **CaF_2:** Rule 4 says $O.N._{Ca} + 2 \times O.N._F = 0$; Rule 3a says $O.N._F = -1$, so $O.N._{Ca} = +2$. **CO_2:** Rule 4 says $O.N._C + 2 \times O.N._O = 0$, and Rule 3d says $O.N._O = -2$, so $O.N._C = +4$. Since none of the atoms changed oxidation states ($O.N._{Ca} = +2$, $O.N._O = -2$, $O.N._C = +4$, and $O.N._F = -1$), this is not a redox reaction.

(c) **Na_2S:** Rule 4 says $2 \times O.N._{Na} + O.N._S = 0$; Rule 3b says $O.N._{Na} = +1$, so $O.N._S = -2$. **HBr:** Rule 4 says $O.N._H + 4 \times O.N._{Br} = 0$; Rule 3a says $O.N._{Br} = -1$, so $O.N._H = +1$. **NaBr:** Rule 4 says $O.N._{Na} + O.N._{Br} = 0$; Rule 3a says $O.N._{Br} = -1$, so $O.N._{Na} = +1$. **H_2S:** Rule 4 says $2 \times O.N._H + 2 \times O.N._S = 0$, and Rule 3c says $O.N._H = +1$, so $O.N._S = -2$. Since none of the atoms changed oxidation states ($O.N._{Na} = +1$, $O.N._S = -2$, $O.N._H = +1$, and $O.N._{Br} = -1$), this is not a redox reaction.

(d) You may have difficulty determining the oxidation numbers in $(NH_4)_2Cr_2O_7$ unless you write it in its ion form: $2\,NH_4^+ + Cr_2O_7^{2-}$. **NH_4^+:** Rule 4 says $O.N._N + 4 \times O.N._H = +1$; Rule 3c says $O.N._H = +1$, so $O.N._N = -3$. **$Cr_2O_7^{2-}$:** Rule 4 says that $2 \times O.N._{Cr} + 7 \times O.N._O = -2$, and Rule 3d says $O.N._O = -2$, so $2 \times O.N._{Cr} = +12$, and $O.N._{Cr} = +6$. **Cr_2O_3:** Rule 4 says $2 \times O.N._{Cr} + 3 \times O.N._O = 0$, and Rule 3d says $O.N._O = -2$, so $2 \times O.N._{Cr} = +6$, and $O.N._{Cr} = +3$. **N_2:** Rule 4 says that $2 \times O.N._N = 0$, so $O.N._N = 0$. **H_2O:** From question 7a, $O.N._H = +1$ and $O.N._O = -2$. Since the oxidation number of nitrogen increases from -3 to 0, the N atoms are oxidized; since the oxidation number of chromium decreases from $+6$ to $+3$, the Cr atoms are reduced.

(e) **N_2H_4:** Rule 4 says $2 \times O.N._N + 4 \times O.N._H = 0$; Rule 3c says $O.N._H = +1$, so $2 \times O.N._N = -4$ and $O.N._N = -2$. **NH_3:** Rule 4 says $O.N._N + 3 \times O.N._H = 0$; Rule 3c says $O.N._H = +1$, so $O.N._N = -3$. **N_2:** Rule 4 says $2 \times O.N._N = 0$ and $O.N._N = 0$. In both N_2H_4 and NH_3, hydrogen has an oxidation state of $+1$. The nitrogen atoms in N_2H_4 have an oxidation state of -2, and in the products nitrogen has an oxidation state of -3 in NH_3 and 0 in N_2. So, the N atoms in N_2H_4 that become N_2 are oxidized and the N atoms in N_2H_4 that become NH_3 are reduced.

8. (a) Gallium metal is oxidized to Ga^{3+}, and the Fe^{2+} ions are reduced to iron metal. Although the Ga and Fe atoms were already balanced, the coefficients are needed to balance charge: $2\,Ga(s) + 3\,Fe^{2+}(aq) \rightarrow 2\,Ga^{3+}(aq) + 3\,Fe(s)$. For gallium metal and Zn^{2+}, there is no reaction: $Ga(s) + Zn^{2+}(aq) \rightarrow N.R.$

(b) Since gallium metal reacts with the Fe^{2+} ions, gallium is a more reactive metal than iron metal; since gallium metal does not react with the Zn^{2+} ions, it is less reactive than zinc metal. So, it should appear between Fe and Zn in the chart. Unfortunately, there is already a metal in between these two metals (nickel) and you can't tell where gallium goes compared to nickel unless you are given data about the reactivities of gallium metal with Ni^{2+} ions or nickel metal with Ga^{3+} ions.

(c)   Since gallium metal is farther away from the reactive metals than aluminum, no displacement reaction will occur: $Ga(s) + Al^{3+}(aq) \rightarrow$ N.R.

(d)   Since potassium metal is closer to the reactive metals than gallium, a displacement reaction will occur. So, $Ga^{3+}(aq)$ becomes $Ga(s)$, and $K(s)$ becomes $K^{+}(aq)$ because it is in column 1A (alkali metal):
$3\,K(s) + Ga^{3+}(aq) \rightarrow 3\,K^{+}(aq) + Ga(s)$.

9.   (a)   There is 25.0 L of water, so the water will make 25.0 L of solution. For the potassium permanganate, the volume of solution is calculated from its molar mass and concentration of the solution.

$$582.5 \text{ g KMnO}_4 \times \frac{\text{mol KMnO}_4}{158.04 \text{ g KMnO}_4} \times \frac{\text{L solution}}{0.235 \text{ mol KMnO}_4} = 15.7 \text{ L solution}$$

So, the student can make 15.7 L of solution, and the $KMnO_4$ is the limiting reactant.

(b)   Calculate the number of moles of $KMnO_4$ from its molar mass, and divide by 25.0 L.

$$582.5 \text{ g KMnO}_4 \times \frac{\text{mol KMnO}_4}{158.04 \text{ g KMnO}_4} = 3.68578 \text{ mol KMnO}_4 \text{ (unrounded)}$$

$$[\text{KMnO}_4] = \frac{3.68578 \text{ mol KMnO}_4}{25.0 \text{ L solution}} = 0.147 \text{ M}$$

10.   One way to solve this problem is to calculate the volume of 0.160 M HCl you can make from 415.2 mL (0.4152 L) of the 12.6 M HCl:

$$0.4152 \text{ L conc. solution} \times \frac{12.6 \text{ mol HCl}}{\text{L conc. solution}} \times \frac{\text{L dil. solution}}{0.160 \text{ mol HCl}} = 32.7 \text{ L dil. solution}$$

If you used all of the concentrated HCl, you could make 32.7 L of 0.450 M HCl, so 35.0 L is not possible.

11.   For each reactant, calculate the amount of solid magnesium hydroxide that can be made.

$$0.7680 \text{ L solution} \times \frac{0.215 \text{ mol Mg(NO}_3)_2}{\text{L solution}} \times \frac{1 \text{ mol Mg(OH)}_2}{1 \text{ mol Mg(NO}_3)_2} \times \frac{58.33 \text{ g Mg(OH)}_2}{\text{mol Mg(OH)}_2} = 9.63 \text{ g Mg(OH)}_2$$

$$0.2420 \text{ L solution} \times \frac{0.594 \text{ mol LiOH}}{\text{L solution}} \times \frac{1 \text{ mol Mg(OH)}_2}{2 \text{ mol LiOH}} \times \frac{58.33 \text{ g Mg(OH)}_2}{\text{mol Mg(OH)}_2} = 4.19 \text{ g Mg(OH)}_2$$

So, LiOH is the limiting reactant, and 4.19 g $Mg(OH)_2$ can be made.

12.   (a)   This is an acid-base exchange reaction between HNO_3 (H^+ and NO_3^-) and $Ca(OH)_2$ (Ca^{2+} and OH^-):
$2\,HNO_3(aq) + Ca(OH)_2(aq) \rightarrow Ca(NO_3)_2(aq) + 2\,H_2O(\ell)$.

(b)   At the equivalence point, the acid-base indicator would have a color intermediate between the acid color (red) and the base color (yellow), so the solution containing the indicator would be orange at the equivalence point.

(c)   Calculate the moles of $Ca(OH)_2$ from the volume and concentration of HNO_3, then divide by the volume of $Ca(OH)_2$ solution in the experiment.

$$0.0335 \text{ L solution} \times \frac{0.0746 \text{ mol HNO}_3}{\text{L solution}} \times \frac{1 \text{ mol Ca(OH)}_2}{2 \text{ mol HNO}_3} = 1.24955 \times 10^{-3} \text{ mol Ca(OH)}_2 \text{ (unrounded)}$$

$$[\text{Ca(OH)}_2] = \frac{1.24955 \times 10^{-3} \text{ mol Ca(OH)}_2}{0.2000 \text{ L solution}} = 6.25 \times 10^{-3} \text{ M}$$

(d)   To convert concentration of $Ca(OH)_2$ to concentration of OH^-, write a mole ratio from the formula of $Ca(OH)_2(aq)$.

$$[\text{OH}^-] = \frac{6.25 \times 10^{-3} \text{ mol Ca(OH)}_2}{\text{L solution}} \times \frac{2 \text{ mol OH}^-}{1 \text{ mol Ca(OH)}_2} = 0.0125 \text{ M}$$

ANSWERS — PRACTICE TEST

Chapter 6

1. (a) The mass is 263 g or 0.263 kg; the gravitational constant is 9.81 m/s^2, and the height of the cliff is 593 m. Plugging into the equation and remembering the 1 J = 1 kg m^2/s^2, the potential energy is:

$$PE = mgh = (0.263 \text{ kg})\left(9.81 \ \frac{\text{m}}{\text{s}^2}\right)(593 \text{ m}) = 1{,}530 \ \frac{\text{kg m}^2}{\text{s}^2} = 1{,}530 \text{ J}$$

   (b) When the ball hits the ground, all of the potential energy has been converted to kinetic energy. Using the equation for kinetic energy ($KE = 1/2 \ mv^2$) and solving for v, the velocity is:

$$KE = \frac{1}{2} \ mv^2, \ v = \sqrt{\frac{2 \ KE}{m}} = \sqrt{\frac{2 \times 1{,}530 \text{ kg m}^2/\text{s}^2}{0.263 \text{ kg}}} = 108 \ \frac{\text{m}}{\text{s}}$$

2. Multiply the mass of the egg and the caloric value to get its kJ value, then convert kJ to J and Cal.

$$5.86 \ \frac{\text{kJ}}{\text{g}} \times 45 \text{ g} = 260 \text{ kJ} \times \frac{1{,}000 \text{ J}}{\text{k J}} = 260{,}000 \text{ J} \times \frac{1 \text{ cal}}{4.184 \text{ J}} \times \frac{1 \text{ Cal}}{1000 \text{ cal}} = 63 \text{ Cal}$$

   So, the 45 g egg contains 260 kJ, 260,000 J, or 63 Cal of energy.

3. For the glass mug, you are given the heat capacity, so use the formula $q = C\Delta T$. You are given the specific heat capacity of water, so use the formula $q = mc\Delta T$, and since you have the molar heat capacity of iron use $q = nc_m\Delta T$. For all of the objects $\Delta T = T_f - T_i = 15.2°C - 76.5°C = -61.3°C$.

$$q_{mug} = C_{mug}\Delta T_{mug} = \left(417 \ \frac{\text{J}}{\text{C}}\right)(-61.3 \text{ C}) = -25{,}600 \text{ J}$$

$$q_{water} = m_{water}c_{water}\Delta T_{water} = (163 \text{ g})\left(4.184 \ \frac{\text{J}}{\text{g C}}\right)(-61.3 \text{ C}) = -41{,}800 \text{ J}$$

$$q_{spoon} = n_{spoon}c_m\Delta T_{spoon} = \left(218.9 \text{ g Fe} \times \frac{\text{mol Fe}}{55.85 \text{ g Fe}}\right)\left(25.2 \ \frac{\text{J}}{\text{mol C}}\right)(-61.3 \text{ C}) = -6{,}050 \text{ J}$$

   Even though the water weighs less than the mug or the iron, it released the most heat (largest magnitude). These values are all negative because the objects are losing (releasing) heat to the surroundings as they cool.

4. (a) Endothermic reactions absorb heat, so the final energy of the system is greater than its initial energy. Exothermic reactions release heat, so the final energy of the system is less than its initial energy.

   (b) The signs of the energy terms for endothermic reactions are positive because the energy of the system has increased; the signs of the energy terms for exothermic reactions are negative because the energy of the system has decreased.

   (c) Because endothermic reactions absorb heat from the surroundings, the surroundings lose heat and their temperature will drop (decrease) and feel colder; because exothermic reactions release heat to the surroundings, the surroundings gain heat and their temperature will rise (increase) and feel warmer.

   (d) In general, endothermic reactions will not happen spontaneously, but can be made to happen if energy is added to the system; exothermic reactions, on the other hand, tend to be spontaneous and release energy to the surroundings (It should be noted that this is a *general* trend, and is not always true).

5. (a) This reaction is endothermic because $\Delta H°$ is positive.

   (b) $$25.0 \text{ g NaHCO}_3 \times \frac{1 \text{ mol NaHCO}_3}{84.01 \text{ g NaHCO}_3} \times \frac{62.8 \text{ kJ}}{2 \text{ mol NaHCO}_3} = 9.34 \text{ kJ}$$

   $$0.2000 \text{ L solution} \times \frac{0.749 \text{ mol H}_2\text{SO}_4}{\text{L solution}} \times \frac{62.8 \text{ kJ}}{1 \text{ mol H}_2\text{SO}_4} = 9.41 \text{ kJ}$$

   So, $NaHCO_3$ is the limiting reactant, and 9.34 kJ of energy would be absorbed by this reaction.

(c)   You can use the percent yield formula with $\Delta H°$ values since it is also a product of the reaction, but first you must determine the amount of energy absorbed by the solution (*soln*).

$$q_{soln} = m_{soln}c_{soln}\Delta T_{soln} = (25.0 \text{ g} + 200.0 \text{ g})\left(4.184 \ \frac{\text{J}}{\text{g C}}\right)(-9.5 \text{ C}) = -8,943.3 \text{ J (unrounded)}$$

$$\% \ yield = \frac{actual \ yield}{theoretical \ yield} \times 100\% = \frac{8,943.3 \text{ J}}{9,344.1 \text{ J}} \times 100\% = 96\%.$$

The energy released by the solution was –8,943.3 J, so the reaction must have gained the same amount of energy (+8,943.3 J, now positive because the energy was gained).

6.   First, sum all q values to zero. You are given the specific heat capacity (c) for water, the enthalpy of reaction ($\Delta H°_{rxn}$) for benzoic acid, and you want to find the heat capacity (C) for the steel container (cont).

$$q_{water} + q_{rxn} + q_{cont} = 0$$

$$q_{water} = m_{water}c_{water}\Delta T_{water} = \left(2.5 \text{ L} \times \frac{1 \text{ mL}}{10^{-3} \text{ L}} \times \frac{1.00 \text{ g}}{\text{mL}}\right)\left(4.184 \ \frac{\text{J}}{\text{g C}}\right)(30.2 \text{ C} - 22.7 \text{ C}) = 78,450 \text{ J}$$

$$q_{rxn} = n_{rxn}\Delta H_{rxn} = \left(3.05 \text{ g C}_6\text{H}_5\text{COOH} \times \frac{\text{mol C}_6\text{H}_5\text{COOH}}{122.12 \text{ g C}_6\text{H}_5\text{COOH}}\right)\left(\frac{-6,453,700 \text{ J}}{2 \text{ mol C}_6\text{H}_5\text{COOH}}\right) = -80,592 \text{ J}$$

$$q_{cont} = C_{cont}\Delta T_{cont} = (C_{cont})(30.2 \text{ C} - 22.7 \text{ C}) = 7.5 \text{ C} \times C_{cont}$$

$$(78,450 \text{ J}) + (-80,592 \text{ J}) + 7.5 \text{ C} \times C_{cont} = 0$$

$$7.5 \text{ C} \times C_{cont} = 2,142 \text{ J, and } C_{cont} = \frac{2,142 \text{ J}}{7.5 \text{ C}} = 290 \ \frac{\text{J}}{\text{C}}$$

7.   First, sum all q values to zero. Since Styrofoam is a good thermal insulator, you may ignore it ($q_{cup} = 0$) in this calculation. You are given the specific heat capacity (c) for water and you want the specific heat capacity for the metal, so use $q = mc\Delta T$ for both calculations.

$$q_{water} + q_{metal} = 0$$

$$q_{water} = m_{water}c_{water}\Delta T_{water} = (100.0 \text{ g})\left(4.184 \ \frac{\text{J}}{\text{g C}}\right)(27.5 \text{ C} - 24.7 \text{ C}) = 1,171.52 \text{ J}$$

$$q_{metal} = m_{metal}c_{metal}\Delta T_{metal} = (48.2 \text{ g})(c_{metal})(27.5 \text{ C} - 97.3 \text{ C}) = -3,364.36 \text{ g C} \times c_{metal}$$

$$1,171.52 \text{ J} + (-3,364.36 \text{ g C} \times c_{metal}) = 0, \ c_{metal} = \frac{-1,171.52 \text{ J}}{-3,364.36 \text{ g C}} = 0.35 \ \frac{\text{J}}{\text{g C}}$$

The specific heat capacity of the metal more closely matches zinc's value of 0.38 J/g °C, so it is zinc.

8.   Use the target equation to decide how to use the four equations provided to you. Since you need 1 molecule of $TiCl_4$ as a reactant, use Equation 3 written backwards (changing the sign of $\Delta H°_3$).

$$\textbf{TiCl}_4\textbf{(g)} + 2 \text{ H}_2(g) \rightarrow \text{Ti(s)} + 4 \text{ HCl(g)} \qquad\qquad -\Delta H_3° = +435.0 \text{ kJ}$$

Since $COCl_2(g)$ appears in Equations 1 and 2, skip it and come back to it. To get 1 formula unit of $TiO_2(s)$ as a product, use Equation 4 as written. Notice that the 2 $H_2(g)$ and the Ti(s) can be canceled right now.

$$\text{Ti(s)} + 2 \text{ H}_2\text{O(g)} \rightarrow \textbf{TiO}_2\textbf{(s)} + 2 \text{ H}_2(g) \qquad\qquad \Delta H_4° = -456.1 \text{ kJ}$$

To get 2 molecules of $CCl_4(\ell)$ as a product, use Equation 1 and double it. This equation also gives you 4 molecules of $COCl_2(g)$, but you only want 2 molecules, so the other two must be canceled.

$$2 \text{ COCl}_2\text{(g)} + 2 \text{ COCl}_2(g) \rightarrow 2 \text{ CO}_2(g) + \textbf{2 CCl}_4\textbf{(}\ell\textbf{)} \qquad\qquad 2 \times \Delta H_1° = -182.6 \text{ kJ}$$

Even though you have all of the chemicals needed for the target equation, you also have molecules you do not need: 2 $COCl_2(g)$ and 2 $H_2O(g)$ as reactants, and 4 HCl(g) and 2 $CO_2(g)$ as products. If you use Equation 2, written backwards and doubled, it will cancel all of the extra molecules.

$$2 \text{ CO}_2\text{(g)} + 4 \text{ HCl(g)} \rightarrow 2 \text{ COCl}_2(g) + 2 \text{ H}_2\text{O(g)} \qquad\qquad -2 \times \Delta H_2° = +235.0 \text{ kJ}$$

To calculate $\Delta H°$, add up the $\Delta H°$ values. Since this value is positive, this reaction is endothermic.

$$\Delta H° = -\Delta H_3° + \Delta H_4° + 2 \times \Delta H_1° + -2 \times \Delta H_2° = 435.0 \text{ kJ} - 456.1 \text{ kJ} - 182.6 \text{ kJ} + 235.0 \text{ kJ} = 31.3 \text{ kJ}$$

9. $$\Delta H = \left\{ \left(1 \text{ mol N}_2\text{O}_4 \times 9.16 \frac{\text{kJ}}{\text{mol N}_2\text{O}_4} \right) + \left(10 \text{ mol H}_2\text{O} \times -285.83 \frac{\text{kJ}}{\text{mol H}_2\text{O}} \right) \right\} -$$

$$\left\{ \left(2 \text{ mol NH}_3 \times -80.29 \frac{\text{kJ}}{\text{mol NH}_3} \right) + \left(7 \text{ mol H}_2\text{O}_2 \times -191.17 \frac{\text{kJ}}{\text{mol H}_2\text{O}_2} \right) \right\} =$$

$$\left\{ (9.16 \text{ kJ}) + (-2852.30 \text{ kJ}) \right\} - \left\{ (-160.58 \text{ kJ}) + (-1338.19 \text{ kJ}) \right\} = -2,843.14 \text{ kJ} - (-1,498.77 \text{ kJ}) = -1,344.37 \text{ kJ}$$

10. (a) The product must be solid benzoic acid with a coefficient of 1; the reactants must be elemental carbon, hydrogen, and oxygen in their standard states. The reactants must have 7 C atoms, 6 H atoms, and 2 O atoms: $7 \text{ C(s)} + 3 \text{ H}_2\text{(g)} + \text{O}_2\text{(g)} \rightarrow \text{C}_6\text{H}_5\text{COOH(s)}$.

(b) Standard molar enthalpy of formation values for all elements in their standard states are defined to be zero, so you do not have to be given this number.

(c) $$\Delta H = \left\{ \left(14 \text{ mol CO}_2 \times -393.5 \frac{\text{kJ}}{\text{mol CO}_2} \right) + \left(6 \text{ mol H}_2\text{O} \times -285.8 \frac{\text{kJ}}{\text{mol H}_2\text{O}} \right) \right\} -$$

$$\left\{ \left(2 \text{ mol C}_6\text{H}_5\text{COOH} \times \Delta H_f\{\text{C}_6\text{H}_5\text{COOH(s)}\} \right) + \left(15 \text{ mol O}_2 \times 0.0 \frac{\text{kJ}}{\text{mol O}_2} \right) \right\} = -6,453.7 \text{ kJ}$$

$$\left\{ (-5,509.0 \text{ kJ}) + (-1,714.8 \text{ kJ}) \right\} - \left\{ \left(2 \text{ mol} \times \Delta H_f\{\text{C}_6\text{H}_5\text{COOH(s)}\} \right) + 0.0 \text{ kJ} \right\} = -6,453.7 \text{ kJ}$$

$$2 \text{ mol} \times \Delta H_f\{\text{C}_6\text{H}_5\text{COOH(s)}\} = \left\{ (-5,509.0 \text{ kJ}) + (-1,714.8 \text{ kJ}) \right\} - (-6,453.7 \text{ kJ}) = -770.1 \text{ kJ}$$

$$\Delta H_f\{\text{C}_6\text{H}_5\text{COOH(s)}\} = \frac{-770.1 \text{ kJ}}{2 \text{ mol}} = -385.1 \frac{\text{kJ}}{\text{mol}}$$

11. (a) $236 \text{ g grapes} \times \dfrac{2.9 \text{ kJ}}{\text{g grapes}} = 680 \text{ kJ} \times \dfrac{10^3 \text{ J}}{1 \text{ kJ}} \times \dfrac{1 \text{ cal}}{4.184 \text{ J}} \times \dfrac{1 \text{ Cal}}{10^3 \text{ cal}} = 160 \text{ Cal}$. So, 236 g of grapes contain 680 kJ or 160 Cal of energy.

(b) $163.58 \text{ Cal} \times \dfrac{\text{min}}{9.0 \text{ Cal}} = 18 \text{ min}$.

So, it takes 18 minutes to use the up the energy stored in 236 g of grapes.

ANSWERS — PRACTICE TEST

Chapter 7

1. (a) $260 \text{ nm} \times \dfrac{10^{-9} \text{ m}}{1 \text{ nm}} = 2.6 \times 10^{-7} \text{ m}$. A wavelength of 260 nm falls outside of the visible spectrum (see the first example on p. 57 of this study guide), so this light is in the ultraviolet (UV) region.

   (b) $c = \lambda \nu, \ \nu = \dfrac{c}{\lambda} = \dfrac{2.998 \times 10^8 \text{ m/s}}{2.6 \times 10^{-7} \text{ m}} = 1.2 \times 10^{15} \text{ s}^{-1}$.

   (c) $E_{quantum} = h\nu = (6.626 \times 10^{-34} \text{ J} \times \text{s})(1.2 \times 10^{15} \text{ s}^{-1}) = 7.6 \times 10^{-19} \text{ J}$.

   (d)   To get the energy, convert the energy of one molecule into moles, then convert to energy in J to kJ.

$$\frac{7.6 \times 10^{-19} \text{ J}}{1 \text{ O–O bond}} \times \frac{6.022 \times 10^{23} \text{ O–O bond}}{1 \text{ mol}} \times \frac{1 \text{ kJ}}{10^3 \text{ J}} = 460 \ \frac{\text{kJ}}{\text{mol}}$$

2. (a)   Since the value of n decreases, the electron moves closer to the nucleus and is in a smaller orbit.
   (b)   Moving closer to the nucleus is exothermic, so the electron loses energy and ΔE is negative.
   (c)   Use Rydberg's equation, where $n_1 = 2$ and $n_2 = 6$ $(n_1 < n_2)$, and $R_H = 1.097 \times 10^7 \text{ m}^{-1}$.

$$\frac{1}{\lambda} = R_H \left(\frac{1}{n_1^2} - \frac{1}{n_2^2} \right) = 1.097 \times 10^7 \text{ m}^{-1} \left(\frac{1}{2^2} - \frac{1}{6^2} \right) = 2.438 \times 10^6 \text{ m}^{-1}$$

$$\frac{1}{\lambda} = 2.438 \times 10^6 \text{ m}^{-1}, \ \lambda = \frac{1}{2.438 \times 10^6 \text{ m}^{-1}} = 4.102 \times 10^{-7} \text{ m} \times \frac{1 \text{ nm}}{10^{-9} \text{ m}} = 410.2 \text{ nm}$$

   A wavelength of 410.2 nm falls in the visible (violet) region.

3.   Some advantages of Rydberg's equation are that this formula shows why some wavelength values are acceptable and most are not, and that it can be used to predict new lines in the infrared and ultraviolet regions that had not been previously known. The major disadvantage of this formula is that it cannot explain why the formula has $1/n^2$ (instead of n or n^2 or $1/n$) in it.

4.   The major advantage of Bohr's theory is that it explains why Rydberg's equation works and why the value of $1/n^2$ is important (it was part of the energy term), and it is also a simple and intuitive theory that chemists and nonchemists can understand. One disadvantage of Bohr's theory is that it can only explain the line emission spectrum of hydrogen. The spectra of other elements have more lines that cannot be explained using this theory. Another disadvantage is that it assumes that the orbits of electrons are orderly and predictable, which is wrong. A third disadvantage is that this (incorrect, or at least incomplete) theory is a simple and intuitive theory that chemists and nonchemists can understand, and (unfortunately) is much easier to understand that the more complete and more accurate quantum mechanical model of the atom.

5. (a)   The distance of the electron from the nucleus is determined by n (the larger n is, the farther the electron is from the nucleus).
   (b)   The shape of the orbital is determined by ℓ. When $\ell = 0$ (the s orbital) it is spherical, when $\ell = 1$ (p orbitals) they are dumbbell-shaped, and when $\ell = 2$ (d orbitals) they are cloverleaf-shaped.
   (c)   The energy of the electron is determined by both n and ℓ. Bohr's theory states that an electron's energy is: $E = -b/n^2$, where b is a constant. Within the same shell (same n value), the energy increases as ℓ increases (so the $3s$ is the lowest in energy, then the $3p$ orbitals, and finally the $3d$ orbitals).
   (d)   The region of space occupied by the electron is determined by both n and m_ℓ. As n increases, the electron is further from the nucleus. While the ℓ value tells you what type of orbital the electron is in (s, p, d, f), the m_ℓ value tells you which regions of space these are orbitals occupy $(p_x, p_y, p_z,$ etc.).

6. (a)   This set of quantum numbers is possible. When n = 2, ℓ can be 0 or 1 (so 1 is acceptable); when $\ell = 1$, m_ℓ can be –1, 0, or +1 (OK), and m_s can be –1/2 or +1/2 (OK). This is a $2p$ ($n = 2$, $\ell = 1$) orbital.

(b)   This set of quantum numbers is not possible. When $n = 3$, ℓ can be 0, 1, or 2 (OK); however, when $\ell = 2$, m_ℓ can be $-2, -1, 0, +1,$ or $+2$ (so -3 is not acceptable),. This is a $3d(n = 3, \ell = 2)$ orbital, which can exist, just not with $m_\ell = -3$.

(c)   This set of quantum numbers is not possible. When $n = 4$, ℓ can be 0, 1, 2, or 3 (so 4 isn't acceptable). This would be a $4g$ ($n = 4$, $\ell = 4$) orbital, which does not exist.

(d)   This set of quantum numbers is possible. When $n = 5$, ℓ can be 0, 1, 2, 3, or 4 (so 3 is acceptable); when $\ell = 3$, m_ℓ can be $-3, -2, -1, 0, +1, +2,$ or $+3$ (so -2 is OK), and m_s can be $-1/2$ or $+1/2$ (OK). This is a $5f$ ($n = 5$, $\ell = 3$) orbital.

7.   (a)   The cloverleaf-shaped orbitals are d orbitals. Since the lobes lie on the x- and y-axes, this is the $d_{x^2-y^2}$ orbital. This is similar to the d_{xy} orbital, which has its lobes *between* the x- and y-axes.

(b)   Although this orbital is not shaped like a cloverleaf, it is the d_{z^2} orbital (with the two lobes on the z-axis and the donut in the xy plane).

(c)   The dumbbell-shaped orbitals are p orbitals. Since the orbital lies on the y-axis, this is the p_y orbital.

(d)   The cloverleaf-shaped orbitals are d orbitals. Since the lobes of this orbital lie between the x- and z-axes, this is the d_{xz} orbital.

8.   (a)   To write electron configurations using the periodic table, start with hydrogen at the top and add electrons until you reach the target atom. Remember that the s-block starts at $1s$, the p-block starts at $2p$, the transition elements start at $3d$, and the f-block starts at $4f$. For nitrogen, 2 e$^-$ go in the $1s$ with opposite spins, 2 e$^-$ go in the $2s$, and 3 e$^-$ go in each of the $2p$ orbitals with parallel spin:

(b)   For a neutral silicon atom you have 14 e$^-$, and the configuration is $1s^22s^22p^63s^23p^2$. To make Si$^-$, simply add an electron to the outermost orbital (the empty $3p$ orbital):

(c)   The condensed notation for I ($Z = 53$) is $1s^22s^22p^63s^23p^64s^23d^{10}4p^65s^24d^{10}5p^5$. If you add up all of the superscripted numbers ($2 + 2 + 6 + 2 + 6 + 2 + 10 + 6 + 2 + 10 + 5 = 53$), this should equal the total number of electrons in the atom or ion and is a good way to make sure you haven't made a mistake.

(d)   The condensed notation for Co ($Z = 27$) is $1s^22s^22p^63s^23p^64s^23d^7$. When you remove electrons from a transition element to make cation, remember that the electrons must be removed from the $4s$ orbital before removing electrons from the $3d$ orbital—Co^{2+}: $1s^22s^22p^63s^23p^63d^7$.

(e)   The largest noble gas within the platinum atom ($Z = 78$) is xenon [Xe]. The next 2 e$^-$ go in an s orbital (Cs and Ba); to determine its number, count down from the top starting at 1 (this is the $6s$ orbital). The next e$^-$ goes in a d orbital; counting down from the top of the d-block starting with 3, this is the $5d$ orbital. The next 14 e$^-$ go in the $4f$ orbitals (f orbitals start at $n = 4$, and this is the first row of them). The last 7 e$^-$ go in the $5d$ orbitals along with the first one (for a total of 8 e$^-$ in the $5d$ orbitals)—Pt: [Xe] $6s^24f^{14}5d^8$; you can also write these in order of increasing n values: [Xe] $4f^{14}5d^86s^2$. To double check, add up the total number of electrons, starting with 54 for [Xe]—$54 + 2 + 14 + 8 = 78$.

(f)   The noble gas notation for the selenium atom ($Z = 34$) is [Ar] $4s^23d^{10}4p^4$. You must add 2 e$^-$ to Se to get Se^{2-}, and these electrons fill in the vacancies in the $4p$ orbitals. The electron configuration for Se^{2-} is [Ar] $4s^23d^{10}4p^6$, which also equals [Kr].

9.   (a)   The atom must be carbon, because only carbon atoms have six electrons (and six protons).

(b)   Orbital box diagram II is correct.

(c)   Diagram I is incorrect because it violates Hund's rule. The 2 e$^-$ in the $2p$ orbitals would be more stable if they occupied different orbitals with parallel spins. Diagram III is incorrect because it violates the Pauli exclusion principle (electrons in the same atom can have the same four quantum numbers). The 2 e$^-$ in the $2s$ orbital cannot have the same spin (both $+1/2$ or both $-1/2$) because being in the $2s$ orbital already requires $n = 2$, $\ell = 0$, and $m_\ell = 0$. Diagram IV is wrong because it violates the aufbau principle (fill lower-energy orbitals before filling higher-energy orbitals). One of the electrons in the $2p$ orbitals should be in the lower-energy $2s$ orbital.

10. (a) The order is F (smallest) < Cl < Br < I (largest). These atoms are in the same column, and their noble gas notations show that the outer electrons are in the $n = 2$ shell for F ([He] $2s^2 2p^5$), the $n = 3$ shell for Cl ([Ne] $3s^2 3p^5$), the $n = 4$ shell for Br ([Ar] $4s^2 3d^{10} 4p^5$), and the $n = 5$ shell for I ([Ar] $5s^2 4d^{10} 5p^5$). Electrons in shells with larger n values have larger radii, so F is the smallest and I is the largest.

(b) The order is Ar (smallest) < S < Si < Mg (largest). These atoms are in the same row, and noble gas notations for these atoms show that the outer electrons are all in the $n = 3$ shell for Mg ([Ne] $3s^2$), Si ([Ne] $3s^2 3p^2$), S ([Ne] $3s^2 3p^4$), and Ar ([Ne] $3s^2 3p^6$). Since the outer electrons in these atoms are in the same shell, the number of electrons does not affect the atom sizes. However, each of these atoms have different $\#p^+$ in the nucleus (12 for Mg, 14 for Si, 16 for S, and 18 for Ar), so the atom with the most protons (Ar) has the smallest radius because the electrons are pulled closely to the nucleus (small radius) while the atom with the least number of protons (Mg) has the largest radius because these electrons are not held as closely to the nucleus.

(c) The order is Fe^{3+} (smallest) < Fe^{2+} < Fe < Fe^- (largest). The noble gas notations for these atoms/ions show that Fe^{3+} ([Ar] $3d^5$) and Fe^{2+} ([Ar] $3d^6$) have their outermost electrons in the $n = 3$ shell ($3d$), while Fe ([Ar] $4s^2 3d^6$) and Fe^- ([Ar] $4s^2 3d^7$) have their outermost electrons in the $n = 4$ shell ($4s$). So, Fe^{3+} and Fe^{2+} are smaller than Fe and Fe^-. You can see that Fe^{3+} would be smaller than Fe^{2+} because even though the have the same $\#p^+$, Fe^{3+} has fewer electrons than Fe^{2+}. The ion with fewer electrons has fewer electron-electron repulsions pushing the electrons away from each other and the nucleus, so it has the smaller radii. The same argument explains why Fe is smaller than Fe^-. In general, ionic radii get smaller as the positive charge increases and larger as the negative charge increases.

(d) The order is Sc^{3+} (smallest) < K^+ < Cl^- < P^{3-} (largest). The noble gas notations for these ions are the same (isoelectronic)—$1s^2 2s^2 2p^6 3s^2 3p^6$ = [Ar]. Because they have the same $\#e^-$, you might expect them to have the same radius, but both $\#e^-$ and $\#p^+$ affect ionic radii. Since Sc^{3+} has the most protons (21), its outer electrons are pulled closest to the nucleus and it is the smallest ion; since P^{3-} has the fewest protons (15), its outer electrons are not pulled in closely to the nucleus and it is the largest ion.

11. This observation can be explained by looking at the noble gas notations of P ([Ne] $3s^2 3p^3$), N ([He] $2s^2 2p^5$), and Na ([Ne] $3s^1$). P is bigger than N because its outer electrons are in a larger orbital ($3s/3p$ versus $2s/2p$). P is smaller than Na because even though both atoms have their outer electrons are in the $n = 3$ orbitals, P has 15 protons pulling the electrons closer to the P nucleus compared to the 11 protons in the Na nucleus.

12. (a) B has a larger ionization energy than Al because B has a smaller radius than Al. Outer electrons in atoms with small radii are more strongly attracted to the nucleus than outer electrons in atoms with larger radii, and it takes more energy (larger IE) to remove these electrons from the smaller atom.

(b) F has a larger ionization energy than C because F has a smaller radius than C.

(c) Cl has a larger ionization energy than Cl^- because Cl has a smaller radius than Cl^-.

(d) Mg^{2+} has a larger ionization energy than Na^+ because Mg^{2+} has a smaller radius than Na^+.

13. The balanced equations for the EA of Cl and the IE of Cl^- are listed below. The IE of Cl^- is just the reverse reaction of the EA of Cl, so if the EA of Cl is –349 kJ/mol, then the IE of Cl^- is +349 kJ/mol. Note that the IE of Cl^- is smaller than the IE of the Cl atom, +1,251 kJ/mol (this is question 12c listed above).

$$Cl(g) + e^- \rightarrow Cl^-(g) \qquad \qquad \Delta H = EA \text{ for Cl}$$
$$Cl^-(g) \rightarrow Cl(g) + e^- \qquad \qquad \Delta H = IE \text{ for } Cl^-$$

14. (a) The condensed notation for N is $1s^2 2s^2 2p^3$.

(b) The first 3 e^- are removed from the $2p$ orbital and the next two (electrons #4 and #5) are removed from the $2s$ orbital. The reason the IE jumps dramatically for electron #6 is that it must be removed from the $1s$ orbital which is much lower in energy (core electrons instead of valence electrons) and it takes much more energy to remove an electron in a lower-energy orbital.

(c) The curve stops after electron #7 because nitrogen has only seven electrons.

ANSWERS — PRACTICE TEST

Chapter 8

1. (a)   Sulfur has 6 valence electrons, and Br has 7. The Lewis dot symbols are drawn below. Note that the where the single e^- and the lone pairs are placed is irrelevant; all that matters is S has 2 lone pairs and 2 unpaired e^- and Br has 3 lone pairs and 1 unpaired e^-. The number of bonds made is the same as the number of unpaired e^-—2 for S and 1 for Br.

    (b)   Because S makes 2 bonds and Br makes 1 bond, the simplest molecule can be made with 1 S and 2 Br atoms. Another compound can be made by bonding two S atoms together. Since each S atom needs 1 more bond, a total of 2 Br atoms are needed.

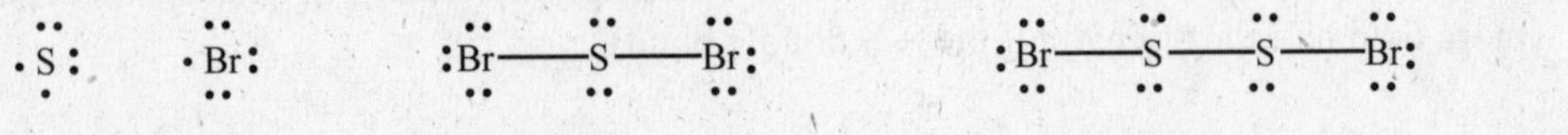

    (c)   The first compound has the formula SBr_2 and is named sulfur dibromide; the second compound has the formula S_2Br_2 and is named disulfur dibromide.

2. (a)   (1) The B atom has 3 e^-, each Br atom has 7 e^-, and the negative charge adds another e^-, which equals $3 + 4(7) + 1 = 32$ e^-. (2) B is the central atom, and bonding the 4 Br atoms to it takes 8 e^-. (3) There are 24 e^- left, and completing the octet takes 6 e^- for each Br atom (24 e^-). (4) There are no e^- left. (5) The B atom has an octet from the four bonds, so no multiple bonds are needed.

    (b)   (1) The S atom has 6 e^-, each Cl atom has 7 e^-, and the positive charge means 1 e^- is removed. So, there are $6 + 3(7) - 1 = 26$ e^-. (2) S is the central atom, and bonding the 3 Cl atoms to it takes 6 e^-. (3) There are 20 e^- left, and the Cl atoms take a total of 18 e^- (6 per Cl atom) to complete their octets. (4) There are 2 e^- left, so they are placed on the S atom as a lone pair. (5) The S atom has an octet from the three bonds and its lone pair, so no multiple bonds are needed.

    (c)   (1) The H atom has 1 e^-, the N has 5 e^-, and the S atom has 6 e^-, for a total of 12 e^-. (2) N is the central atom, and bonding to H and S takes 4 e^-. (3) There are 8 e^- left, six of which are placed on the S atom to complete its octet (H only wants 2 e^-). (4) There are 2 e^- left and are placed on N. (5) Since N has only 6 e^-, it makes a multiple bond with S; the S atom shares one lone pair to make a N-S double bond.

    (d)   (1) The Xe has 8 e^- (noble gas), O has 6 e^-, and each F has 7 e^- for a total of $8 + 6 + 4(7) = 42$ e^-. (2) Xe is the central atom, and bonding the O and the 4 F atoms to it takes 10 e^-. (3) There are 32 e^- left; the O and F atoms each need 6 e^- to complete their octets (30 e^-). (4) There are 2 e^- left, so they are added to Xe as a lone pair. (5) Since Xe already has 12 e^-, there is no need to make multiple bonds.

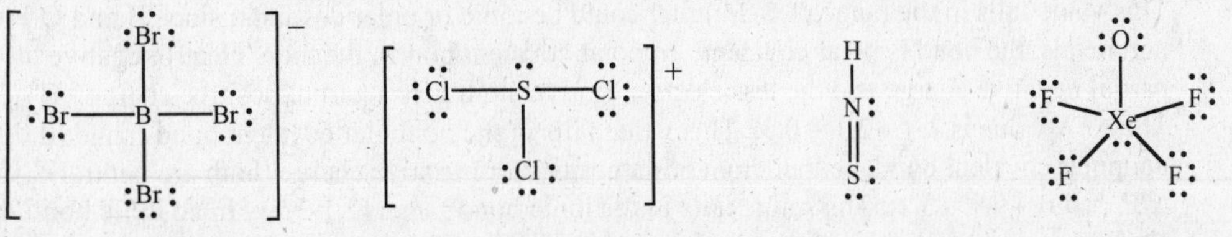

3. (a)   This alkene cannot have *cis-trans* isomers because the substituents on the left C atom are the same (H).

    (b)   This alkene can have *cis-trans* isomers because each C atom has two different substituents on it, and there are common substituents on the two C atoms (H and CH_3). Because the two CH_3 groups (and the two H atoms) are on opposite sides of the double bond, this is the *trans* isomer.

    (c)   This alkene cannot have *cis-trans* isomers because the substituents on the both C atoms are the same (H for the left one and CH_3 for the right one).

    (d)   This alkene can have *cis-trans* isomers because each C atom has two different substituents on it, but there is a common substituent on the two C atoms (CH_3). Because the two CH_3 groups are on the same side of the double bond, this is the *cis* isomer.

4. (a) $$\Delta H = \left\{ 8\ \text{mol}_{S-S} \times \frac{226\ \text{kJ}}{\text{mol}_{S-S}} \right\} - \left\{ 4\ \text{mol}_{S=S} \times \frac{425\ \text{kJ}}{\text{mol}_{S=S}} \right\} = +108\ \text{kJ}.$$

(b)   Because $\Delta H°$ is positive, this is an endothermic reaction. In endothermic reactions, the reactants are at lower energies (more stable) than the products, so S_8 is more stable than S_2 and elemental sulfur exists in its most stable form as S_8 rings.

(c) $\Delta H = \left\{ 8\ mol_{O-O} \times \dfrac{146\ kJ}{mol_{O-O}} \right\} - \left\{ 4\ mol_{O=O} \times \dfrac{498\ kJ}{mol_{O=O}} \right\} = -824\ kJ$.

(d)   This reaction is exothermic for oxygen ($\Delta H°$ negative). In exothermic reactions, the products are at lower energies (more stable) than the reactants, so O_2 is more stable than O_8 and elemental oxygen exists in its most stable form as O_2 molecules. Looking at the double bond strengths, the O-O double bond is slightly stronger than the S-S double bond (498 kJ/mol vs. 425 kJ/mol, respectively); however, the O-O single bond is much weaker than the S-S single bond (146 kJ/mol vs. 226 kJ/mol). So, oxygen exists in its doubly-bonded form because that is a very strong bond (more than twice the strength of its single bond) but sulfur exists in its singly-bonded form because that is a strong bond (more than half the strength of the S-S double bond).

5.   The $\Delta H°$ value for this reaction is simply $2 \times \Delta H°_f\{NOCl(g)\} = +106\ kJ$.

$$\Delta H = \left\{ \left(1\ mol_{N\equiv N} \times \frac{946\ kJ}{mol_{N\equiv N}} \right) + \left(1\ mol_{O=O} \times \frac{498\ kJ}{mol_{O=O}} \right) + \left(1\ mol_{Cl-Cl} \times \frac{242\ kJ}{mol_{Cl-Cl}} \right) \right\} -$$

$$\left\{ \left(2\ mol_{N=O} \times D(N=O) \right) + \left(2\ mol_{N-Cl} \times \frac{193\ kJ}{mol_{N-Cl}} \right) \right\} = +106\ kJ$$

$$\Delta H = 1{,}686\ kJ - 2\ mol_{N=O} \times D(N=O) - 386\ kJ = +106\ kJ$$

$$-2\ mol_{N=O} \times D(N=O) = +106\ kJ - 1{,}686\ kJ + 386\ kJ = -1{,}194\ kJ$$

$$D(N=O) = \frac{-1{,}194\ kJ}{-2\ mol_{N=O}} = 597\ \frac{kJ}{mol}$$

6.   (a)   The trend for electronegativity is that the electronegativity decreases as you look down a column in the periodic table. So, Be is the most electronegative of the three and Ca is the least electronegative (most electropositive); the order is Ca (smallest) < Mg < Be (largest).

(b)   The trend for electronegativity is that the electronegativity increases as you look across a row in the periodic table. So, Al is the least electronegative and Cl is the most electronegative, and the order is Al (smallest) < P < Cl (largest).

7.   (a)   The ΔEN value is $3.4 - 1.9 = 1.5$ (remember that ΔEN is always the larger EN minus the smaller one). This value falls in the range (1.5-2.0) that could be ionic or polar covalent; since B and O are both nonmetals, the bond is polar covalent. In polar covalent bonds, the more electronegative atom gets a partial negative charge and the less electronegative atom gets a partial positive charge: $B^{\delta+}-O^{\delta-}$.

(b)   The ΔEN value is $2.4 - 2.1 = 0.3$. This value falls in the nonpolar covalent bond range (0.0-0.4). In a nonpolar covalent bond, neither atom has a positive or negative charge (both are neutral): C—H.

(c) $\Delta EN = 4.0 - 0.9 = 3.1$. This value falls in the ionic bond range (2.1-3.2). In an ionic bond, the more electronegative atom gets a full negative charge and the less electronegative atom gets a full positive charge: Na^+-F^-.

(d)   The ΔEN value is $3.0 - 3.0 = 0.0$. This value falls in the nonpolar covalent range (0.0-0.4)—the fact that this is a triple bond instead of a single bond does not affect the ΔEN value or how the bond is classified. Neither atom gets a positive or negative charge: N≡N.

(f)   The ΔEN value is $3.4 - 2.3 = 1.1$, which falls in the polar covalent range (0.5-1.4). So, the more electronegative atom gets a partial negative charge and the less electronegative atom gets a partial positive charge: $S^{\delta+}=O^{\delta-}$.

(d) $\Delta EN = 3.4 - 1.6 = 1.8$. This value falls in the range (1.5-2.0) that could be ionic or polar covalent; since Zn is a metal and O is a nonmetal, the bond is ionic: Zn^+-O^- (in reality it is $Zn^{2+}-O^{2-}$).

8.   (a)   Neutral C atoms have 4 e^- and this C atom has 5 e^-, so its formal charge is –1 (one more e^- than neutral); neutral N atoms have 5 e^- and this N atom has 5 e^-, so its formal charge is 0.

A 22

(b)   Neutral N atoms have 5 e^- and this N atom has 4 e^-, so its formal charge is +1. Neutral H atoms have 1 e^- and this H atom has 1 e^-, so its formal charge is 0. Neutral O atoms have 6 e^-; the singly-bonded O has 7 e^- (formal charge of –1), and the doubly-bonded O and the O bonded to the N and H atoms have 6 e^- (formal charges of 0).

(c)   Neutral N atoms have 5 e^- and this N atom has 5 e^-, so its formal charge is 0; neutral O atoms have 6 e^- and this O atom has 5 e^-, so its formal charge is +1 (one e^- less than neutral).

(d)   Neutral S atoms have 6 e^- and this S atom has 5 e^-, so its formal charge is +1. Neutral Cl atoms have 7 e^- and both Cl atoms have 7 e–, so their formal charges are 0. Neutral O atoms have 6 e^- and this O atom O has 7 e^-, so it has a formal charge of –1. Note that in each Lewis structure the sum of formal charges is equal to the overall charge of the molecule or ion.

9.   (a)

	Br_I	C_I	N_I	Br_{II}	C_{II}	N_{II}	Br_{III}	C_{III}	N_{III}
#e^- on neutral atom	7	4	5	7	4	5	7	4	5
#e^- on atom in structure	5	4	7	6	4	6	7	4	5
formal charge	+2	0	–2	+1	0	–1	0	0	0

(b)   Based on the formal charges, structure III (Br—C≡N) is the most stable because it has the minimum formal charges for each atom (all zero).

(c)   Because there is one single most stable structure, that structure is used and the other two are ignored. So, the C–N bond is a triple bond, its length should be 115 pm, and its enthalpy should be 866 kJ/mol.

10.   (a)

	F_I	C_I	top/side O_I	F_{II}	C_{II}	top/side O_{II}	F_{III}	C_{III}	top/side O_{III}
#e^- on neutral atom	7	4	6	7	4	6	7	4	6
#e^- in the structure	6	4	7/7	7	4	6/7	7	4	7/6
formal charge	+1	0	–1/–1	0	0	0/–1	0	0	–1/0

(b)   Based on the formal charges, structure I is the least stable because it has the maximum number of nonzero formal charges. Structures II and II are equally stable since both have a 0 charge on the C atom, the F atom, and one of the O atoms, and a –1 charge on the other O atom.

(c)   Because there are two resonance structures, the C-F bond will be the average of the two bonds in these structures. The average of a single bond and a single bond is a single bond. So, the C-F bond is a single bond, its length should be 141 pm, and its enthalpy should be 486 kJ/mol.

(d)   Because there are two resonance structures, the C-O bonds will be the averages of the two bonds in these structures. For the top C-O bond and for the side C-O bond, this bond is an average of a single and double bond, so its bond length will be 1/2(143 pm + 122 pm) = 133 pm, and its bond enthalpy will be 1/2(336 kJ/mol + 695 kJ/mol) = 516 kJ/mol.

11.   (a)   Because four of the six spots on the ring do not have the same substituents, this compound cannot be *ortho*, *meta*, or *para* (none of these).

(b)   There are two non-H substituents on C atoms separated by another C atom in the ring, so this is the *meta* isomer.

(b)   There are two non-H substituents on neighboring C atoms in the ring, so this is the *ortho* isomer.

(c)   There are two non-H substituents on opposite C atoms in the ring, so this is the *para* isomer.

12.   Molecular orbital theory says that bonding orbitals can exist over more than two atoms in any molecule. Looking at the resonance structures for CO_2F^- ion (in question 10b above), you can see that while the C-F bond is a single bond, the C-O bonds are somewhere in between a single and double bond. Molecular orbital theory says that a pair of electrons in a bond can be shared among the C atom and the two O atoms, resulting in a bond that can be thought of as two 'half-bonds' being shared in the two regions of space between the C atom and each of the O atoms.

ANSWERS — PRACTICE TEST

Chapter 9

1. The general theory behind VSEPR is that the pairs of electrons on the central atom (whether lone pairs or bonding pairs) repel each other and want to get as far from each other as possible while still being around the central atom. The tetrahedral shape is preferred over the square planar shape because the electrons can get farther away from each other in a tetrahedral shape (109.5° apart) than in a square planar shape (90° apart).

2. The AXE notations are determined by the number of bonded atoms and lone pairs on the central atom. The electron-pair geometries are determined by the $X + E$ values from the AXE notation; if the central atom has no lone pairs, the electron-pair geometries and the molecular geometries are the same. The bond angles and the hybridization are determined by the electron-pair geometries; when the central atom has lone pairs, the bond angle is slightly less than predicted.

chemical	AXE notation	electron-pair geometry	molecular geometry	bond angles	hybridization	molecular shape
HCO^+	AX_2E_0	linear	linear	180°	sp	$[H\!-\!C\!\equiv\!O]+$
HOCl	AX_2E_2	tetrahedral	bent	< 109.5°	sp^3	
$TeBr_6$	AX_6E_0	octahedral	octahedral	90°	—	
SO_2	$A\dot{X}_2E_1$	triangular planar	bent	< 120°	sp^2	
SiH_3F	AX_4E_0	tetrahedral	tetrahedral	109.5°	sp^3	
NO_3^-	AX_3E_0	triangular planar	triangular planar	120°	sp^2	

3. Any atom that has four different objects attached to it is chiral. Chiral atoms are usually identified by marking them with an asterisk (as in the pictures of these molecules on the next page).
    In the first molecule (polypropylene oxide, a polymer) there is only 1 chiral atom per repeating unit. The CH_2 group and the CH_3 group are not chiral because they have at least two atoms in common (H atoms). The second C atom in the polymer is bonded to a CH_2 group, and H atom, a CH_3 group, and an O atom (chiral).
    In the second molecule (threonine, an amino acid) there are 2 chiral atoms—the C atom next to the C=O group and the C atom just below that atom.
    The third molecule (glycine, an amino acid) has no chiral atoms, so glycine is an achiral molecule.
    In the fourth molecule (glucose) there are 5 chiral atoms.

4. (a) The rules described in this book for determining molecular polarities only work for neutral molecules. This compound is a cation and must be paired with an anion in a liquid or solid, and the intermolecular forces in an ionic compound are ionic bonds.

(b) ΔEN is $3.4 - 2.1 = 1.3$ (polar covalent) for the O-H bond and $3.4 - 2.7 = 0.7$ (polar covalent) for the O-Cl bond. Since the O atom has lone pairs and the outer atoms are different, this molecule is polar. This polar molecule has dipole-dipole forces (because the O-Cl bond is polar) and hydrogen bonds (because H is bonded to an O atom). It also has London forces because all molecules have them.

(c) The ΔEN value is $2.6 - 1.9 = 0.7$ (polar covalent). Since the Te has no lone pairs and all of the outer atoms are the same (Br), this molecule is nonpolar due to symmetry. So, even though this molecule has polar bonds, it is nonpolar. Therefore, it will have only London forces.

(d) The ΔEN value is $3.4 - 2.3 = 1.1$ (polar covalent). Since the S atom has a lone pair, this molecule is polar. This molecule has dipole-dipole forces but no hydrogen bonds. It also has London forces.

(e) ΔEN for the Si-H bond is $2.1 - 1.6 = 0.5$ (polar covalent) and is $4.0 - 1.6 = 2.4$ (polar covalent) for the Si-F bond. Since the Si atom has a polar bond and the outer atoms are different, this molecule is polar. This molecule has dipole-dipole forces, and it has London forces because all molecules have them.

(f) This compound is an anion and must be paired with a cation in a liquid or solid, and the intermolecular forces in an ionic compound are ionic bonds. Note that even though the N has no lone pairs and all the outer atoms are the same, it is not nonpolar because it is charged.

5. (a) The ΔEN value is $3.4 - 2.4 = 1.0$ (polar covalent). Since this is a diatomic molecule, the bond polarity is the molecular polarity, and CO is polar. The C atom has a net positive charge because it has a smaller EN value and the O atom has a net positive charge because it has a larger EN value.

(b) In the Lewis structure, each atom has $5\ e^-$. Neutral C atoms have $4\ e^-$, so this C atom has a formal charge of -1; neutral O atoms have $6\ e^-$, so this O atom has a formal charge of $+1$.

(c) CO has such a small dipole moment because these two factors work against each other. In the actual molecule, the bond polarity wins ($C^{\delta+}O^{\delta-}$), but the charge separation is much less than is expected.

6. Since melting points are a measure of intermolecular forces, the molecule with a higher melting point (HCl) must have the stronger intermolecular forces. Diatomic fluorine has a molar mass of 38.0 g/mol and has a total of $18\ e^-$, while hydrogen chloride has a molar mass of 36.5 g/mol and has a total of $18\ e^-$. Since these values are close, London forces are probably not responsible for the large difference in melting points. F_2 is nonpolar ($\Delta EN = 4.0 - 4.0 = 0.0$) but HCl is polar ($\Delta EN = 2.6 - 2.1 = 0.5$), so the molecules in solid F_2 will attracted by London forces while those in HCl will be kept in the solid state by London forces and dipole-dipole interactions. HCl has the stronger intermolecular forces due to its dipole-dipole interactions.

7. Since boiling points are a measure of intermolecular forces, the molecule with a highest boiling point (H_2O) has the strongest intermolecular forces, then HF, and NH_3 must have the weakest intermolecular forces. Since they have molar masses from 17-20 g/mol and all have $10\ e^-$, it is unlikely that London forces are responsible for the differences. These molecules all have London forces and hydrogen bonds. Since dipole-dipole interactions increase as bond polarity increases (and hydrogen bonds are dipole-dipole interactions), we might expect the trend to match the ΔEN values (0.9 for NH_3, 1.3 for H_2O, and 1.9 for HF). NH_3 and HF follow the trend, but H_2O does not; it has stronger hydrogen bonds than expected. This can be explained by looking at the Lewis structures of each molecule. The N atom in NH_3 has 3 H atoms and 1 lone pair, the O atom in H_2O has 2 H atoms and 2 lone pairs, and the F atom in HF has 1 H atom and 3 lone pairs. H_2O has the perfect balance of lone pairs and H atoms, so each water molecule can make 4 hydrogen bonds (2 using

its H atoms, and 2 using its lone pairs on O). Although NH_3 and HF could make 4 hydrogen bonds per molecule, NH_3 is short of lone pairs on N to make that many bonds, and HF is short of H atoms to make that many bonds (they make a total of 2 hydrogen bonds per molecules). The additional hydrogen bonds possible in water make its total intermolecular forces greater than HF even though its ΔEN is smaller than in HF.

8. (a) Both HF and H_2O are polar and have hydrogen bonds (because they have H atoms bonded to F or O atoms). When HF molecules interact, the H atoms ($\delta+$) are attracted to F atoms ($\delta-$) in other HF molecules; similarly, when H_2O molecules interact, the H atoms ($\delta+$) are attracted to O atoms ($\delta-$) in other H_2O molecules. When HF and H_2O molecules interact the H atoms ($\delta+$) in both molecules are attracted to the F and O atoms ($\delta-$) in HF and H_2O, respectively.

(b) The hydrogen bonds are very strong attractions as far as intermolecular forces are concerned. HF molecules are attracted to other HF molecules by strong hydrogen bonds, and H_2O molecules are attracted to each other by strong hydrogen bonds. These liquids will mix to form a homogeneous mixture (and the molecules will interact with each other) only if the attractions between the HF and H_2O molecules are strong as well. Since these attractions are also strong hydrogen bonds, these molecules have just as strong a preference for each other as they do for themselves, and they will mix. These results can be generalized to dissolving any polar or ionic compound in water. Water mixes with these molecules because the attractions between the water molecules and these polar or ionic compounds are comparable in strength to the attractions between water molecules.

9. (a) Hydrogen (H_2) molecules are nonpolar ($\Delta EN = 2.1 - 2.1 = 0.0$, nonpolar covalent), so they only have rather weak London forces holding them together. Water, on the other hand, is polar ($\Delta EN = 3.4 - 2.1 = 1.3$, polar covalent) and has very strong hydrogen bonds holding water molecules together. When H_2 and H_2O molecules interact, they are weakly attracted by London forces—when a water molecule (with its permanent dipole of $H^{\delta+}O^{\delta-}H^{\delta+}$) approaches a neutral H_2 molecule, it induces a dipole in this molecule so that the atoms near the O atom in water are made to be positive and the atoms near the H atoms in water are made to be negative. In general, these London forces are much weaker than the hydrogen bonds between water molecules.

(weak London forces are not shown in these pictures).

(b) The hydrogen bonds between water molecules are very strong intermolecular forces. On the other hand, the H_2 molecules are only attracted to each other and to the H_2O molecules by relatively weak London forces. Because the water molecules are more attracted to each other than they are to the H_2 molecules, they crowd closely together forcing the nonpolar H_2 molecules out of the way. In the last picture, you can see that the water molecules are strongly attracted to each other and the H_2 atoms are 'on their own'. This is an unstable configuration, and once the H_2 molecules are expelled, the strong attractions in the water will be strengthened because the H_2 molecules will not be getting in the way of these attractions. These results can be generalized to dissolving any nonpolar compound in water (hydrocarbons, grease, oil, etc.). Water does not mix with these molecules because the attractions between the water molecules are so much stronger than the attractions between water molecules and these nonpolar compounds that the water molecules would prefer to mix with themselves than with the nonpolar molecules.

ANSWERS — PRACTICE TEST

Chapter 10

1. The difference between a sharp knife and a dull one is that the sharp knife has a smaller cutting surface (i.e., thin and pointed) than the dull knife. Assuming that it takes the same pressure to cut through an object whether the knife is sharp or dull (and remembering that $P = F/A$), you must exert a greater force (larger F) when the cutting surface is dull (large A) than when using a knife with a smaller cutting surface (small A).

2. To convert in Hg to mm Hg, use the length conversions introduced in Chapter 1 (1 in = 2.54 cm, 1 cm = 10 mm), then convert mm Hg to atm using the definition of atm (1 atm = 760 mm Hg).

$$30.00 \text{ in Hg} \times \frac{2.54 \text{ cm Hg}}{1 \text{ in Hg}} \times \frac{10 \text{ mm Hg}}{1 \text{ cm Hg}} = 762 \text{ mm Hg} \times \frac{1 \text{ atm}}{760 \text{ mm Hg}} = 1.00 \text{ atm}$$

3. (a) Because gas molecules move randomly in all directions of the container, if they are given a larger volume to occupy, these random motions will eventually result in the gas molecules being everywhere inside the container. Liquid and solid molecules are held in place by their intermolecular forces.

   (b) The two types of gas molecules do not attract or repel other each other. So, their random motions allow them to expand to the entire volume of the container, independent of each other. Since both types of gas molecules occupy the container's entire volume, this is a homogeneous mixture.

4. (a) Since you are given two temperatures, you should use the combined gas law. In this question, you are not told anything about the pressure or moles (the tire is assumed not to have a leak), so you should assume these values are constant. Also, remember to convert temperatures to the Kelvin scale.

$$\frac{PV_1}{PV_2} = \frac{nT_1}{nT_2}, \frac{V_1}{V_2} = \frac{T_1}{T_2}, V_2 = \frac{T_2}{T_1} \times V_1 = \frac{(273.15 + -3.2 \text{ K})}{(273.15 + 34.7 \text{ K})} \times 24.0 \text{ L} = 21.0 \text{ L}$$

   (b) As the gas sample inside the tire cools, the gas particles move slower and hit the inner walls of the tire less often. The gas particles outside the tire are hitting the outer walls of the tire more often, so they push the walls of the tire inward (decreasing the tire volume) until the number of collisions inside and outside the tire are equal (Charles's law).

   (c) The initial volume (the winter volume before the tire was reinflated) is 21.0 L, and the final volume is 24.0 L. You must assume that the air pressure and temperature do not change while the tire is being reinflated in the winter.

$$\frac{PV_1}{PV_2} = \frac{n_1 T}{n_2 T}, \frac{V_1}{V_2} = \frac{n_1}{n_2}, n_2 = \frac{V_2}{V_1} \times n_1 = \frac{24.0 \text{ L}}{21.0 \text{ L}} \times 1.09 \text{ mol} = 1.25 \text{ mol}$$

   (d) When more gas particles were added to the inside of the tire, this led to more collisions between the gas particles inside the tire and the inner walls of the tire. These increased collisions pushed the walls of the tire outward (increasing the tire volume) until the number of collisions inside and outside the tire are equal (Avogadro's law).

5. Since you are given only one volume and temperature, you should use the ideal gas law. Solve for pressure (recognizing that the mass of I_2 will give you the number of moles of I_2).

$$P = \frac{nRT}{V} = \frac{\left(312.8 \text{ g } I_2 \times \dfrac{\text{mol } I_2}{253.81 \text{ g } I_2}\right)\left(0.082057 \dfrac{\text{L atm}}{\text{mol K}}\right)(273.15 + 253.4 \text{ K})}{10.00 \text{ L}} = 5.325 \text{ atm}$$

6. Since you are given the mass and volume of the gas, you can use the formula $d = m/V$.

$$d = \frac{m}{V} = \frac{312.8 \text{ g}}{10.00 \text{ L}} = 31.28 \frac{\text{g}}{\text{L}}$$

7. Since you are only given two of the four gas parameters (T and P) and you are asked to solve for the volume of the gas, perform a stoichiometry first (solving for n for H_2), then perform an ideal gas law calculation.

$$47.47 \text{ g B}_2\text{H}_6 \times \frac{\text{mol B}_2\text{H}_6}{27.67 \text{ g B}_2\text{H}_6} \times \frac{6 \text{ mol H}_2}{1 \text{ mol B}_2\text{H}_6} = 10.29346 \text{ mol H}_2 \text{ (unrounded)}$$

$$V = \frac{nRT}{P} = \frac{(10.29346 \text{ mol})(0.082057 \text{ L atm/mol K})(273.15 + 18.8 \text{ K})}{789.2 \text{ mm Hg} \times \frac{1 \text{ atm}}{760 \text{ mm Hg}}} = 237.5 \text{ L}$$

8.   (a)   Since you are given three of the four gas parameters (V, T, and n calculated from the mass), you should use the ideal gas law. This answer should be greater than 12.14 atm (the pressure after O_2 has reacted).

$$P = \frac{nRT}{V} = \frac{\left(500.0 \text{ g O}_2 \times \frac{\text{mol O}_2}{32.00 \text{ g O}_2}\right)\left(0.082057 \frac{\text{L atm}}{\text{mol K}}\right)(273.15 + 174.3 \text{ K})}{30.0 \text{ L}} = 19.12 \text{ atm}$$

(b)   Since the pressure would have started at 19.12 atm but dropped to 12.14 atm as a result of the reaction, the amount of oxygen that reacted must have exerted a pressure of 19.12 atm − 12.14 atm = 6.98 atm.

$$n = \frac{PV}{RT} = \frac{(6.98 \text{ atm})(30.0 \text{ L})}{(0.082057 \text{ L atm/mol K})(273.15 + 174.3 \text{ K})} = 5.7032 \text{ mol O}_2 \text{ (unrounded)}$$

(c) $n_M = 5.7032 \text{ mol O}_2 \times \frac{2 \text{ mol M}}{1 \text{ mol O}_2} = 11.40634 \text{ mol M}$. $M(M) = \frac{500.0 \text{ g M}}{11.40634 \text{ mol M}} = 43.8 \frac{\text{g}}{\text{mol}}$.

Since this metal must be an alkali earth metal, it is most likely calcium, Ca ($M = 40.08$ g/mol).

9.   Since you are given m, T, V, and P, use the equation $M = mRT/PV$.

$$M = \frac{mRT}{PV} = \frac{(100.0 \text{ g})(0.082057 \text{ L atm/mol K})(273.15 + 374 \text{ K})}{(1.71 \text{ atm})(25.0 \text{ L})} = 124.2 \frac{\text{g}}{\text{mol}}$$

$$n = \frac{M \text{ molecular}}{M \text{ empirical}} = \frac{124.2 \text{ g/mol}}{30.97 \text{ g/mol}} = 4.01 = 4. \text{ So, white phosphorus is P}_4.$$

10.   Calculate P for water vapor using the ideal gas law and the van der Waals equation.

$$PV = nRT, \; P = \frac{nRT}{V} = \frac{(15.00 \text{ mol})(0.082057 \text{ L atm/mol K})(273.15 + 180 \text{ K})}{89.5 \text{ L}} = 6.23 \text{ atm}$$

$$\left(P + \frac{n^2 a}{V^2}\right)(V - nb) = nRT, \; P = \frac{nRT}{(V - nb)} - \frac{n^2 a}{V^2} =$$

$$\frac{(15.00 \text{ mol})(0.082057 \text{ L atm/mol K})(273.15 + 180 \text{ K})}{89.5 \text{ L} - (15.00 \text{ mol})(0.0305 \text{ L/mol})} - \frac{(15.00 \text{ mol})^2(5.46 \text{ L}^2 \text{ atm/mol}^2)}{(89.5 \text{ L})^2} =$$

$$6.26 \text{ atm} - 0.15 \text{ atm} = 6.47 \text{ atm}$$

The first term (6.26 atm) represents the correction using b. This value closely matches the ideal gas law value (a correction of 0.03 atm); the correction using a is 0.15 atm, which is bigger. So, the correction using a (which represents the attractions of gas particles) is bigger than the correction using b (which represents the nonzero volume of the gas particles). This makes sense for a very small molecule that has very strong noncovalent interactions (hydrogen bonds).

11.   The reason C-Cl bonds are broken instead of C-F bonds is that the C-Cl bonds (327 kJ/mol) are weaker than the C-F bonds (486 kJ/mol), so they are easier to break.

12.   No, the scientist should not do this. First of all, plants need carbon dioxide to exist so removing it from the air would kill all plants on earth. Second, removing all of the carbon dioxide does not remove all of the greenhouse gases present in earth's atmosphere (water vapor is a greenhouse gas), and removing all of the carbon dioxide could cause the planet's average temperature to drop too much (possibly freezing the earth's oceans and lakes).

A 28

ANSWERS — PRACTICE TEST

Chapter 11

1. (a) When liquid water is heated, water vapor, $H_2O(g)$, is produced. As you may know, hydrogen gas ignites in the presence of air (if you have ever seen a video of the *Hindenberg* burning, that was an example of combustion of hydrogen gas). If boiling water released hydrogen gas, then any cook using an open flame (like a gas cook-top) would experience the combustion of hydrogen.

    (b) When liquid water is converted to $H_2O(g)$, the added energy goes to breaking the intermolecular forces (hydrogen bonds) holding liquid molecules together. Chapter 9 says that the strength of hydrogen bonds is usually 10-40 kJ/mol. When liquid water is converted to $H_2(g)$ and $O_2(g)$, the covalent O-H bonds in water must be broken; Table 8.2 in the textbook (p. 345) shows that breaking O-H bonds takes 467 kJ/mol. Since it takes much less energy to convert liquid water into gaseous water than it does to break the covalent O-H bonds in water, it makes sense that $H_2O(g)$ will be formed.

2. (a) The trend is that the vapor pressure decreases as the molecules get bigger (more atoms, larger molar mass). Since these compounds are hydrocarbons (nonpolar molecules), the only intermolecular forces present are London forces. In general, London forces are stronger in larger molecules because it is easier to have a temporary imbalance (induced dipole) of the electrons around larger molecules. Stronger London forces mean fewer molecules will have enough energy to escape the liquid and become gas molecules (which causes a lower vapor pressure).

    (b) The trend is that the vapor pressure decreases as the molecules get bigger (more atoms, larger molar mass). Since these compounds have polar O-H bonds, the intermolecular forces present are London forces and hydrogen bonds (which are dipole-dipole attractions). Even though the alcohols have hydrogen bonds in them, the London forces are responsible for the observed trend. The effect of the hydrogen bonds is the same because each alcohol has one O-H per molecule. The difference in these molecules is the number of CH_2 groups present, and increasing the number of CH_2 groups strengthens the London forces, which make it more difficult for the molecules to escape the liquid as gases.

    (c) In order to determine the effect of hydrogen bonds, you need to control for London forces by having two molecules with the same approximate size (molar mass). In this case, you can compare pentane (M = 72.15 g/mol, no hydrogen bonds) with 1-butanol (M = 74.12 g/mol, hydrogen bonds). The difference here is very large. Pentane's vapor pressure is 651 mm Hg while 1-butanol's vapor pressure is 18 mm Hg. The hydrogen bonds make the 1-butanol molecules much more strongly attracted to each other, so many fewer 1-butanol molecules will have enough energy to escape the liquid as a gas, and its vapor pressure is very low; pentane molecules are held in the liquid by relatively weak London forces, so many of them can escape the liquid, resulting in a high vapor pressure.

    (d) The alcohols and water all have hydrogen bonds in them. The reason water has a much lower vapor pressure than the alcohols (even though it has a much lower molar mass) is that water has *two* O-H groups per molecule. This means that water can make many more hydrogen bonds with each other, increasing the strength of the intermolecular forces holding water molecules in the liquid, lowering the number of water molecules having enough energy to escape the liquid, and lowering its vapor pressure.

3. (a) Since water has polar O-H bonds, its intermolecular forces include London forces and hydrogen bonds (which are dipole-dipole attractions). Since phosphorus has nonpolar bonds ($\Delta EN = 2.1 - 2.1 = 0.0$), its intermolecular forces are limited to London forces.

    (b) The boiling point of a liquid is a measure of its intermolecular forces. Since phosphorus has a much higher boiling point, it has stronger intermolecular forces than water. Even though water has strong hydrogen bonding forces, P_4 is much bigger than H_2O (more atoms and more electrons) so its overall London forces must be stronger than the hydrogen bonding and London forces present in water.

    (c) Molecules with weaker intermolecular forces have higher vapor pressures, so water should have the higher vapor pressure at 60°C (149 mm Hg for water, less than 1 mm Hg for phosphorus).

4. (a) In region 2 the solid is melting into a liquid, and in region 4 the liquid is boiling into a gas.

    (b) Regions 1 and 2 contain solids—region 1 represents the solid being heated until it reaches its melting point; region 2 represents the solid melting (at its melting point) until there is no solid left.

(c)   Regions 2, 3, and 4 contain liquids. Region 2 represents the solid melting into a liquid, region 3 represents the liquid being heated from its melting point to its boiling point, and region 4 represents the liquid boiling away as a gas.

(d)   Regions 4 and 5 contain gases—region 4 represents the liquid boiling (at its boiling point) into a gas and region 5 represents the gas being heated above its boiling point.

(e)   In regions 1, 3, and 5, the added energy is going to raise the temperature of the solid, the liquid, and the gas (respectively). You can see this because these are the regions where the temperature is changing as heat is added (the flat lines show the temperature staying constant as heat is added).

5.   (a)   Steam burns are more dangerous because they transfer more energy to the burn victim than boiling water. Boiling water transfer energy as it cools from 100°C to 37°C (q_{cool}), but steam transfers energy as it condenses into liquid water ($q_{cond.}$) plus the energy transferred as the newly formed liquid at 100°C cools to 37°C (q_{cool}).

(b) $q_{cool} = mc\Delta T = (10.0 \text{ g})\left(4.184 \dfrac{\text{J}}{\text{g C}}\right)(37 \text{ C} - 100 \text{ C}) = -2,600 \text{ J} \times \dfrac{1 \text{ kJ}}{10^3 \text{ J}} = -2.6 \text{ kJ}$.

(c) $q_{cond.} = n\Delta H = \left(10.0 \text{ g} \times \dfrac{\text{mol H}_2\text{O}}{18.02 \text{ g H}_2\text{O}}\right)\left(-40.7 \dfrac{\text{kJ}}{\text{mol H}_2\text{O}}\right) = -22.6 \text{ kJ}$. The total heat lost by the steam is –2.6 kJ + –22.6 kJ = –25.2 kJ (almost ten times the energy lost by the boiling water).

6.   In order to use the Clausius-Clapeyron equation, you need to calculate $\Delta H°_{vap}$ using the $\Delta H°_f$ values.

$$CS_2(\ell) \rightarrow CS_2(g) \hspace{4cm} \Delta H°_{vap}$$

$$\Delta H_{vap} = \left\{1 \text{ mol CS}_2(g) \times 117.36 \dfrac{\text{kJ}}{\text{mol CS}_2(g)}\right\} - \left\{1 \text{ mol CS}_2(\ell) \times 89.70 \dfrac{\text{kJ}}{\text{mol CS}_2(\ell)}\right\} = +27.66 \text{ kJ}$$

This is the value for 1 mol of CS_2, so $\Delta H°_{vap} = 27.66$ kJ/mol. This equation requires you to know $\Delta H°_{vap}$ and the vapor pressure of the substance at a known temperature. Using the boiling point of CS_2, you know that at 46.2°C ($T_1 = 319.35$ K) the vapor pressure of CS_2 (P_1) must be 760.0 mm Hg (the definition of the normal boiling point). Using this information, you can solve for the vapor pressure (P_2) at 25.0°C ($T_2 = 298.15$ K).

$$\ln\left(\dfrac{P_2}{760.0 \text{ mm Hg}}\right) = -\dfrac{27,660 \text{ J/mol}}{8.314 \text{ J/mol K}}\left(\dfrac{1}{298.15 \text{ K}} - \dfrac{1}{319.35 \text{ K}}\right)$$

$$\ln\left(\dfrac{P_2}{760.0 \text{ mm Hg}}\right) = -3,327 \text{ K}\left(0.003354 \text{ K}^{-1} - 0.003131 \text{ K}^{-1}\right) = -0.7408$$

$$\dfrac{P_2}{760.0 \text{ mm Hg}} = e^{-0.7408} = 0.47675, \quad P_2 = 0.47675 \times 760.0 \text{ mm Hg} = 362 \text{ mm Hg}$$

7.   (a)   The phase diagram for oxygen appears to the right.

(b)   At room temperature (25°C, or 298 K) and room pressure (760 mm Hg), oxygen is a gas.

(c)   At 63 K and 1,000 mm Hg, oxygen is a liquid—the phase in between the solid and gas phases.

(d)   At 7 K and 350 mm Hg, oxygen is a solid—the phase to the right (low T) is the solid phase.

(e)   At 170 K and 40,000 mm Hg, oxygen is a supercritical liquid (fluid) since this is above the critical T (154 K) and critical P (37,823 mm Hg) values.

(f)   Oxygen will only sublime (convert directly from solid to gas) at pressures below the triple point (below 1 mm Hg).

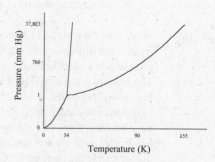

8.   (a)   Since CO_2 never conducts electricity, it is either a molecular or a covalent network solid. Molecular solids have very low melting points and covalent network solids have very high melting points. Since CO_2 has a very low melting (sublimation) point of –78.5°C, it is a molecular solid.

A 30

(b)   Since SiO_2 never conducts electricity, it is either a molecular or a covalent network solid. Molecular solids have very low melting points and covalent network solids have very high melting points. Since SiO_2 has a very high melting point (1723°C), it is a covalent network solid.

(c)   Since SnO_2 does not conduct electricity as a solid, but does when it is melted, SnO_2 is an ionic solid.

9. (a)   Since gold (Au) appears in the transition elements, you would expect it to be a metallic solid.

    (b)   Since iron pyrite (FeS_2) contains iron (a metal) and sulfur (a nonmetal), you should expect it to be an ionic solid (iron cations and sulfur anions).

    (c)   Melting point is not really going to help you distinguish between a metallic solid (gold) and an ionic solid (fool's gold) because both are expected to have strong forces holding them together, which means high melting points. In fact, the melting points of these two compounds are similar (1064°C for gold and 1171°C for fool's gold). Electrical conductivity can help you distinguish between a metallic solid and an ionic solid—gold (a metallic solid) should conduct electricity as a solid but fool's gold (an ionic solid) should not conduct electricity as a solid (but would if it were melted). If your solid conducts electricity, it is probably gold; if it does not, it is probably fool's gold.

10. (a)   For the bcc unit cell $4r = \sqrt{3}\ell$, and for the fcc unit cell $4r = \sqrt{2}\ell$. Solving for the atomic radius:

$$4r_{bcc} = \sqrt{3}\ell, \; r_{bcc} = \frac{\sqrt{3}\ell}{4} = \frac{\sqrt{3}(0.287\text{ nm})}{4} = 0.124 \text{ nm}$$

$$4r_{fcc} = \sqrt{2}\ell, \; r_{fcc} = \frac{\sqrt{2}\ell}{4} = \frac{\sqrt{2}(0.363\text{ nm})}{4} = 0.128 \text{ nm}$$

(b)   In order to determine the metal's density, you need to determine the mass and volume of the unit cells. The volumes will be $V_{cube} = \ell^3$. The masses are determined by the weight of an iron atom (in g) and the number of iron atoms in each unit cell (2 for bcc, 4 for fcc).

$$m_{bcc} = 2 \text{ Fe atoms} \times \frac{1 \text{ mol Fe atoms}}{6.022 \times 10^{23} \text{ Fe atoms}} \times \frac{55.85 \text{ g Fe atoms}}{\text{mol Fe atoms}} = 1.8549 \times 10^{-22} \text{ g Fe atoms}$$

$$V_{bcc} = \ell^3 = \left(0.287 \text{ nm} \times \frac{10^{-9} \text{ m}}{1 \text{ nm}} \times \frac{1 \text{ cm}}{10^{-2} \text{ m}} \right)^3 = 2.3640 \times 10^{-23} \text{ cm}^3$$

$$d_{bcc} = \frac{m}{V} = \frac{1.8549 \times 10^{-22} \text{ g}}{2.3640 \times 10^{-23} \text{ cm}^3} = 7.85 \text{ g/cm}^3$$

$$m_{fcc} = 4 \text{ Fe atoms} \times \frac{1 \text{ mol Fe atoms}}{6.022 \times 10^{23} \text{ Fe atoms}} \times \frac{55.85 \text{ g Fe atoms}}{\text{mol Fe atoms}} = 3.7097 \times 10^{-22} \text{ g Fe atoms}$$

$$V_{fcc} = \ell^3 = \left(0.363 \text{ nm} \times \frac{10^{-9} \text{ m}}{1 \text{ nm}} \times \frac{1 \text{ cm}}{10^{-2} \text{ m}} \right)^3 = 4.7832 \times 10^{-23} \text{ cm}^3$$

$$d_{fcc} = \frac{m}{V} = \frac{3.7097 \times 10^{-22} \text{ g}}{4.7832 \times 10^{-23} \text{ cm}^3} = 7.76 \text{ g/cm}^3$$

(c)   Ultimately, the density for the bcc unit cell is slightly larger because the atomic radii for the iron atoms in the bcc unit cell are slightly smaller. The iron atoms still have the same mass upon heating, but their radii increase from 0.124 nm in the bcc unit cell to 0.128 nm in the fcc cell. This means the same mass of iron atoms is now taking up more space (a larger volume), so the density of solid iron decreases when it changes from the bcc to the fcc unit cell.

11. (a)   Since Zn has two fewer e^- than Ge and Se has two more e^- than Ge, the combination of ZnSe has the same number of valence electrons as GeGe (elemental germanium). Since Ge is a semimetal (metalloid), it should behave like a semiconductor (which it does).

    (b)   In order to make a p-type semiconductor, you need to dope the semiconductor with atoms containing fewer electrons that the semiconductor. Since Zn has fewer e^- than Ge (and GaAs is isoelectronic with Ge), it makes sense to have an excess of Zn atoms if you want to make a p-type semiconductor.

    (c)   In order to make an n-type semiconductor, you need to dope add atoms with more electrons that the semiconductor. Since Se has more e^- than Ge, extra Se atoms would make an n-type semiconductor.

A 31

ANSWERS — PRACTICE TEST

Chapter 12

1. Since the molecules in petroleum feedstocks (crude oil) tend to be nonpolar hydrocarbons, the intermolecular forces present in these compounds are only London forces. London forces are stronger for bigger molecules because they have more electrons that can become temporarily unbalanced. So, the larger molecules have stronger intermolecular forces, and it takes more energy (a higher temperature) to break these forces holding the molecules in the liquid state so they can boil and become gases.

2. The assignments of octane number for heptane (0) and isooctane (100) were completely arbitrary. Heptane does not burn very efficiently, but there are molecules that can burn even less efficiently than heptane; similarly, while isooctane burns rather efficiently, there are other molecules that burn even more efficiently.

3. All three of these resources are fossil fuels (derived from the remains of previous animal and plant life). The major difference between the three is their states of matter. Coal represents the solid form of fossil fuels, crude oil represents its liquid form, and natural gas represents its gaseous form.

4. (a) Aldehydes (molecules with a -C(=O)H group) are made from primary alcohols (RCH_2OH). So, the primary alcohol you should use is $ClCH_2CH_2CH_2CH_2OH$. When a primary alcohol is oxidized, an aldehyde is initially produced, but it quickly reacts with the oxidizing agent to make the carboxylic acid ($ClCH_2CH_2CH_2COOH$) so you may not actually be able to collect the aldehyde product.

    (b) Ketones (molecules of the general formula $RC(=O)R'$) are made from secondary alcohols ($RR'CHOH$) so you should oxidize the secondary alcohol $(CH_3)_2CHOH$ (isopropyl or rubbing alcohol).

    (c) Carboxylic acids (molecules with a -C(=O)OH group) are made from primary alcohols (RCH_2OH), so you should oxidize the primary alcohol CF_3CH_2OH.

    (d) Esters (molecules with the general formula $RC(=O)OR'$) are made from alcohols and carboxylic acids. In order to make this ester compound, you need the alcohol CH_3OH (methanol) to get the $-OCH_3$ group of the ester and the carboxylic acid CH_3COOH (acetic acid) to get the $CH_3C(=O)-$ group. Since you are only allowed to use alcohols as reactants, you must make the acetic acid from the primary alcohol CH_3CH_2OH (ethanol). Once you oxidize the ethanol to acetic acid, remove the oxidizing agent and add the methanol to make the ester.

5. A triglyceride is simply a triester made by the condensation reaction of the triol glycerol ($C_3H_5(OH)_3$) and three carboxylic acids (fatty acids). The first acid is $CH_3(CH_2)_5CH=CH(CH_2)_7COOH$, a monounsaturated fat (because it has only one C-C double bond); the second acid is $CH_3(CH_2)_{14}COOH$, a saturated fat with no C-C double bonds; and the third acid is $CH_3CH_2CH=CHCH_2CH=CHCH_2CH=CH(CH_2)_7COOH$, a polyunsaturated fat containing three C-C double bonds (the other reactant, glycerol, is none of the above since it isn't an acid).

6. When an alkene monomer forms an addition polymer, each C atom in the monomer trades a C-C double bond for two C-C single bonds (still a total of two bonds), so when an alkyne monomer forms an addition polymer each C atom in the monomer trades a C-C triple bond for a C-C double bond and a C-C single bond (for a total of three bonds). Because the alkyne becomes an alkene in the polymer chain, there are three types of polyalkene chains that can be formed —all *cis*, all *trans*, or a mixture of both *cis* and *trans*. The all *cis* and all *trans* are drawn below.

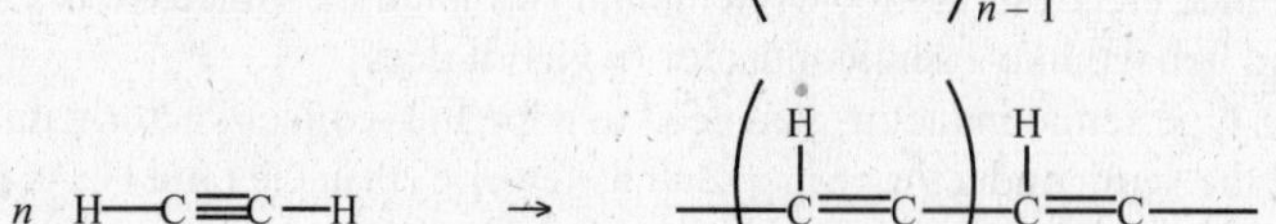

7.   (a)   Since this polymer has no O or N atoms in the chain, it must be an addition polymer made from an alkene. Since the repeat pattern has 3 H atoms and 1 CH_3 group, the alkene must be $CH_2=CHCH_3$.

   (b)   Since this polymer only has C atoms in the chain, it is also an addition polymer made from alkenes. Since the repeat pattern contains four C atoms, it is probably made from two alkenes. The first alkene is $CF_2=CF_2$ and the second is $CH_2=CH_2$ (50-50 mix). This chain could also be made from $CH_2=CF_2$ groups in an alternating pattern (CF_2 groups bond to CF_2 groups and CH_2 groups bond to CH_2 groups).

   (c)   This polymer has N atoms and -C(=O)- groups in the chain, so it must be a condensation polymer (polyamide) made from amines and carboxylic acids. Since the repeat pattern has 2 N atoms and 2 -C(=O)- groups, it is likely that this chain is made from two different monomers. In order to determine the two monomers, all amide bonds (bonds between the N atom and the C atom of the -C(=O)- group) are hydrolyzed so that N gets an H atom from water and the -C(=O)- group gets the OH from water. This results in two compounds—$NH_2CH_2CH_2NH_2$ (a diamine) and $HOC(=O)CH_2C(=O)OH$ (a diacid).

   (d)   This chain has an O atom and a -C(=O)- group in the backbone, so it must be a condensation polymer (polyester) made from an alcohol and a carboxylic acid. Since the repeat pattern has only 1 O atom and 1 -C(=O)- group in the polymer chain, it is likely that this chain is made from a single monomer (acid alcohol). In order to determine the monomer, the ester bonds (bonds between the O atom and the C atom of the -C(=O)- group) are hydrolyzed so that the O gets an H atom from water and the -C(=O)- group gets the OH from water. This results in the compound $HOC(=O)CH_2CH_2OH$ as the monomer.

8.   (a)   This molecule is salicylic acid (the active ingredient in Compound W and other wart removers). Salicylic acid has an aromatic (benzene) ring, an alcohol (-OH group) attached to the ring, and a carboxylic acid (-C(=O)OH group) attached to the ring.

   (b)   This molecule is cinnamaldehyde (the chemical responsible for the smell and taste of cinnamon). Cinnamaldehyde contains an aromatic (benzene) ring, and a side chain that contains an alkene (in the *trans* position) and an aldehyde (-C(=O)H group).

9.   If you write alanine as A, methionine as M, and serine as S, then the possibilities including one of each are AMS, ASM, MAS, MSA, SAM, and SMA (you can make more trimers if you do not require one of each amino acid—AAA, MMM, SSS, AAS, MMA, etc.). Two of these trimers (AMS and SAM) are drawn below.

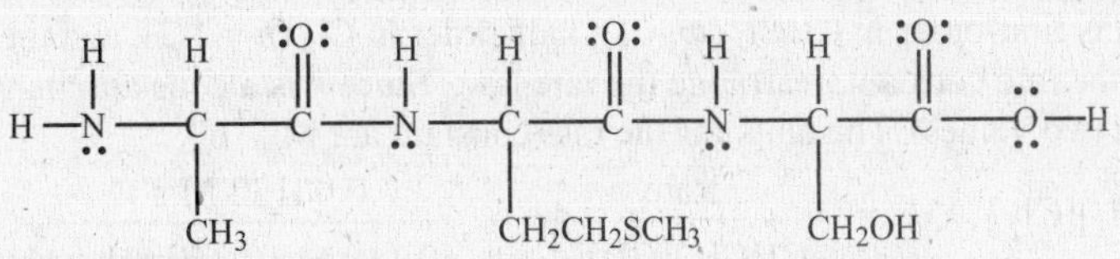

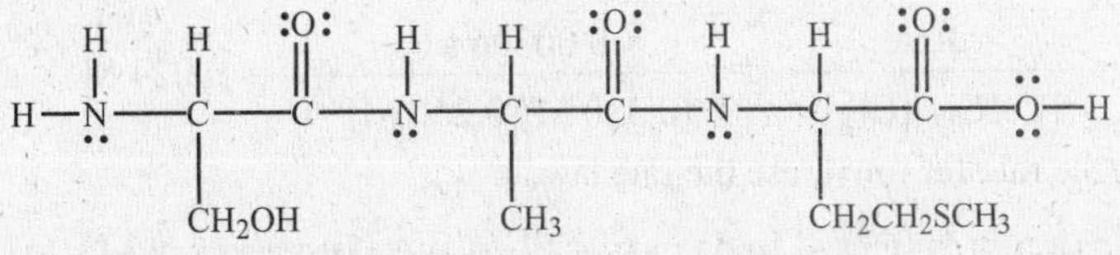

10.   The R group in phenylalanine (-$CH_2C_6H_5$) has only C and H atoms in it, so all of the bonds in this R group must be nonpolar, making this group nonpolar. The R group in tyrosine (-$CH_2C_6H_4OH$) has an -OH group in it, which is very polar and can undergo hydrogen-bonding attractions with other polar group.

ANSWERS — PRACTICE TEST

Chapter 13

1. (a) The rate of N_2O_5 decomposition is simply $\Delta[N_2O_5]/\Delta t$. These values are negative because N_2O_5 is reacting (and disappearing) as the reaction occurs.

$$\frac{\Delta[N_2O_5]_{0\text{-}15}}{\Delta t} = \frac{[N_2O_5]_f - [N_2O_5]_i}{t_f - t_i} = \frac{0.430\ M - 0.500\ M}{15\ s - 0\ s} = -0.0047\ \frac{M}{s}$$

$$\frac{\Delta[N_2O_5]_{45\text{-}60}}{\Delta t} = \frac{0.274\ M - 0.318\ M}{60\ s - 45\ s} = -0.0029\ \frac{M}{s}$$

(b) The reason the magnitude of the rate is larger from 0 to 15 s than it is from 45s to 60 s is that the N_2O_5 concentration is higher at $t = 0$ than at $t = 45$. As the concentration of N_2O_5 decreases over time, there are fewer N_2O_5 molecules present that can react and the rate of decomposition decreases.

(c) To determine the rates for NO_2 and O_2, use the rates calculated above for N_2O_5 and the balanced equation. These rates are positive because NO_2 and O_2 are being produced during the reaction.

$$\text{Rate}_{NO_2} = -0.0047\ \frac{\text{mol }N_2O_5/L}{s} \times \frac{4\text{ mol }NO_2}{-2\text{ mol }N_2O_5} = +0.0093\ \frac{\text{mol }NO_2/L}{s} = +0.0093\ \frac{M}{s}$$

$$\text{Rate}_{O_2} = -0.0029\ \frac{\text{mol }N_2O_5/L}{s} \times \frac{1\text{ mol }O_2}{-2\text{ mol }N_2O_5} = +0.0015\ \frac{\text{mol }O_2/L}{s} = +0.0015\ \frac{M}{s}$$

2. (a) The general formula for the rate law is: Rate $= k[CHCl_3]^m[Cl_2]^n$. To determine the value of m, use two runs where only $[CHCl_3]$ changes (runs 1 and 2). When $[CHCl_3]$ is tripled from run 1 to run 2, the rate also triples. Since $[CHCl_3]^m$ is proportional to the rate, then $[3]^m = 3$ and $m = 1$. To determine the value of n, look at two runs where only $[Cl_2]$ changes (runs 1 and 3). When $[Cl_2]$ is quadrupled from run 1 to run 3, the rate doubles. Since $[Cl_2]^n$ is proportional to the rate, then $[4]^n = 2$ and $n = 1/2$ (since the square root of 4 is 2). The specific rate law is Rate $= k[CHCl_3][Cl_2]^{1/2}$. Note that the exponents of the rate law do not match the reactant coefficients (Cl_2 has a coefficient of 1, but an exponent of 1/2).

(b) This reaction is first-order in $CHCl_3$ ($m = 1$), half-order in Cl_2 ($n = 1/2$), and the overall order is 3/2.

(c) To solve for the rate constant, rearrange the rate law. Since k is a constant, its value will be the same if you calculate two values. The units for the rate constant are $M^{-1/2}\ s^{-1}$.

$$\text{Rate} = k[CHCl_3][Cl_2]^{1/2},\ k = \frac{\text{Rate}}{[CHCl_3][Cl_2]^{1/2}} = \frac{0.00132\ M/s}{[0.110\ M][0.250\ M]^{1/2}} = 0.024\ M^{-1/2}\ s^{-1}$$

$$k = \frac{\text{Rate}}{[CHCl_3][Cl_2]^{1/2}} = \frac{0.00396\ M/s}{[0.330\ M][0.250\ M]^{1/2}} = 0.024\ M^{-1/2}\ s^{-1}$$

(d) To determine the rate for run 4, use the rate law.

$$\text{Rate} = k[CHCl_3][Cl_2]^{1/2} = (0.024\ M^{-1/2}\ s^{-1})[0.180\ M][0.360\ M]^{1/2} = 0.00259\ \frac{M}{s}$$

3. (a) Since $m(NaCl)$ vs. t is linear, this reaction is zero-order and the rate law is Rate $= -\Delta m(NaCl)/\Delta t = k$. Since NaCl is a solid, it does not have a concentration and its amount must be measured by its mass.

(b) To solve for the rate constant, plug two sets of data into the integrated rate law.

$$m(NaCl)_t = m(NaCl)_0 - kt,\ k = \frac{m(NaCl)_0 - m(NaCl)_t}{t} = \frac{2.25\ g - 2.00\ g}{12.0\ s} = 0.021\ \frac{g}{s}$$

(c) To solve for concentration after 50.0 s, use the $t = 0$ data and the rate constant.

$$m(NaCl)_t = m(NaCl)_0 - kt = 2.25\ g - \left(0.021\ \frac{g}{s}\right)(50.0\ s) = 1.21\ g$$

(d) To find the time needed to have a mass of 0.00 g, use the $t = 0$ data and the rate constant.

$$m(NaCl)_t = m(NaCl)_0 - kt,\ t = \frac{m(NaCl)_0 - m(NaCl)_t}{k} = \frac{2.25\ g - 0.00\ g}{0.0208\ g/s} = 108\ s$$

(e) After every 12.0 s, the mass lost is the same (2.25 g – 2.00 g = 0.25 g; 2.00 g – 1.75 g = 0.25 g, etc.). For zero-order reactions, the rate (the amount disappearing over the same time frame) is a constant.

4. (a) Since $\ln[O_3]$ versus t is linear, this reaction is first-order and the rate law is Rate $= -\Delta[O_3]/\Delta t = k[O_3]$.

   (b) To solve for the rate constant, plug two sets of data into the integrated rate law.

$$\ln\frac{[O_3]_{16}}{[O_3]_0} = \ln\left(\frac{1.92\ M}{3.00\ M}\right) = -0.4463 = -kt,\ k = \frac{-0.4463}{-16.0\ s} = 0.0279\ s^{-1}$$

   (c) To solve for concentration after 10.0 s, use the $t = 0.0$ data and the rate constant.

$$\ln\frac{[O_3]_{10}}{[O_3]_0} = -kt = -(0.0279\ s^{-1})(10.0\ s) = -0.279,\ \frac{[O_3]_{10}}{[O_3]_0} = e^{-0.279} = 0.757$$

$$[O_3]_{10} = 0.757 \times [O_3]_0 = 0.757\ (3.00\ M) = 2.27\ M$$

   This value is reasonable since it falls between the $t = 8.0$ and the $t = 16.0$ values (2.40 M and 1.92 M).

   (d) To find the time needed to have $[O_3] = 1.00$ M, use the $t = 0.0$ data and the rate constant.

$$\ln\frac{[O_3]_t}{[O_3]_0} = \ln\left(\frac{1.00\ M}{3.00\ M}\right) = -1.099 = -kt,\ t = \frac{-1.099}{-0.0279\ s^{-1}} = 39.4\ s$$

   (e) After every 8.0 s, the proportion of $[O_3]$ remaining is the same (3.00 M/2.40 M = 0.80; 1.92 M/2.40 M = 0.80, etc.). For first-order reactions, the proportion of reactants disappearing over the same time frame is constant.

5. If the concentration at $t = 0$ is $[A]_0$, then the concentration at $t = t_{1/2}$ will be $1/2[A]_0$.

   <u>zero-order:</u> $\frac{1}{2}[A]_0 = [A]_0 - kt_{1/2},\ kt_{1/2} = [A]_0 - \frac{1}{2}[A]_0 = \frac{1}{2}[A]_0,\ t_{1/2} = \frac{[A]_0}{2k}$

   <u>second-order:</u> $\frac{1}{1/2[A]_0} = \frac{2}{[A]_0} = \frac{1}{[A]_0} + kt_{1/2},\ kt_{1/2} = \frac{2}{[A]_0} - \frac{1}{[A]_0} = \frac{1}{[A]_0},\ t_{1/2} = \frac{1}{k[A]_0}$

   The reason these values are not tabulated is that they are not constants. In other words, the half-lives of zero- and second-order reactions depend on the initial concentration of the reactants.

6. (a) The value for E_a is calculated from the two-point Arrhenius equation, and the A value is determined from the (one-point) Arrhenius equation.

$$\ln\frac{k_{425}}{k_{333}} = \frac{E_a}{R}\left(\frac{1}{T_{333}} - \frac{1}{T_{425}}\right),\ \ln\frac{k_{425}}{k_{333}} = \ln\frac{0.112\ M^{-1/2}\ s^{-1}}{2.01 \times 10^{-4}\ M^{-1/2}\ s^{-1}} = \ln(557) = 6.3229$$

$$\frac{E_a}{R}\left(\frac{1}{606.15\ K} - \frac{1}{698.15\ K}\right) = \frac{E_a}{R}(2.17 \times 10^{-4}\ K^{-1}),\ \frac{E_a}{R} = \frac{6.3229}{2.17 \times 10^{-4}\ K^{-1}} = 29{,}084\ K$$

$$E_a = 29{,}084\ K \times R = (29{,}084\ K)\left(8.314\ \frac{J}{mol\ K}\right) = 2.42 \times 10^5\ \frac{J}{mol}$$

$$A = ke^{+E_a/RT} = (0.112\ M^{-1/2}\ s^{-1})e^{\frac{2.42 \times 10^5\ J/mol}{(8.314\ J/mol\ K)(698.15\ K)}} = 1.39 \times 10^{17}\ M^{-1/2}\ s^{-1}$$

   (b) The reason the reaction rate increases when the temperature increases is that more of the reactants ($CHCl_3$ and Cl_2) have enough energy to reach the top of the 'energy hill', and have enough energy to react and form the products (CCl_4 and HCl).

   (c) To determine the value of k at 400°C, you can use the one-point or two-point Arrhenius equation.

$$k = Ae^{-E_a/RT} = (1.39 \times 10^{17}\ M^{-1/2}\ s^{-1})e^{-\frac{2.42 \times 10^5\ J/mol}{(8.314\ J/mol\ K)(673.15\ K)}} = 0.0239\ M^{-1/2}\ s^{-1}$$

7. (a) The four steps in the proposed mechanism must add to the overall equation $CHCl_3(g) + Cl_2(g) \rightarrow CCl_4(g) + HCl(g)$. So, the final step must be $2\ Cl(g) \rightarrow Cl_2(g)$ to cancel the unwanted chemicals.

   (b) The intermediates are the chemicals that are produced in one step of the mechanism and then used in a subsequent step of the mechanism. The chemical species that first appear as products, and then appear as reactants are $Cl(g)$ and $CCl_3(g)$.

   (c) If step 1 was rate-limiting, the rate law would be Rate $= k[Cl_2]$, which does not match the actual rate law. So, step 1 cannot be the rate-limiting step. If step 2 was rate-limiting, the rate law would be

Rate = $k[CHCl_3][Cl·]$. Since Cl· is an intermediate it shouldn't appear in the rate law. If step 2 is the slow step, then step 1 represents an *equilibrium reaction* (where Rate$_f$= Rate$_r$).

$$\text{Step 1':} \quad Cl_2(g) \underset{k_{-1}}{\overset{k_1}{\rightleftharpoons}} 2\, Cl·(g)$$

$$\text{Rate}_f = k_1[Cl_2] = \text{Rate}_r = k_{-1}[Cl·]^2, \quad [Cl·] = \sqrt{\frac{k_1}{k_{-1}}}[Cl_2]^{1/2}$$

$$\text{Rate} = k_2[CHCl_3][Cl·] = k_2\sqrt{\frac{k_1}{k_{-1}}}[CHCl_3][Cl_2]^{1/2} = k'[CHCl_3][Cl_2]^{1/2}$$

Since this is consistent with the actual rate law, the rate-limiting step in this mechanism is step 2.

8. (a) Given the rate law Rate = $k[NO]^2[CO]$, this reaction is second-order in NO ($m = 2$), first-order in CO ($n = 1$), and third-order overall ($m + n = 2 + 1 = 3$).

   (b) Since the units of Rate are M/s and the units for concentrations are M, the units on k are:

$$\text{Rate} = k[NO]^2[CO], \quad k = \frac{\text{Rate}}{[NO]^2[H_2]} = \frac{M/s}{M^2 \times M} = M^{-2}\,s^{-1}$$

   (c) The intermediates in mechanism I are $N_2O_2(g)$ and $N_2O(g)$, the intermediates in mechanism II are C_2O_2 and NCO, and the intermediates in mechanism III are CNO_2 and CO_3.

   (d) For mechanism I, if the first step was rate-limiting the rate law would be Rate = $k_1[NO]^2$. If the second step was rate-limiting, the rate law would be Rate = $k_2[N_2O_2][CO] = k'[NO]^2[CO]$, which matches the actual rate law. For mechanism II, if the first step was rate-limiting the rate law would be Rate = $k_1[CO]^2$. If the second was rate-limiting, the rate law would be Rate = $k_2[C_2O_2][NO] = k'[NO][CO]^2$. If the third step was rate-limiting, the rate law would be Rate = $k_3[NCO][NO] = k"[NO]^2[CO]^2/[CO_2]$. Since none of the steps in mechanism II matches the actual rate law, mechanism II cannot be the correct mechanism. For mechanism III, if the first step was rate-limiting the rate law would be Rate = $k_1[NO][CO]$. If the second step was rate-limiting, the rate law would be Rate = $k_2[CNO_2][NO]$ = $k'[NO]^2[CO]$, which also matches the actual rate law. Since the second step in mechanisms I and III are consistent with the actual rate law, either mechanism I or III could be the actual mechanism.

9. If step 1 is rate-limiting, the E_a value for step 1 will be greater than E_a for step 2 (the first energy hill is taller than the second energy hill, as seen in the graph on the left). If step 2 is rate-limiting, the E_a value for step 2 will be greater than E_a for step 1 (the second energy hill is taller than the first energy hill, as seen in the graph on the right).

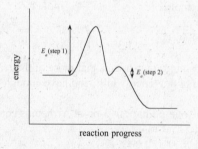
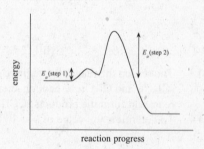

10. A homogeneous catalyst is a chemical that that speeds up the rate of a reaction and has the same state of matter as the reactants; a heterogeneous catalyst is a chemical that speeds up the rate of a reaction but has a state of matter that is different from the reactants. Industrial chemists prefer to use heterogeneous catalysts because they are easier to remove from the reactants and products (to be used again with another batch of reactants) than homogeneous catalysts.

ANSWERS — PRACTICE TEST

Chapter 14

1. (a) This system has reached equilibrium by 27 seconds. One way to determine that the system is at equilibrium is that the concentrations stop changing.

   (b) The values for the forward and reverse rates are shown below. The fact that these two values are not zero tells you that the reaction continues to happen even after the system has reached equilibrium.

   $$\text{Rate}_f = k_f[N_2O_4] = 0.013 \text{ s}^{-1}(4.915 \text{ M}) = 0.064 \text{ M/s}$$

   $$\text{Rate}_r = k_r[NO_2]^2 = 2.2 \text{ M}^{-1}\text{ s}^{-1}(0.170 \text{ M})^2 = 0.064 \text{ M/s}$$

   (c) $K_c = \dfrac{[NO_2]^2}{[N_2O_4]} = \dfrac{(0.170 \text{ M})^2}{4.915 \text{ M}}$, or $K_c = \dfrac{k_f}{k_r} = \dfrac{0.013 \text{ s}^{-1}}{2.2 \text{ M}^{-1}\text{ s}^{-1}}$. Both answers are 0.0059 $M = 0.0059$.

   (d) $Q = \dfrac{(\text{conc. }NO_2)^2}{(\text{conc. }N_2O_4)} = \dfrac{(0.154 \text{ M})^2}{4.923 \text{ M}} = 0.0048 \text{ M} = 0.0048$. Since $Q < K_c$ the reaction will shift to the right until $Q = K_c$. This shift can be seen as the reaction proceeds to equilibrium, since the equilibrium concentrations is larger for NO_2 and smaller for N_2O_4 (shifted to the right).

   (e) Because the K_c value is less than 1 (10^{-x}), this reaction is reactant-favored. You can also see that it is reactant-favored because most of the reactant did not react (about 2% reacted).

   (f) $K_P = \dfrac{P^2_{NO_2}}{P_{N_2O_4}} = K_c(RT)^{\Delta n} = (0.0059 \text{ M})\left(0.082057 \dfrac{\text{atm}}{\text{M K}} \times 298.15 \text{ K}\right)^1 = 0.14 \text{ atm} = 0.14$.

2. (a) Perform a Hess's law calculation to get the target equation, then calculate K_c from these values. Use equation 1 as written to get $NH_3(aq)$ and $NH_4^+(aq)$ and use equation 2 as written to get $HF(aq)$ and $F^-(aq)$. Finally, use equation 3 written backwards to get rid of the H_2O, $H_3O^+(aq)$, and $OH^-(aq)$.

   $$\textbf{NH}_3\textbf{(aq)} + H_2O(\ell) \rightleftharpoons \textbf{NH}_4^+\textbf{(aq)} + OH^-(aq)$$

   $$\textbf{HF(aq)} + H_2O(\ell) \rightleftharpoons H_3O^+(aq) + \textbf{F}^-\textbf{(aq)}$$

   $$H_3O^+(aq) + OH^-(aq) \rightleftharpoons 2 H_2O(\ell)$$

   $$K_c = \dfrac{[NH_4^+][F^-]}{[NH_3][HF]} = K_{c_1} \times K_{c_2} \times \dfrac{1}{K_{c_3}} = \left(1.8 \times 10^{-5}\right)\left(7.2 \times 10^{-4}\right)\left(\dfrac{1}{1.0 \times 10^{-14}}\right) = 1.3 \times 10^6$$

   (b) Since only gases exert pressures on their surroundings, aqueous compounds will not have partial pressures, and K_P would be meaningless.

   (c) Since K_c is greater than 1 (10^{+x}), this reaction is product-favored.

3. (a) $K_c = \dfrac{[CO_2][H_2]^2}{[H_2O]^2}$. Carbon is omitted from this equation because it is a solid. Solids and liquids have constant concentrations, so they are left out of the K_c expression.

   (b) The initial concentrations are $[H_2O] = 3.00$ mol/0.500 L $= 6.00$ M and $[CO_2] = [H_2] = 0.00$ M. You cannot calculate a concentration for C(s) since the solid will not expand to fill the entire container (i.e., its volume is not 0.500 L). Since there is no CO_2 or H_2 under the initial conditions, some C and H_2O will react to form CO_2 and H_2 (the reaction shifts to the right), so the signs for the C row are positive for the products and negative for the reactants.

	2 H₂O(g)	+	C(s)	⇌	CO₂(g)	+	2 H₂(g)
I	6.00 M				0.00 M		0.00 M
C	−2x				+x		+2x
E	6.00 M − 2x				x		2x

   At equilibrium, $[H_2] = 0.30$ M $= x$. So, $x = 0.15$ M. Using this value of x, $[CO_2] = x = 0.15$ M and $[H_2O] = 6.00$ M $- 2x = 6.00$ M $- 2(0.15$ M$) = 5.70$ M.

(c) $K_c = \dfrac{[CO_2][H_2]^2}{[H_2O]^2} = \dfrac{[0.15\ M][0.30\ M]^2}{[5.70\ M]^2} = 4.2 \times 10^{-4}\ M = 4.2 \times 10^{-4}.$

(d)   Because the K_c value is less than 1 (10^{-x}), this reaction is reactant-favored. You can also see that it is reactant-favored because most of the reactant did not react (about 5% reacted).

(e) $K_P = \dfrac{P_{CO_2} P_{H_2}^2}{P_{H_2O}^2} = K_c(RT)^{\Delta n} = (4.2 \times 10^{-4}\ M)\left(0.082057\ \dfrac{atm}{M\ K} \times 573.15\ K\right)^1 = 0.020\ atm = 0.020.$

4. (a) $K_c = \dfrac{[CO_2]}{[SO_3]}$. $CoCO_3$ and $CoSO_4$ are omitted from this equation because they are solids.

(b)   The initial concentrations are $[CO_2]$ = 18.0 mol/1.00 L = 18.0 M and $[SO_3]$ = 0.0 M. You cannot calculate concentrations for $CoCO_3$(s) or $CoSO_4$(s) since solids do not fill the entire container. Since there is no SO_3 under the initial conditions, some $CoSO_4$ and CO_2 will react to form $CoCO_3$ and SO_3 (the reaction shifts to the left), so the values in C are positive for reactants and negative for products.

	$CoCO_3$(s)	+	SO_3(g)	$\rightleftharpoons$	$CoSO_4$(s)	+	CO_2(g)
I			0.0 M				18.0 M
C			+x				−x
E			x				18.0 M − x

Since you don't know the equilibrium concentration of any chemical in this reaction, you must solve for x and use it to find the equilibrium concentrations.

$$K_c = \frac{[CO_2]}{[SO_3]} = \frac{[18.0\ M - x]}{[x]} = 61.2,\ 61.2x = 18.0\ M - x$$

$$x + 61.2x = 62.2x = 18.0\ M,\ x = \frac{18.0\ M}{62.2} = 0.289\ M = [SO_3]$$

$$[CO_2] = 18.0\ M - x = 18.0\ M - 0.289\ M = 17.7\ M$$

$$\underline{Check:}\ K_c = \frac{[CO_2]}{[SO_3]} = \frac{[17.7\ M]}{[0.289\ M]} = 61.2$$

(c)   Because the K_c value is greater than 1 (10^{+x}), this reaction is product-favored. You can see that it is product-favored because very little of the products were transformed into reactants (about 2% reacted).

(d) $K_P = \dfrac{P_{CO_2}}{P_{SO_3}} = K_c(RT)^{\Delta n} = (61.2)\left(0.082057\ \dfrac{atm}{M\ K} \times 2773.15\ K\right)^0 = 61.2.$

Note that when the number of gas molecules on both sides of the equation are the same, $K_c = K_P$.

5. (a) $K_c = \dfrac{[PCl_3][Cl_2]}{[PCl_5]}.$

(b) $Q = \dfrac{(conc.\ PCl_3)(conc.\ Cl_2)}{(conc.\ PCl_5)} = \dfrac{(3.00\ M)(1.00\ M)}{2.00\ M} = 1.50\ M = 1.50.$

Since $Q > K_c$ (1.50 > 1.21), the reaction will shift to the left, using up PCl_3 and Cl_2 to make PCl_5.

(c)   The initial concentrations are $[PCl_5]$ = 2.00 M, $[PCl_3]$ = 3.00 M, and $[Cl_2]$ = 1.00 M. Since the reaction shifts to the left, the signs for the C row are positive for the reactants and negative for the products.

	PCl_5(g)	$\rightleftharpoons$	PCl_3(g)	+	Cl_2(g)
I	2.00 M		3.00 M		1.00 M
C	+x		−x		−x
E	2.00 M + x		3.00 M − x		1.00 M − x

$$K_c = \frac{[PCl_3][Cl_2]}{[PCl_5]} = \frac{[3.00\ M - x][1.00\ M - x]}{[2.00\ M + x]} = 1.21 = 1.21\ M$$

$$3.00\ M^2 - 4.00\ M\,x + x^2 = (1.21\ M)(2.00\ M + x) = 2.42\ M^2 + 1.21\ M\,x$$

$$x^2 - 5.21\text{ M } x + 0.58\text{ M}^2 = 0, \quad x = \frac{5.21\text{ M} \pm \sqrt{(5.21\text{ M})^2 - 4(1)(0.58\text{ M}^2)}}{2(1)} = 0.11\text{ M or } 5.10\text{ M}$$

Although the quadratic formula gives you two values, only one of them (0.11 M) is valid. The value of 5.10 M is not possible because it would lead to negative equilibrium concentrations for PCl_3 and Cl_2. The equilibrium concentrations are $[PCl_5] = 2.00\text{ M} + x = 2.00\text{ M} + 0.11\text{ M} = 2.11\text{ M}$, $[PCl_3] = 3.00\text{ M} - x = 3.00\text{ M} - 0.11\text{ M} = 2.89\text{ M}$, and $[Cl_2] = 1.00\text{ M} - x = 1.00\text{ M} - 0.11\text{ M} = 0.89\text{ M}$.

$$\underline{\text{Check:}} \quad K_c = \frac{[PCl_3][Cl_2]}{[PCl_5]} = \frac{[2.89\text{ M}][0.89\text{ M}]}{[2.11\text{ M}]} = 1.22\ M = 1.22$$

(d)   Because the K_c value is close to 1, this reaction is not really product-favored or reactant-favored. You can see that in this reaction both the products and reactants are present at comparable concentrations.

6.   (a)   When H_2 is added, $[H_2]$ increases and so does Rate$_f$ (but Rate$_r$ stays the same). Since Rate$_f$ > Rate$_r$, more HBr is being made from H_2 and Br_2 (forward reaction) than is reacting in the reverse reaction, and the reaction is shifted to the right. Since Br_2 is reacting, the brown color will get lighter (fade).

        (b)   When HBr is added, [HBr] increases and Rate$_r$ increases (Rate$_f$ stays the same). Since Rate$_r$ > Rate$_f$, more H_2 and Br_2 are being made from HBr (reverse reaction) than is reacting in the forward reaction and the reaction is shifted to the left until Rate$_f$ = Rate$_r$. Since Br_2 is made, the color gets darker.

        (c)   Since there is the same number of gas particles on both sides of the equation, decreasing the volume will not shift the reaction. The decreased volume will result in more collisions, but since both of the reactions require one bimolecular reaction, the increased collisions affect both reactions in the same way. Even though the amount of Br_2 in the system has not changed, its concentration is now higher since it is now in a smaller volume and the color of the system would look darker brown.

        (d)   Since this is an exothermic reaction, heat is a product in the system. Adding heat by increasing the temperature will shift the reaction to the left since the 2 HBr will absorb some of this heat as they react. Since Br_2 is being made, the brown color in the system will get darker.

7.   (a)   When Cl^- ions are added, $[Cl^-]$ increases and so does Rate$_f$ (Rate$_r$ stays the same). Since Rate$_f$ > Rate$_r$, more $Co(H_2O)_6^{2+}$ react with Cl^- (forward reaction) than is produced from $CoCl_4^{2-}$ and H_2O (reverse reaction), and the reaction is shifted to the left until Rate$_f$ = Rate$_r$. Since the red ion is disappearing and the blue ion is being produced, the solution becomes more blue (less red).

        (b)   Normally, adding solids or liquids to an equilibrium system will not shift the reaction since they do not appear in the K_c expression. However, adding water to the system increases the system volume and this will cause a shift in the reaction. Increasing V decreases the number of collisions between $Co(H_2O)_6^{2+}$ and Cl^- ions, slowing Rate$_f$ dramatically (since it requires at least five bimolecular reactions); however, the reaction of $CoCl_4^{2-}$ and H_2O does change dramatically because $CoCl_4^{2-}$ can still collide with water molecules. This results in a shift in the reaction to the left, causing the solution to become less blue and more red.

        (c)   Since this reaction shifts to the right (more blue/less red) when the reaction is heated, heat must be used by $Co(H_2O)_6^{2+}$ and Cl^- when they react. That means that this reaction is endothermic.

8.   (a)   The K_c values do not change when reactant or product concentrations are changed.

        (b)   The K_c values do not change when the volume of a system is changed.

        (c)   The reaction in question 6 is exothermic, so heat is a product. Adding heat by increasing the temperature will shift the reaction to the left, making it more reactant-favored. So, the K_c value will be smaller. Since the reaction in question 7 is endothermic, the reverse is true and K_c will be larger.

        (d)   For the reaction in question 6, the number of gaseous molecules is the same on both sides of the equation. So, the dispersal of energy among reactants or products would be about the same, and ΔS° should be close to zero. For the reaction in question 7, the number of aqueous and liquid molecules is larger on the left side of the equation (seven molecules, versus five on the left side), so energy can be dispersed more efficiently among the products and ΔS° should be positive.

        (e)   At low T, ΔH° determines the favored direction of a reaction. So, reaction 6 will be product-favored at low T and reaction 7 will be reactant-favored at low T. At high T, ΔS° determines the favored direction of a reaction. Since ΔS° is close to zero for the reaction in question 6, it has little effect. So, ΔH° is still important and this reaction is likely to be product-favored at high T as well. Since ΔS° is positive for the reaction in question 7, that reaction is product-favored at high T.

ANSWERS — PRACTICE TEST

Chapter 15

1.   (a)   Since solid I_2 sank to the bottom of the container and the water did not turn purple, iodine is insoluble in water. Water is a polar substance but iodine (I—I, $\Delta EN = 0.0$) is nonpolar. Water molecules are attracted to each other by very strong hydrogen bonds, but iodine molecules are attracted to H_2O and other I_2 molecules by rather weak London forces. Since the water molecules are more attracted to each other (hydrogen bonds) than they are to the I_2 molecules (London forces), the water molecules prefer to interact with each other and will not mix with the I_2 molecules.

      (b)   Since the hexane solution turned bright purple, the purple solid I_2 must be soluble in hexane. Hexane and iodine are nonpolar, so hexane molecules attract other hexane and I_2 molecules by London forces; similarly, I_2 molecules are attracted to hexane and other I_2 molecules by London forces. Since all of these molecules are attracted to each other by London forces, the energy of the two pure samples and the solution would be about the same, and they will mix together because it is entropy-favored.

2.   (a)   Since the water solution turned pale green, the green solid $FeCl_2$ must be soluble in water. Water is a polar substance and $FeCl_2$ is ionic (Fe^{2+} and Cl^- ions). Water molecules are attracted to each other by strong hydrogen bonds, water molecules are attracted to the ions by strong ion-dipole forces, and the ions in $FeCl_2$ are attracted to each other by strong ion-ion forces. Since all of these particles are attracted to each other by strong forces, the energy of the two pure samples and the solution would be about the same, and they will mix together because it is entropy-favored.

      (b)   Since solid $FeCl_2$ sank to the bottom of the container and the hexane did not turn green, $FeCl_2$ is insoluble in hexane. $FeCl_2$ is ionic, but hexane is nonpolar. The Fe^{2+} and Cl^- ions are attracted to each other by very strong ion-ion forces, but hexane molecules are attracted to the Fe^{2+} and Cl^- ions and the other hexane molecules by rather weak London forces. Since the ions are more attracted to each other than they are to the hexane molecules, the Fe^{2+} and Cl^- ions prefer to interact with each other and will not mix with the hexane molecules, remaining as solid $FeCl_2$.

3.   (a)   Since the mixing of the molecules within the two liquids into a larger volume leads to a greater dispersal of energy, the $\Delta S°_{soln}$ value for mixing is positive.

      (b)   Since C_6H_6 is a nonpolar molecule with no net positive or negative charges, they will be weakly attracted to all molecules and ions (including H_2O molecules) by weak London forces.

      (c)   Since the polar water molecules are more attracted to each other than they are to the C_6H_6 molecules, the enthalpy of separating the water molecules is much larger than the energy released when the H_2O and C_6H_6 molecules are attracted to each other, and this reaction will be endothermic ($\Delta H°_{soln} > 0$).

      (d)   The positive $\Delta S°_{soln}$ value favors the mixing process, but the positive $\Delta H°_{soln}$ value does not favor the mixing process. Since you are told that the two liquids do not mix, the enthalpy factor must be more important (stronger) than the entropy factor.

4.   (a)   When the gas molecules are dissolved in liquid water, the gas molecules are forced into the much smaller liquid volume. This results in a negative $\Delta S°_{soln}$ value for both of the gases.

      (b)   Water is a very polar liquid but methane is nonpolar. Since the polar water molecules are more attracted to each other (by hydrogen bonding forces) than they are to the CH_4 molecules (by London forces), the enthalpy of separating the water molecules from each other is much larger than the energy released when the H_2O and CH_4 molecules are attracted to each other, and this reaction will be endothermic ($\Delta H°_{soln} > 0$). On the other hand, ammonia is a polar molecule that can undergo hydrogen bonding attractions (because it has H atoms attached to an N atom). Since water molecules are attracted to other H_2O molecules and the NH_3 molecules by hydrogen bonding forces (and the NH_3 molecules are attracted to both types of molecules by hydrogen bonds), you might expect $\Delta H°_{soln}$ to be close to zero. However, $\Delta H°_{soln}$ is actually negative (–34.1 kJ/mol).

      (c)   Since CH_4 has a positive $\Delta H°_{soln}$ value and a negative $\Delta S°_{soln}$ value, it will be insoluble in water at low temperatures and at high temperatures. Since NH_3 has a negative $\Delta H°_{soln}$ value and a negative $\Delta S°_{soln}$ value, it will be soluble in water at low temperatures but insoluble in water at high temperatures.

5. If the solid completely dissolves and there is none left at the bottom of the container, you can't be sure if the solution is saturated (because there was just enough solid present to make the solution saturated) or if it is unsaturated (because there wasn't enough solid present to make a saturated solution). On the other hand, if some solid remains undissolved after the solution is made, you can be reasonably sure that this solution is saturated—if more solid needed to dissolve to make the solution saturated, it would have dissolved instead of remaining undissolved.

6. The mass fraction is determined by multiplying by the ratio of 1 divided by 10^6 ppm; the mass percent can be found by multiplying the mass fraction by 100%.

$$mass\ fraction\ HCN = 3,404\ ppm \times \frac{1}{10^6\ ppm} = 0.003404$$

$$mass\ percent\ HCN = 0.003404 \times \frac{100\%}{1} = 0.3404$$

7. (a) First, you need to calculate the concentration of CO_2 by converting the mass of CO_2 into moles, and determining the solution volume. Since you were not told otherwise, you should assume that the density of the solution is the same as water (1.00 g/mL, and m_{soln} = 100.0 g water + 0.169 g CO_2 = 100.169 g soln). Then, use Henry's law to solve for the Henry's law constant k_H.

$$[CO_2] = \frac{n_{CO_2}}{V_{soln}} = \frac{0.169\ g\ CO_2 \times \dfrac{mol\ CO_2}{44.01\ g\ CO_2}}{100.169\ g\ soln \times \dfrac{mL}{1.00\ g\ soln} \times \dfrac{10^{-3}\ L}{1\ mL}} = \frac{0.00384\ mol\ CO_2}{0.100169\ L} = 0.0383\ M$$

$$[CO_2] = k_H P_{CO_2},\ k_H = \frac{[CO_2]}{P_{CO_2}} = \frac{0.0383\ M}{1.00\ atm \times \dfrac{760\ mm\ Hg}{1\ atm}} = 5.04 \times 10^{-5}\ \frac{M}{mm\ Hg}$$

(b) $$[CO_2] = k_H P_{CO_2} = \left(5.04 \times 10^{-5}\ \frac{M}{mm\ Hg}\right)(8,400\ mm\ Hg) = 0.42\ M.$$

(c) $$[CO_2] = k_H P_{CO_2},\ P_{CO_2} = \frac{[CO_2]}{k_H} = \frac{1.00\ M}{5.04 \times 10^{-5}\ \dfrac{M}{mm\ Hg}} \times \frac{1\ atm}{760\ mm\ Hg} = 26.1\ atm.$$

8. (a) In order to calculate the mass fraction and weight percent, you need to determine the mass of water in the soda (assuming that the mass of the solution is the same as the mass of the sugar and water).

$$m_{H_2O} = m_{soln} - m_{C_6H_{12}O_6} = \left(355\ mL \times 1.025\ \frac{g}{mL}\right) - 44.0\ g = 319.9\ g\ H_2O$$

$$mass\ fraction\ C_6H_{12}O_6 = \frac{m_{C_6H_{12}O_6}}{m_{total}} = \frac{44.0\ g\ C_6H_{12}O_6}{319.9\ g\ H_2O + 44.0\ g\ C_6H_{12}O_6} = 0.121 \times \frac{100\%}{1} = 12.1\%$$

(b) First, calculate the number of moles of sugar present using its mass and its molar mass. Remember that molarity is moles of solute per volume of solution (total volume), and that molality is moles of solute per kilogram of solvent (mass of the water only).

$$n_{C_6H_{12}O_6} = 44.0\ g\ C_6H_{12}O_6 \times \frac{mol\ C_6H_{12}O_6}{180.16\ g\ C_6H_{12}O_6} = 0.24423\ mol\ C_6H_{12}O_6\ (unrounded)$$

$$[C_6H_{12}O_6] = \frac{n_{C_6H_{12}O_6}}{V_{soln}} = \frac{0.24423\ mol\ C_6H_{12}O_6}{355\ mL \times \dfrac{10^{-3}\ L}{1\ mL}} = 0.688\ \frac{mol}{L} = 0.688\ M$$

$$m_{C_6H_{12}O_6} = \frac{n_{C_6H_{12}O_6}}{m_{solvent}} = \frac{0.24423 \text{ mol } C_6H_{12}O_6}{319.9 \text{ g } H_2O \times \dfrac{1 \text{ kg } H_2O}{10^3 \text{ g } H_2O}} = 0.764 \ \frac{\text{mol}}{\text{kg}} = 0.764 \text{ m}$$

(c)   First, calculate the mole fraction of the water, then use Raoult's law to find the solution vapor pressure.

$$n_{H_2O} = 319.9 \text{ g } H_2O \times \frac{\text{mol } H_2O}{18.02 \text{ g } H_2O} = 17.751 \text{ mol } H_2O \text{ (unrounded)}$$

$$X_{C_6H_{12}O_6} = \frac{n_{C_6H_{12}O_6}}{n_{total}} = \frac{17.751 \text{ mol } H_2O}{17.751 \text{ mol } H_2O + 0.24423 \text{ mol } C_6H_{12}O_6} = 0.9864 \text{ (unrounded)}$$

$$P_{soln} = X_{solvent}P = (0.9864)(433.6 \text{ mm Hg}) = 427.7 \text{ mm Hg}$$

(d)   Use the Δbp and Δfp formulas to determine the boiling and freezing points of this solution.

$$bp(\text{solution}) - bp\ (\text{solvent}) = K_b m_{solute} i_{solute} = \left(0.52 \ \frac{C}{m}\right)(0.764 \text{ m})(1) = 0.40 \text{ C}$$

$$bp(\text{solution}) = bp\ (\text{solvent}) + 0.40 \text{ C} = 100.0 \text{ C} + 0.40 \text{ C} = 100.4 \text{ C}$$

$$fp(\text{solution}) - fp\ (\text{solvent}) = -K_f m_{solute} i_{solute} = -\left(1.86 \ \frac{C}{m}\right)(0.764 \text{ m})(1) = -1.42 \text{ C}$$

$$fp(\text{solution}) = fp\ (\text{solvent}) - 1.42 \text{ C} = 0.0 \text{ C} - 1.42 \text{ C} = -1.4 \text{ C}$$

(e) $\Pi_{C_6H_{12}O_6} = [C_6H_{12}O_6]RTi_{solute} = \left(0.688 \ \dfrac{\text{mol}}{L}\right)\left(0.082057 \ \dfrac{L \text{ atm}}{\text{mol K}}\right)(310.15 \text{ K})(1) = 17.5 \text{ atm.}$

9.   (a) $\Delta fp = fp(\text{solution}) - fp\ (\text{solvent}) = 5.44 \text{ C} - 6.50 \text{ C} = -1.06 \text{ C}.$

(b)   Use the Δmp formula to solve for m_{solute} (remembering that $i_{solute} = 1$ for nonelectrolytes).

$$\Delta fp = -K_f m_{solute} i_{solute}, \ m_{solute} = \frac{\Delta fp}{-K_f i_{solute}} = \frac{5.44 \text{ C} - 6.50 \text{ C}}{-20.2 \ \dfrac{C}{m} \times 1} = 0.0525 \text{ m} = 0.0525 \ \frac{\text{mol S}}{\text{kg}}$$

(c)   Use the molality of the solution and the mass of the solvent to find the moles of sulfur present.

$$m_S = \frac{n_S}{m_{solvent}}, \ n_S = m_S m_{solvent} = \left(0.0525 \ \frac{\text{mol S}}{\text{kg } C_6H_{12}}\right)\left(152.00 \text{ g } C_6H_{12} \times \frac{1 \text{ kg } C_6H_{12}}{10^3 \text{ g } C_6H_{12}}\right) = 0.00798 \text{ mol S}$$

(d)   To determine the molar mass of sulfur, divide the mass of sulfur by the number of moles of sulfur.

$$M_S = \frac{m_S}{n_S} = \frac{2.05 \text{ g S}}{0.00798 \text{ mol S}} = 257 \ \frac{\text{g S}}{\text{mol S}}$$

$$n = \frac{M \text{ molecular}}{M \text{ empirical}} = \frac{257 \text{ g/mol}}{32.06 \text{ g/mol}} = 8.02 =. 8$$

So, the molecular formula of sulfur is S_8.

10.   Since boiling point elevation (and freezing point lowering) are colligative properties, you need to determine the total molality of molecules and ions in the solution. Pure water has a total molality of 0.0 m (nothing dissolved in it), 0.75 m Li_2SO_4 has 3×0.75 m = 2.25 m ions ($i_{solute} = 3$), 1.0 m CH_2O has 1.0 m molecules ($i_{solute} = 1$), 1.0 m $NaCH_3COO$ has 2×1.0 m = 2.0 m ions ($i_{solute} = 2$), 1.5 m $Al(NO_3)_3$ has 4×1.5 m = 6.0 m ions ($i_{solute} = 4$), and 2.25 m $C_{12}H_{22}O_{11}$ has 2.25 m molecules ($i_{solute} = 1$). The greater the number of particles dissolved in the water, the higher the boiling point will be. So, the solutions in order of increasing boiling points are (1) pure water, (2) 1.0 m CH_2O, (3) 1.0 m $NaCH_3COO$, (4) 0.75 m Li_2SO_4, (5) 2.25 m $C_{12}H_{22}O_{11}$, and (6) 1.5 m $Al(NO_3)_3$. Although the 0.75 m Li_2SO_4 and 2.25 m $C_{12}H_{22}O_{11}$ have the same number of dissolved particles, the 2.25 m $C_{12}H_{22}O_{11}$ probably has a higher boiling point since the i_{solute} for Li_2SO_4 is probably a little lower than 3. The freezing points of these solutions would follow the same trend, but the pure water would have the highest freezing point and 1.5 m $Al(NO_3)_3$ would have the lowest freezing point.

11. (a) Assume that you have exactly 1.00 L of this solution, and determine the mass of the solution, the mass of NH_4F, and then mass of C_2H_5OH. Then, calculate the mass fraction and ppm values.

$$m_{soln} = 1.00 \text{ L} \times \frac{1 \text{ mL}}{10^{-3} \text{ L}} \times 0.789 \frac{\text{g}}{\text{mL}} = 789 \text{ g}$$

$$m_{NH_4F} = 1.00 \text{ L solution} \times 0.0052 \frac{\text{mol } NH_4F}{\text{L solution}} \times \frac{37.04 \text{ g } NH_4F}{\text{mol } NH_4F} = 0.193 \text{ g } NH_4F$$

$$m_{C_2H_5OH} = m_{soln} - m_{NH_4F} = 789 \text{ g} - 0.193 \text{ g} = 788.807 \text{ g } C_2H_5OH$$

$$mass\ fraction\ NH_4F = \frac{m_{NH_4F}}{m_{total}} = \frac{0.193 \text{ g } NH_4F}{789 \text{ g solution}} = 2.45 \times 10^{-4} \times \frac{10^6 \text{ ppm}}{1} = 245 \text{ ppm}$$

(b) To calculate the molality, divide the moles of NH_4F by the mass of C_2H_5OH.

$$m_{NH_4F} = \frac{n_{NH_4F}}{m_{solvent}} = \frac{0.0052 \text{ mol } NH_4F}{788.807 \text{ g } C_2H_5OH \times \frac{1 \text{ kg } C_2H_5OH}{10^3 \text{ g } C_2H_5OH}} = 0.0066 \frac{\text{mol}}{\text{kg}} = 0.0066 \text{ m}$$

(c) $$\Pi_{NH_4F} = [NH_4F]RTi_{solute} = \left(0.0052 \frac{\text{mol}}{\text{L}}\right)\left(0.082057 \frac{\text{L atm}}{\text{mol K}}\right)(295.15 \text{ K})(2) = 0.25 \text{ atm}.$$

(d) In part c, you assumed that the NH_4^+ ions and the F^- ions are completely independent of each other, and that $i_{solute} = 2$. However, this leads to a theoretical value of the osmotic pressure that is too large. So, i_{solute} must be lower than 2, which means that some of the NH_4^+ ions and the F^- ions are staying together instead of moving independently (i.e., they are forming ion pairs in the solution).

12. Salty ocean water is placed in a container with a semi-permeable membrane. A pressure greater than the osmotic pressure of the solution is exerted on the solution, and pure water comes through the membrane, but very little (if any) Na^+ or Cl^- ions will pass through the membrane.

13. (a)

(b) The hydrophilic (ionic) ends of the soap molecules are very strongly attracted to each other (more than they are attracted to the nonpolar heptane molecules) and will cluster together in the soap-heptane solution. The hydrophobic (nonpolar) ends of the soap molecules are weakly attracted to each other and the heptane molecules, so the soap molecules will orient themselves so that their hydrophobic ends are in the heptane and their polar ends do not interact with the heptane molecules. They can form a monolayer (where the ionic ends are sticking out of the solution and the tails are in the heptane) or a micelle (where the soap molecules cluster together with the tails pointing into the heptane and the ionic heads clustering together in an ionic region).

14. (a) The 0.32 M Na_3AsO_4 solution has 0.96 M Na^+ ions and 0.32 M AsO_4^{3-} ions. After going through the cation-exchange resin, the solution has 0.96 M H_3O^+ ions and 0.32 M AsO_4^{3-} ions (0.32 M H_3AsO_4). After going through the anion-exchange resin, the solution has 0.75 M H_3O^+ ions and 0.75 M OH^- ions (the anion-exchange resin must replace each AsO_4^{3-} ion with 3 OH^- ions).

(b) The H_3O^+ ions and the OH^- ions will neutralize each other to make H_2O molecules. Therefore, all of the ions in the solution are now gone and the solution is 'deionized'.

(c) No, this solution is not guaranteed to be pure water. These ion-exchange resins remove cation and anion impurities from the water, but do not remove any neutral impurities that may be in the water sample.

ANSWERS — PRACTICE TEST

Chapter 16

1. (a) In the forward reaction, CH_3NH_2 gains an H^+ ion and H_2O loses an H^+ ion, so CH_3NH_2 is the base and H_2O is the acid. In the reverse reaction, $CH_3NH_3^+$ loses an H^+ ion and OH^- gains an H^+ ion, so $CH_3NH_3^+$ is the conjugate acid and OH^- is the conjugate base.

   (b) In the forward reaction, $H_2PO_4^-$ loses H^+ (acid) and CO_3^{2-} loses H^+ (base); in the reverse reaction, HCO_3^- loses H^+ (conjugate acid) and HPO_4^{2-} gains H^+ (conjugate base).

   (c) In the forward reaction, the $NH_3(g)$ molecule gains H^+ (base) and the $HCl(g)$ molecule loses H^+ (acid); in the reverse reaction, NH_4^+ in $NH_4Cl(s)$ loses H^+ (conjugate acid) and Cl^- gains H^+ (conjugate base).

   (d) In the forward reaction, one of the HSO_3^- gains H^+ (base) and the other HSO_3^- loses H^+ (acid); in the reverse reaction, H_2SO_3 loses H^+ (conjugate acid) and SO_3^{2-} gains H^+ (conjugate base).

2. (a) Compound I has no ionized molecules, so it is a nonacid (0% ionization). Compound II has all of its molecules ionized (100% ionization), which makes it a strong acid. In compound III, some of its molecules are ionized so it is an acid, but only 1 out of 5 (20%) are ionized; therefore, it is a weak acid.

   (b) Since compound II has the most free H^+ (H_3O^+) ions in solution, it will have the lowest pH value.

   (c) Compound II and III have the same total acid concentration, so they will react with equal amounts of OH^- ions; since compound I is not an acid, it will not react with any OH^- ions.

3. (a) A solution with pOH less than 7.00 would be basic (its pH is $14.00 - 2.43 = 11.57$).

   (b) A solution with $[OH^-] = 1.0 \times 10^{-7}$ M also has $[H_3O^+] = 1.0 \times 10^{-7}$ M. Since $[H_3O^+] = [OH^-]$, this solution would be neutral.

   (c) A solution that has $[H_3O^+] > 1.0 \times 10^{-7}$ M would be acidic (its pH is 4.15).

   (d) A solution with pH = 7.00 also has pOH = 7.00. Since pH = pOH, this solution would be neutral.

   (e) A solution that has $[H_3O^+] < 1.0 \times 10^{-7}$ M would be basic (its pH is 11.46).

   (f) A solution with pH less than 7.00 would be acidic.

4. Use the K_w formula to determine $[H_3O^+]$, and the pOH formula to find pOH. The pH value can be calculated using the pH-pOH relationship or by using the pH formula, and you should get the same value either way.

$$K_w = [H_3O^+][OH^-], \quad [H_3O^+] = \frac{K_w}{[OH^-]} = \frac{1.0 \times 10^{-14}\ M^2}{3.4 \times 10^{-5}\ M} = 2.9 \times 10^{-10}\ M$$

$$pOH = -\log[OH^-] = -\log[3.4 \times 10^{-5}] = 4.47$$

$$pH = -\log[H_3O^+] = -\log[2.9 \times 10^{-10}] = 9.53 = 14.00 - 4.47$$

5. (a) You are given K_a values, so they must be weak acids. Write the balanced equation for the acid reactions, showing the chemical and water as reactants and H_3O^+ and the conjugate base as products.

$$HOCl(aq) + H_2O(\ell) \rightleftharpoons H_3O^+(aq) + OCl^-(aq) \qquad K_a = \frac{[H_3O^+][OCl^-]}{[HOCl]}$$

$$HOBr(aq) + H_2O(\ell) \rightleftharpoons H_3O^+(aq) + OBr^-(aq) \qquad K_a = \frac{[H_3O^+][OBr^-]}{[HOBr]}$$

   (b) Since $K_a(HOCl) > K_a(HOBr)$, HOCl is a stronger acid than HOBr (the reaction above is more product-favored for HOCl than HOBr). Therefore, the pH of the HOCl solution should be lower.

   (c) For 0.0015 M HOCl, the initial concentrations are [HOCl] = 0.0015 M, and $[H_3O^+] = [OCl^-] = 0.0$ M. Some HOCl and H_2O will react, producing H_3O^+ and OCl^- ions.

	$HOCl(aq)$	$+$	$H_2O(\ell)$	$\rightleftharpoons$	$H_3O^+(aq)$	$+$	$OCl^-(aq)$
I	0.0015 M				0.0000 M		0.0000 M
C	$-x$				$+x$		$+x$
E	$0.0015\ M - x$				x		x

Since K_a is very small, very little HOCl will ionize and $0.0015\text{ M} - x \approx 0.0015\text{ M}$.

$$K_a = \frac{[H_3O^+][OCl^-]}{[HOCl]} = \frac{[x][x]}{[0.0015\text{ M} - x]} = \frac{x^2}{0.0015\text{ M} - x} \approx \frac{x^2}{0.0015\text{ M}}$$

$$\frac{x^2}{0.0015\text{ M}} = 6.8 \times 10^{-8}\ M, \quad x^2 = (0.0015\text{ M})(6.8 \times 10^{-8}\text{ M}) = 1.0 \times 10^{-10}\text{ M}^2$$

$$x = \sqrt{1.0 \times 10^{-10}\text{ M}^2} = 1.0 \times 10^{-5}\text{ M} = [H_3O^+]$$

$$\text{pH} = -\log[H_3O^+] = -\log[1.0 \times 10^{-5}] = 5.00$$

5% rule: $\dfrac{x}{0.0015\text{ M}} = \dfrac{1.0 \times 10^{-5}\text{ M}}{0.0015\text{ M}} \times 100\% = 0.67\% < 5\%$ (good)

	HOBr(aq)	+	H₂O(ℓ)	⇌	H₃O⁺(aq)	+	OBr⁻(aq)
I	0.0015 M				0.0000 M		0.0000 M
C	$-x$				$+x$		$+x$
E	0.0015 M $- x$				x		x

Since K_a is very small, very little HOBr will ionize and $0.0015\text{ M} - x \approx 0.0015\text{ M}$.

$$K_a = \frac{[H_3O^+][OBr^-]}{[HOBr]} = \frac{[x][x]}{[0.0015\text{ M} - x]} = \frac{x^2}{0.0015\text{ M} - x} \approx \frac{x^2}{0.0015\text{ M}}$$

$$\frac{x^2}{0.0015\text{ M}} = 2.5 \times 10^{-9}\ M, \quad x^2 = (0.0015\text{ M})(2.5 \times 10^{-9}\text{ M}) = 3.8 \times 10^{-12}\text{ M}^2$$

$$x = \sqrt{3.8 \times 10^{-12}\text{ M}^2} = 1.9 \times 10^{-6}\text{ M} = [H_3O^+]$$

$$\text{pH} = -\log[H_3O^+] = -\log[1.9 \times 10^{-6}] = 5.71$$

5% rule: $\dfrac{x}{0.0075\text{ M}} = \dfrac{1.9 \times 10^{-6}\text{ M}}{0.0015\text{ M}} \times 100\% = 0.13\% < 5\%$ (good)

(d) The percent ionization for each acid is the same as the 5% rule, so 0.67% of the HOCl molecules ionize and 0.13% of the HOBr molecules ionize.

(e) Because Cl has a higher electronegativity than Br (2.7 and 2.6, respectively), Cl pulls electrons away from the O and H atoms better than Br does, so the H atom in HOCl is slightly more positive (and therefore more acidic) than the H atom in HOBr.

6. (a) The reactants are the basic form of alanine and H_3O^+, and one of the products is water.

N atom: $NH_2CH(CH_3)COO^-(aq) + H_3O^+(aq) \rightleftharpoons H_2O(\ell) + NH_3^+CH(CH_3)COO^-(aq)$

O atom: $NH_2CH(CH_3)COO^-(aq) + H_3O^+(aq) \rightleftharpoons H_2O(\ell) + NH_2CH(CH_3)COOH(aq)$

(b) The K values for the reactions above are $K_b/K_w = 1/K_a$. The K value for the first reaction (7.4×10^9) is larger than the K value for the second reaction (1.0×10^4), so the first reaction is more likely to occur. This is consistent with the idea that 'neutral' amino acids exist as zwitterions.

7. (a) Since the solution is clearly acidic, H_3O^+ must be one of the products.

$$HSeO_4^-(aq) + H_2O(\ell) \rightleftharpoons H_3O^+(aq) + SeO_4^{2-}(aq) \qquad K_a = \frac{[H_3O^+][SeO_4^{2-}]}{[HSeO_4^-]}$$

(b) Set up the *ICE* table for this acid-base reaction.

	HSeO₄⁻(aq)	+	H₂O(ℓ)	⇌	H₃O⁺(aq)	+	SeO₄²⁻(aq)
I	0.20 M				0.000 M		0.000 M
C	$-x$				$+x$		$+x$
E	0.20 M $- x$				x		x

You can calculate the value of $[H_3O^+] = x$ from the pH of the solution.

$$[H_3O^+] = 10^{-pH} \text{ M} = 10^{-1.26} \text{ M} = 0.055 \text{ M}$$

$$K_a = \frac{[H_3O^+][OBr^-]}{[HOBr]} = \frac{[x][x]}{[0.20 \text{ M} - x]} = \frac{[0.055 \text{ M}][0.055 \text{ M}]}{[0.20 \text{ M} - 0.055 \text{ M}]} = 0.021 \ M = 0.021$$

(c)   A strong acid that has a concentration of 0.20 M should have a pH $= -\log[0.20] = 0.70$. Since this solution has a pH that is higher than 0.70, $HSeO_4^-$ is a weak acid. The fact that the K_a value is less than 1 also suggests that this reaction is reactant-favored (i.e., a weak acid). Finally, you can calculate the percent ionization of the acid—if it is 100%, the acid is strong; if it is less than 100%, it is a weak acid.

$$\text{% ionization:} \quad \frac{0.055 \text{ M}}{0.20 \text{ M}} \times 100\% = 27\%$$

8.   (a)   Since ammonia can also be called ammonium hydroxide, ammonia must be a base.

$$NH_3(aq) + H_2O(\ell) \rightleftharpoons NH_4^+(aq) + OH^-(aq) \qquad K_b = \frac{[NH_4^+][OH^-]}{[NH_3]}$$

(b)   Because this reaction has OH^- as a product, this is a K_b reaction (the K_b reaction for NH_3, a weak base). Looking in Table 16.2 of the textbook (p. 772), $K_b(NH_3) = 1.7 \times 10^{-5}$. Since this value is less than 1, this reaction is reactant-favored (mostly NH_3 and H_2O, very little NH_4^+ and OH^-).

(c)   For 14.8 M NH_3, the initial concentrations are $[NH_3] = 14.8$ M, and $[NH_4^+] = [OH^-] = 0.0$ M. Some NH_3 and H_2O will react, producing NH_4^+ and OH^- ions.

	$NH_3(aq)$	$+$	$H_2O(\ell)$	$\rightleftharpoons$	$NH_4^+(aq)$	$+$	$OH^-(aq)$
I	14.8 M				0.0 M		0.0 M
C	$-x$				$+x$		$+x$
E	14.8 M $- x$				x		x

Since K_b is very small, very little NH_3 will ionize and 14.8 M $- x \approx 14.8$ M.

$$K_b = \frac{[NH_4^+][OH^-]}{[NH_3]} = \frac{[x][x]}{[14.8 \text{ M} - x]} = \frac{x^2}{14.8 \text{ M} - x} \approx \frac{x^2}{14.8 \text{ M}}$$

$$\frac{x^2}{14.8 \text{ M}} = 1.7 \times 10^{-5} \ M, \ x^2 = (14.8 \text{ M})(1.7 \times 10^{-5} \ M) = 2.5 \times 10^{-4} \text{ M}^2$$

$$x = \sqrt{2.5 \times 10^{-4} \text{ M}^2} = 0.016 \text{ M} = [OH^-]$$

$$pOH = -\log[OH^-] = -\log[0.016] = 1.80, \ pH = 14.00 - pOH = 14.00 - 1.80 = 12.20$$

$$\underline{\text{5% rule:}} \quad \frac{x}{14.8 \text{ M}} = \frac{0.016 \text{ M}}{14.8 \text{ M}} \times 100\% = 0.11\% \ < \ 5\% \text{ (good)}$$

(d)   The percent ionization is the same as the 5% rule, so 0.11% of the NH_3 molecules ionize.

(e)   There are at least two ways to determine whether reaction is mostly NH_3 and H_2O or NH_4^+ and OH^-. The fact that the K_b value is less than 1 suggests that this reaction is reactant-favored (a weak base) that exists mostly as its neutral base form (NH_3). Also, the percent ionization of NH_3 is only 0.11%, so 99.89% of the NH_3 molecules are unionized (mostly neutral NH_3).

9.   (a)   For the first reaction, write the solid $Al(NO_3)_3$ dissociating into Al^{3+} and NO_3^- ions, then write the hydrolysis reactions for these ions; $Al(H_2O)_6^{3+}$ will make the solution acidic but NO_3^- will not react with water (it is the anion of a strong acid), so this solution will be acidic.

$$Al(NO_3)_3(s) \rightarrow Al^{3+}(aq) + 3 \ NO_3^-(aq)$$

$$Al(H_2O)_6^{3+}(aq) + H_2O(\ell) \rightleftharpoons H_3O^+(aq) + Al(H_2O)_5(OH)^{2+}(aq)$$

$$NO_3^-(aq) + H_2O(\ell) \rightarrow N.R.$$

(b)   First, write the reaction of solid $CaBr_2$ dissociating into Ca^{2+} and Br^- ions, then write the hydrolysis reactions for these ions; Ca^{2+} will not react with water (it is the cation of a strong base) and Br^- will not react with water (it is the anion of a strong acid), so this solution will be neutral.

$$CaBr_2(s) \rightarrow Ca^{2+}(aq) + 2\ Br^-(aq)$$

$$Ca^{2+}(aq) + H_2O(\ell) \rightarrow N.R.$$

$$Br^-(aq) + H_2O(\ell) \rightarrow N.R.$$

(c)   First, write the reaction of solid $Fe(CH_3COO)_3$ dissociating into Fe^{3+} and CH_3COO^- ions, then write the hydrolysis reactions for these ions; since neither of these ions are the salts of strong acids or bases, they will both react with water—$Fe(H_2O)_6^{3+}$ will react with water, making it acidic, and CH_3COO^- will react with water, making it basic. To determine which effect is stronger, you need to compare the K_c values. The second reaction is $K_a(Fe(H_2O)_6^{3+}) = 6.3 \times 10^{-3}$, and the third reaction is $K_b(CH_3COO^-) = 5.6 \times 10^{-10}$. Since $K_a > K_b$, the second reaction is more important and the solution will be acidic.

$$Fe(CH_3COO)_3(s) \rightarrow Fe^{3+}(aq) + 3\ CH_3COO^-(aq)$$

$$Fe(H_2O)_6^{3+}(aq) + H_2O(\ell) \rightleftharpoons H_3O^+(aq) + Fe(H_2O)_5(OH)^{2+}(aq)$$

$$CH_3COO^-(aq) + H_2O(\ell) \rightleftharpoons CH_3COOH(aq) + OH^-(aq)$$

(d)   First, write the reaction of solid $Mg(CN)_2$ dissociating into Mg^{2+} and CN^- ions, then write the hydrolysis reactions for these ions; Mg^{2+} will not react with water (it is the cation of a strong base), but CN^- will react with water, making it basic. So, this solution will be basic.

$$Mg(CN)_2(s) \rightarrow Mg^{2+}(aq) + 2\ CN^-(aq)$$

$$Mg^{2+}(aq) + H_2O(\ell) \rightarrow N.R.$$

$$CN^-(aq) + H_2O(\ell) \rightleftharpoons HCN(aq) + OH^-(aq)$$

10.   Using the answer and third reaction from question 9d above, this solution will be basic. You were not given $K_b(CN^-)$, but you can calculate it from K_w and the K_a value for its conjugate acid, $K_a(HCN)$. Also, note that the initial concentration of CN^- is 0.076 M (twice the value of the initial $Mg(CN)_2$ concentration).

$$K_a(HCN)K_b(CN^-) = K_w,\ K_b(CN^-) = \frac{K_w}{K_a(HCN)} = \frac{1.0 \times 10^{-14}}{3.3 \times 10^{-10}} = 3.0 \times 10^{-5}$$

	$CN^-(aq)$	+	$H_2O(\ell)$	$\rightleftharpoons$	$HCN(aq)$	+	$OH^-(aq)$
I	0.076 M				0.000 M		0.000 M
C	$-x$				$+x$		$+x$
E	0.076 M $- x$				x		x

Since K_b is very small, very little N_3^- will react with water and 0.076 M $- x \approx 0.076$ M.

$$K_b = \frac{[HCN][OH^-]}{[CN^-]} = \frac{[x][x]}{[0.076\ M - x]} = \frac{x^2}{0.076\ M - x} \approx \frac{x^2}{0.076\ M}$$

$$\frac{x^2}{0.076\ M} = 3.0 \times 10^{-5}\ M,\ x^2 = (0.076\ M)(3.0 \times 10^{-6}\ M) = 2.3 \times 10^{-6}\ M^2$$

$$x = \sqrt{2.3 \times 10^{-6}\ M^2} = 1.5 \times 10^{-3}\ M = [OH^-]$$

$$pOH = -\log[OH^-] = -\log[1.5 \times 10^{-3}] = 2.82,\ pH = 14.00 - pOH = 14.00 - 2.82 = 11.18$$

$$\underline{5\%\ \text{rule}}:\ \frac{x}{0.076\ M} = \frac{1.5 \times 10^{-3}\ M}{0.076\ M} \times 100\% = 2.0\% < 5\%\ (\text{good})$$

11.   (a)   Since the B atom in BH_3 ends with one more bond, it must have accepted a pair of electrons and is the Lewis acid. F^- ends with one less lone pair, so it must have donated an electron pair (Lewis base).

(b)   Since each C atom in CO ends with one less lone pair, each CO molecule must have donated an electron pair to Ni, making them the Lewis bases. Since the Ni atom ends with four more bonds, it gained an electron pair from each CO molecule, making it the Lewis acid.

ANSWERS — PRACTICE TEST

Chapter 17

1. Acid-base buffers must have a conjugate acid present to react with any added OH^- ions and must have a conjugate base present to react with any added H_3O^+ ions. Buffers with $[A^-]/[HA]$ ratios greater than 10.0 have a lot of A^- ions but very few HA molecules present. These buffers would be very effective at neutralizing added H_3O^+ ions but would fail once a few OH^- ions are added. Similarly, buffers with $[A^-]/[HA]$ ratios less than 0.10 have a lot of HA molecules but very few A^- ions present. These buffers would be effective at neutralizing added OH^- ions but would fail once a few H_3O^+ ions are added.

2. (a) The pK_a values are calculated as the negative logarithm of K_a. Buffers are most effective when the target pH is within 1 pH unit of the pK_a value for the conjugate acid. Since the pK_a of HCN is the only value within 1 pH unit of the target pH (9.25), HCN/CN^- is the only effective buffer system present.

$$pK_a = -\log K_a; \quad pK_a(HF) = -\log(6.8 \times 10^{-4}) = 3.17,$$

$$pK_a(HOCl) = -\log(6.8 \times 10^{-8}) = 7.17, \quad pK_a(HCN) = -\log(3.3 \times 10^{-10}) = 9.48$$

(b) Using the generic K_a expression, the ratio of $[A^-]/[HA] = K_a/[H_3O^+]$. This ratio is 1.2 million for HF (1.2 million F^- ions for every HF molecule), 120 for HOCl (120 OCl^- ions for every HOCl molecule), and 0.59 (0.59 CN^- ions for every HCN molecule). Both the HF/F^- buffer and $HOCl/OCl^-$ buffer would have so few HA molecules that these buffers would not be effective against added OH^- ions.

$$K_a = \frac{[H_3O^+][A^-]}{[HA]}, \quad \frac{[A^-]}{[HA]} = \frac{K_a}{[H_3O^+]}; \quad [H_3O^+] = 10^{-pH} \, M = 10^{-9.25} \, M = 5.6 \times 10^{-10} \, M$$

$$\frac{[F^-]}{[HF]} = \frac{6.8 \times 10^{-4} \, M}{5.6 \times 10^{-10} \, M} = 1.2 \times 10^6, \quad \frac{[OCl^-]}{[HOCl]} = \frac{6.8 \times 10^{-8} \, M}{5.6 \times 10^{-10} \, M} = 120, \quad \frac{[CN^-]}{[HCN]} = \frac{3.3 \times 10^{-10} \, M}{5.6 \times 10^{-10} \, M} = 0.59$$

(c) Since you want to make 50.0 mL of a buffer solution, take 50.0 mL of the 0.10 M HCN solution and dissolve solid KCN in this solution. Now, all you have to do is perform a stoichiometric calculation to determine how much solid KCN to add to the 50.0 mL (0.0500 L) of 0.10 M HCN.

$$\frac{[CN^-]}{[HCN]} = 0.59, \quad [CN^-] = 0.59 \times [HCN] = 0.59 \times 0.10 \, M = 0.059 \, M$$

$$m_{KCN} = 0.0500 \, \text{L solution} \times \frac{0.059 \, \text{mol KCN}}{\text{L solution}} \times \frac{65.12 \, \text{g KCN}}{\text{mol KCN}} = 0.19 \, \text{g KCN}$$

3. (a) The H_3O^+ ions from the added HCl solution react with the conjugate base of the buffer (CN^- ions).

$$CN^-(aq) + H_3O^+(aq) \rightarrow HCN(aq) + H_2O(\ell)$$

(b) First, you need to determine the number of moles of HCN, CN^-, and H_3O^+ ions before the reaction occurs. Then, recognize that CN^- will completely react with H_3O^+ to make HCN (decreasing $[CN^-]$ and increasing [HCN]). Finally, calculate the new $[H_3O^+]$ from the K_a expression.

$$n_{H_3O^+} = 0.0020 \, \text{L solution} \times \frac{0.10 \, \text{mol HCl}}{\text{L solution}} \times \frac{1 \, \text{mol } H_3O^+}{1 \, \text{mol HCl}} = 0.00020 \, \text{mol } H_3O^+$$

$$n_{CN^-} = 0.0500 \, \text{L solution} \times \frac{0.059 \, \text{mol KCN}}{\text{L solution}} \times \frac{1 \, \text{mol } CN^-}{1 \, \text{mol KCN}} = 0.00295 \, \text{mol } CN^-$$

$$n_{HCN} = 0.0500 \, \text{L solution} \times \frac{0.10 \, \text{mol HCN}}{\text{L solution}} = 0.00500 \, \text{mol HCN}$$

$$[CN^-] = \frac{(0.00295 \, \text{mol } CN^- - 0.00020 \, \text{mol } CN^-)}{0.0500 \, \text{L solution} + 0.0020 \, \text{L solution}} = 0.0528 \, M$$

$$[HCN] = \frac{(0.005 \, \text{mol HCN} + 0.00020 \, \text{mol HCN})}{0.0500 \, \text{L solution} + 0.0020 \, \text{L solution}} = 0.100 \, M$$

A 48

$$K_a = \frac{[H_3O^+][CN^-]}{[HCN]}, \ [H_3O^+] = K_a \times \frac{[HCN]}{[CN^-]} = 3.3 \times 10^{-10} \ M \times \frac{[0.100 \ M]}{[0.0528 \ M]} = 6.2 \times 10^{-10} \ M$$

$$pH = -\log[H_3O^+] = -\log[6.2 \times 10^{-10}] = 9.20, \ \Delta pH = 9.20 - 9.25 = -0.05$$

(c) The amount (n) of H_3O^+ added was calculated above (0.00020 mol H_3O^+). The amount of OH^- in a pH = 9.25 solution is determined below. Since there is more H_3O^+ than OH^-, the OH^- will be neutralized and the pH will be defined by $[H_3O^+]$.

$$K_w = [H_3O^+][OH^-], \ [OH^-] = \frac{K_w}{[H_3O^+]} = \frac{1.0 \times 10^{-14} \ M^2}{5.6 \times 10^{-10} \ M} = 1.8 \times 10^{-5} \ M$$

$$n_{OH^-} = 0.0500 \ L \ solution \times \frac{1.8 \times 10^{-5} \ mol \ OH^-}{L \ solution} = 8.9 \times 10^{-7} \ mol \ OH^-$$

$$[H_3O^+] = \frac{(2.0 \times 10^{-4} \ mol \ H_3O^+ - 8.9 \times 10^{-7} \ mol \ H_3O^+)}{0.0500 \ L \ solution + 0.0020 \ L \ solution} = 0.00383 \ M$$

$$pH = -\log[H_3O^+] = -\log[0.00383] = 2.42, \ \Delta pH = 2.42 - 9.25 = -6.83$$

4. The pH at the equivalence point in this curve is 7.0, which suggests that this is the titration curve of a strong acid with a strong base. This curve looks different than the other strong acid/strong base titration curve presented earlier in this book because the strong acid and strong base in this titration curve are much less concentrated (the initial pH of 4.0 suggests that the strong acid had an initial concentration of 0.00010 M).

5. Before you can calculate the pH of the solution, you need to know the amounts (n) of HOBr and KOH present at each part. The amount of HOBr is the same in each question; the amount of KOH depends on its volume.

$$n_{HOBr} = 0.0300 \ L \ solution \times \frac{0.200 \ mol \ HOBr}{L \ solution} = 0.00600 \ mol \ HOBr$$

$$n_{OH^-}(23) = 0.0230 \ L \ solution \times \frac{0.200 \ mol \ OH^-}{L \ solution} = 0.00460 \ mol \ OH^-$$

$$n_{OH^-}(30) = 0.0300 \ L \ solution \times \frac{0.200 \ mol \ OH^-}{L \ solution} = 0.00600 \ mol \ OH^-$$

$$n_{OH^-}(42) = 0.0420 \ L \ solution \times \frac{0.200 \ mol \ OH^-}{L \ solution} = 0.00840 \ mol \ OH^-$$

(a) For this question, you need to find the pH of a 0.200 M HOBr solution using the K_a expression.

	HOBr(aq)	+	$H_2O(\ell)$	$\rightleftharpoons$	H_3O^+(aq)	+	OBr^-(aq)
I	0.200 M				0.000 M		0.000 M
C	$-x$				$+x$		$+x$
E	0.200 M $- x$				x		x

Since K_a is very small, very little HOBr will ionize and 0.200 M $- x \approx$ 0.200 M.

$$K_a = \frac{[H_3O^+][OBr^-]}{[HOBr]} = \frac{[x][x]}{[0.200 \ M - x]} = \frac{x^2}{0.200 \ M - x} \approx \frac{x^2}{0.200 \ M}$$

$$\frac{x^2}{0.200 \ M} = 2.5 \times 10^{-9} \ M, \ x^2 = (0.200 \ M)(2.5 \times 10^{-9} \ M) = 5.0 \times 10^{-10} \ M^2$$

$$x = \sqrt{5.0 \times 10^{-10} \ M^2} = 2.2 \times 10^{-5} \ M = [H_3O^+], \ pH = -\log[H_3O^+] = -\log[2.2 \times 10^{-5}] = 4.65$$

$$\underline{5\% \ rule:} \ \frac{x}{0.0075 \ M} = \frac{2.2 \times 10^{-5} \ M}{0.200 \ M} \times 100\% = 0.011\% < 5\% \ (good)$$

(b) Since the amount of HOBr present is more than the amount of OH^-, all of the OH^- ions react with the HOBr, producing OBr^- ions. Then, you can solve for $[H_3O^+]$ and pH using the K_a expression.

$$[HOBr] = \frac{(0.00600 \text{ mol HOBr} - 0.00460 \text{ mol HOBr})}{0.0300 \text{ L solution} + 0.0230 \text{ L solution}} = 0.02642 \text{ M}$$

$$[OBr^-] = \frac{0.004600 \text{ mol OBr}^-}{0.0300 \text{ L solution} + 0.0230 \text{ L solution}} = 0.08679 \text{ M}$$

$$K_a = \frac{[H_3O^+][OBr^-]}{[HOBr]}, \ [H_3O^+] = K_a \times \frac{[HOBr]}{[OBr^-]} = 2.5 \times 10^{-9} \ M \times \frac{0.02642 \text{ M}}{0.08679 \text{ M}} = 7.6 \times 10^{-10} \text{ M}$$

$$pH = -\log[H_3O^+] = -\log[7.6 \times 10^{-10}] = 9.12$$

(c) Since the amount of HOBr is the same as the amount of OH^-, the HOBr is converted into OBr^- ions. So, you need to find the pH of a 0.100 M OBr^- solution using K_b for OBr^- (the concentration of OBr^- is halved since the volume of the solution has doubled from 30.0 mL to 60.0 mL).

$$K_a(HOBr)K_b(OBr^-) = K_w, \ K_b(OBr^-) = \frac{K_w}{K_a(HOBr)} = \frac{1.0 \times 10^{-14}}{2.5 \times 10^{-9}} = 4.0 \times 10^{-6}$$

	$OBr^-(aq)$	+	$H_2O(\ell)$	$\rightleftharpoons$	$HOBr(aq)$	+	$OH^-(aq)$
I	0.100 M				0.000 M		0.000 M
C	$-x$				$+x$		$+x$
E	0.100 M $- x$				x		x

Since K_b is very small, very little OBr^- will react with water and 0.100 M $- x \approx 0.100$ M.

$$K_b = \frac{[HOBr][OH^-]}{[OBr^-]} = \frac{[x][x]}{[0.100 \text{ M} - x]} = \frac{x^2}{0.100 \text{ M} - x} \approx \frac{x^2}{0.100 \text{ M}}$$

$$\frac{x^2}{0.100 \text{ M}} = 4.0 \times 10^{-6} \ M, \ x^2 = (0.100 \text{ M})(4.0 \times 10^{-6} \ M) = 4.0 \times 10^{-7} \ M^2$$

$$x = \sqrt{4.0 \times 10^{-7} \ M^2} = 6.3 \times 10^{-4} \text{ M} = [OH^-]$$

$$pOH = -\log[OH^-] = -\log[6.3 \times 10^{-4}] = 3.20, \ pH = 14.00 - pOH = 14.00 - 3.20 = 10.80$$

$$\underline{5\% \text{ rule:}} \ \frac{x}{0.100 \text{ M}} = \frac{6.3 \times 10^{-4} \text{ M}}{0.100 \text{ M}} \times 100\% = 0.63\% < 5\% \text{ (good)}$$

(d) Since the amount of HOBr is the less than the amount of OH^-, the HOBr has been completely converted into OBr^- ions and the unreacted OH^- ions will determine the pH of the solution.

$$[OH^-] = \frac{(0.00840 \text{ mol OH}^- - 0.00600 \text{ mol OH}^-)}{0.0300 \text{ L solution} + 0.0420 \text{ L solution}} = 0.0333 \text{ M}$$

$$pOH = -\log[OH^-] = -\log[0.0333] = 1.48, \ pH = 14.00 - pOH = 14.00 - 1.48 = 12.52$$

6. (a) The ions in K_{sp} are barium ions (Ba^{2+}) and phosphate ions (PO_4^{3-}).

$$Ba_3(PO_4)_2(s) \rightleftharpoons 3 Ba^{2+}(aq) + 2 PO_4^{3-}(aq) \qquad K_{sp} = [Ba^{2+}]^3[PO_4^{3-}]^2$$

(b) The mass solubility is the mass of solid divided by the solution volume; the molar solubility is simply the mass solubility divided by the molar mass of barium phosphate ($M = 601.8$ g/mol).

$$\frac{0.0059 \text{ g Ba}_3(PO_4)_2}{15.0 \text{ L}} = 3.9 \times 10^{-4} \frac{\text{g Ba}_3(PO_4)_2}{\text{L}} \times \frac{\text{mol Ba}_3(PO_4)_2}{601.8 \text{ g Ba}_3(PO_4)_2} = 6.5 \times 10^{-7} \text{ M}$$

(c) Set up the *ICE* table for this reaction, recognizing that the molar solubility is $x = 6.5 \times 10^{-7}$ M.

	$Ba_3(PO_4)_2(s)$	$\rightleftharpoons$	$3 Ca^{2+}(aq)$	+	$2 PO_4^{3-}(aq)$
I			0.00 M		0.00 M
C			$+3x$		$+2x$
E			$3x$		$2x$

$$K_{sp} = [Ba^{2+}]^3[PO_4^{3-}]^2 = [3x]^3[2x]^2 = 108x^5 = 108[6.5 \times 10^{-7}\ M]^5 = 1.2 \times 10^{-29}\ M^5 = 1.3 \times 10^{-29}$$

7. (a) The balanced equation and K_{sp} expression for this reaction is:

$$SrF_2(s) \rightleftharpoons Sr^{2+}(aq) + 2\ F^-(aq) \qquad\qquad K_{sp} = [Sr^{2+}][F^-]^2$$

$SrF_2(s)$	$\rightleftharpoons$	$Sr^{2+}(aq)$	+	$2\ F^-(aq)$
I		0.00 M		0.00 M
C		+x		+2x
E		x		2x

$$K_{sp} = [Sr^{2+}][F^-]^2 = [x][2x]^2 = 4x^3 = 2.8 \times 10^{-9}\ M^3$$

$$x^3 = \frac{K_{sp}}{4},\ x = \sqrt[3]{\frac{K_{sp}}{4}} = \sqrt[3]{\frac{2.8 \times 10^{-9}\ M^3}{4}} = 8.9 \times 10^{-4}\ M$$

(b) The initial concentration of the Sr^{2+} ion is 0.25 M. Since x is very small, assume $0.25\ M + x \approx 0.25\ M$.

$SrF_2(s)$	$\rightleftharpoons$	$Sr^{2+}(aq)$	+	$2\ F^-(aq)$
I		0.25 M		0.00 M
C		+x		+2x
E		0.25 M + x		2x

$$K_{sp} = [Sr^{2+}][F^-]^2 = [0.25\ M + x][2x]^2 = [0.25\ M + x][4x^2] \approx [0.25\ M][4x^2] = 1.00\ M\,x^2$$

$$x^2 = \frac{K_{sp}}{1.00\ M},\ x = \sqrt{\frac{K_{sp}}{1.00\ M}} = \sqrt{\frac{2.8 \times 10^{-9}\ M^3}{1.00\ M}} = 5.3 \times 10^{-5}\ M$$

5% rule: $\dfrac{x}{0.25\ M} = \dfrac{5.3 \times 10^{-5}\ M}{0.25\ M} \times 100\% = 0.021\% < 5\%$ (good)

(c) The initial concentration of the F^- ion is 0.25 M. Since x is very small, assume $0.25\ M + 2x \approx 0.25\ M$.

$SrF_2(s)$	$\rightleftharpoons$	$Sr^{2+}(aq)$	+	$2\ F^-(aq)$
I		0.00 M		0.25 M
C		+x		+2x
E		x		0.25 M + 2x

$$K_{sp} = [Sr^{2+}][F^-]^2 = [x][0.25\ M + 2x]^2 \approx x[0.25\ M]^2 = 0.0625\ M^2\,x$$

$$x = \frac{2.8 \times 10^{-9}\ M^3}{0.0625\ M^2} = 4.5 \times 10^{-8}\ M$$

5% rule: $\dfrac{2x}{0.25\ M} = \dfrac{2(4.5 \times 10^{-8}\ M)}{0.25\ M} \times 100\% = 3.6 \times 10^{-5}\% < 5\%$ (good)

8. Because the precipitation of solid strontium fluoride from the solution (the reverse reaction of the K_{sp} equation) requires the reaction of one Sr^{2+} ion and two F^- ions, the addition of F^- ions affects the solubility of the solid much more than adding Sr^{2+} ions does. This greater effect can also be seen in the K_{sp} equation since the F^- concentration is squared but the Sr^{2+} concentration is not.

9. (a) To determine whether precipitates will form, calculate Q for each solid and compare them to K_{sp}.

$$Q(AgI) = (conc.\ Ag^+)(conc.\ I^-) = (3.0 \times 10^{-6}\ M)(1.7 \times 10^{-6}\ M) = 5.1 \times 10^{-12}\ M^2 > K_{sp}(AgI)$$

$$Q(AuI_3) = (conc.\ Au^{3+})(conc.\ I^-)^3 = (3.0 \times 10^{-6}\ M)(1.7 \times 10^{-6}\ M)^3 = 1.5 \times 10^{-23}\ M^4 > K_{sp}(AuI_3)$$

$$Q(CuI) = (conc.\ Cu^+)(conc.\ I^-) = (3.0 \times 10^{-6}\ M)(1.7 \times 10^{-6}\ M) = 5.1 \times 10^{-12}\ M^2 = K_{sp}(CuI)$$

$Q(HgI_2) = (\text{conc. } Hg^{2+})(\text{conc. } I^-)^2 = (3.0 \times 10^{-6} \text{ M})(1.7 \times 10^{-6} \text{ M})^2 = 8.7 \times 10^{-18} \text{ } M^3 > K_{sp}(HgI_2)$

$Q(PbI_2) = (\text{conc. } Pb^{2+})(\text{conc. } I^-)^2 = (3.0 \times 10^{-6} \text{ M})(1.7 \times 10^{-6} \text{ M})^2 = 8.7 \times 10^{-18} \text{ } M^3 < K_{sp}(PbI_2)$

Since $Q > K_{sp}$ for AgI, AuI$_3$, and HgI$_2$, these solids will precipitate; since $Q < K_{sp}$ for PbI$_2$, it will not precipitate. For CuI, $Q = K_{sp}$ and this solution is saturated (no precipitate will form).

(b) To determine the order in which the precipitates will form, you need to calculate the concentration of I^- ions needed to precipitate each ionic solid.

$$K_{sp}(AgI) = [Ag^+][I^-], \ [I^-] = \frac{K_{sp}(AgI)}{[Ag^+]} = \frac{8.3 \times 10^{-17} \text{ } M^2}{3.0 \times 10^{-6} \text{ M}} = 2.8 \times 10^{-11} \text{ M}$$

$$K_{sp}(AuI_3) = [Au^{3+}][I^-]^3, \ [I^-] = \sqrt[3]{\frac{K_{sp}(AuI_3)}{[Au^{3+}]}} = \sqrt[3]{\frac{1.0 \times 10^{-46} \text{ } M^4}{3.0 \times 10^{-6} \text{ M}}} = 3.2 \times 10^{-14} \text{ M}$$

$$K_{sp}(CuI) = [Cu^+][I^-], \ [I^-] = \frac{K_{sp}(CuI)}{[Cu^+]} = \frac{5.1 \times 10^{-12} \text{ } M^2}{3.0 \times 10^{-6} \text{ M}} = 1.7 \times 10^{-6} \text{ M}$$

$$K_{sp}(HgI_2) = [Hg^{2+}][I^-]^2, \ [I^-] = \sqrt{\frac{K_{sp}(HgI_2)}{[Hg^{2+}]}} = \sqrt{\frac{4.0 \times 10^{-29} \text{ } M^3}{3.0 \times 10^{-6} \text{ M}}} = 3.7 \times 10^{-12} \text{ M}$$

$$K_{sp}(PbI_2) = [Pb^{2+}][I^-]^2, \ [I^-] = \sqrt{\frac{K_{sp}(PbI_2)}{[Pb^{2+}]}} = \sqrt{\frac{7.1 \times 10^{-9} \text{ } M^3}{3.0 \times 10^{-6} \text{ M}}} = 0.049 \text{ M}$$

The solid with the smallest $[I^-]$ will precipitate first and the solid with the largest $[I^-]$ will precipitate last. So, the order of solid precipitation would be—AuI$_3$ (first), HgI$_2$, AgI, CuI, and PbI$_2$ (last).

10. (a) The balanced equations and the K_{sp} formulas are:

$$Ag_2CrO_4(s) \rightleftharpoons 2 \, Ag^+(aq) + CrO_4^{2-}(aq) \qquad K_{sp} = [Ag^+]^2[CrO_4^{2-}]$$

$$PbCrO_4(s) \rightleftharpoons Pb^{2+}(aq) + CrO_4^{2-}(aq) \qquad K_{sp} = [Pb^{2+}][CrO_4^{2-}]$$

(b) For each solid, calculate $[CrO_4^{2-}]$ from the K_{sp} expressions.

$$K_{sp} = [Ag^+]^2[CrO_4^{2-}], \ [CrO_4^{2-}] = \frac{K_{sp}}{[Ag^+]^2} = \frac{9.0 \times 10^{-12} \text{ } M^3}{(0.010 \text{ M})^2} = 9.0 \times 10^{-8} \text{ M}$$

$$K_{sp} = [Pb^{2+}][CrO_4^{2-}], \ [CrO_4^{2-}] = \frac{K_{sp}}{[Pb^{2+}]} = \frac{1.8 \times 10^{-14} \text{ } M^2}{0.010 \text{ M}} = 1.8 \times 10^{-12} \text{ M}$$

(c) As long as the concentration of CrO_4^{2-} remains below 9.0×10^{-8} M, Ag$_2$CrO$_4$ should not precipitate. So, the way to separate the Ag$^+$ and Pb^{2+} ions is to add CrO_4^{2-} ions, making sure its concentration is more than 1.8×10^{-12} M but much less than 9.0×10^{-8} M. Then you can filter the precipitate from the solution; the Pb^{2+} ions should be in the precipitate as solid PbCrO$_4$, and Ag$^+$ should still be dissolved.

11. The addition of NaOH (OH$^-$ ions) to Zn^{2+} ions will result in a precipitate of the slightly soluble Zn(OH)$_2$. The reaction is simply the backwards reaction of K_{sp} (4.5×10^{-17}), so its K_c value is $1/K_{sp}$. Adding more OH$^-$ ions allows the complex ion Zn(OH)$_4^{2-}$ to form ($K_f = 4.5 \times 10^{17}$) from solid Zn(OH)$_2$. This formula is similar to the Ni^{2+}/CN$^-$ example on p. 157 of this book. So, K_c for this reaction will be equal to $K_{sp} \times K_f$. Since the first K_c value is larger than one, it is clearly product-favored. The second K_c value is close to one; according to Le Chatelier's principle, adding more OH$^-$ ions (large [OH$^-$]) will shift this reaction toward the formation of the complex ion.

$$Zn^{2+}(aq) + 2 \, OH^-(aq) \rightleftharpoons Zn(OH)_2(s) \qquad K_c = \frac{1}{K_{sp}} = 2.2 \times 10^{16}$$

$$Zn(OH)_2(s) + 2 \, OH^-(aq) \rightleftharpoons Zn(OH)_4^{2-}(aq) \qquad K_c = K_{sp} \times K_f = 20.$$

ANSWERS — PRACTICE TEST

Chapter 18

1. An exothermic reaction releases <u>energy</u> as a result of the reaction and has a negative $\Delta H°$ value; and exergonic reaction releases <u>free energy</u> as a result of the reaction and has a negative $\Delta G°$ value. An exothermic reaction releases energy to the surroundings, causing the surrounding to become warmer; and exergonic reaction is a product-favored reaction that will occur without outside assistance.

2. Although each of these pictures contains eight atoms, they have very different entropy values. The atoms in III are in the solid state and are forced to stay in a small volume and an organized pattern, so they have a minimal dispersal of energy. Picture I depicts a liquid, in which the atoms have more freedom of movement (and energy dispersion) than atoms in a solid but much less than atoms in a gas. Although II and IV both contain gases, II has eight gas particles (monatomic atoms) and IV has four gas particles (diatomic molecules). Therefore, the order is III (lowest entropy value) < I < IV < II (largest entropy value).

3. (a) Since gaseous carbon dioxide molecules have a larger volume than solid carbon dioxide molecules, they have a greater probability of energy dispersal. So, $CO_2(g)$ has the greater entropy value.

    (b) Solid $MgCl_2$ has Mg^{2+} and $2\ Cl^-$ ions in rigid, organized patterns with very little free motion. Aqueous $MgCl_2$ has Mg^{2+} and $2\ Cl^-$ ions that can move independently throughout the entire volume of the liquid. Since the aqueous ions have a greater volume (and a greater probability of energy dispersal), this means that $MgCl_2(aq)$ has the greater entropy value.

    (c) Since gaseous hydrazine (N_2H_4) has more N and H atoms than ammonia (NH_3), there is a greater chance of energy dispersal across the 6 atoms in hydrazine than across the 4 atoms in ammonia. Therefore, hydrazine gas, $N_2H_4(g)$, has the greater entropy value.

    (d) Aqueous HBr molecules dissolve in water to produce two ions, H^+ (which combines with water to form H_3O^+ ions) and Br^-, and these ions have the freedom to move throughout the entire volume of the liquid. Gaseous HBr molecules, on the other hand, have the freedom to move throughout the entire airspace above the liquid. Since the airspace above an aqueous solution usually has a larger volume than the solution, the gaseous molecules also have a greater probability of energy dispersal, and $HBr(g)$ will have the greater entropy value.

4. The values for $\Delta H°_f$ and $\Delta G°_f$ values are zero for all elements in their standard state at 25°C and 1 bar pressure (for gases); these values are zero because chemists defined them to be zero. The values for $S°$ are zero for all crystalline solids at absolute zero (0 K); these values are zero because these crystalline solids have a minimum dispersal of energy (i.e., this is the 'absence' of entropy which represents no dispersal of energy).

5. (a) Since this reaction converts a gaseous molecule and four liquid/aqueous ions into four liquid/aqueous molecules/ions, the energy would be less dispersed after the reaction and $\Delta S°$ should be negative.

    (b) This reaction converts three gaseous molecules into four gaseous molecules, which results in a dispersal of the energy as a result of the reaction and a positive $\Delta S°$ value.

    (c) Since this reaction converts a solid and gas molecule into a solid and gas molecule, the dispersal of energy before and after the reaction are about the same and the $\Delta S°$ value should be close to zero.

    (d) This reaction converts two types of solids into a solid and a liquid. The liquid product should have a greater potential for energy dispersal and therefore this reaction should have a positive $\Delta S°$ value.

6. (a) $$\Delta S = \left\{\left(1\ \text{mol Na}_2\text{SO}_4 \times 138.1\ \frac{\text{J}}{\text{mol Na}_2\text{SO}_4\ \text{K}}\right) + \left(1\ \text{mol H}_2\text{O} \times 69.9\ \frac{\text{J}}{\text{mol H}_2\text{O K}}\right)\right\} -$$
    $$\left\{\left(1\ \text{mol SO}_3 \times 256.8\ \frac{\text{J}}{\text{mol SO}_3\ \text{K}}\right) + \left(2\ \text{mol NaOH} \times 48.1\ \frac{\text{J}}{\text{mol NaOH K}}\right)\right\} = -145.0\ \frac{\text{J}}{\text{K}}$$

    This matches your prediction of a negative $\Delta S°$ value.

(b) $\Delta S = \left\{ \left(1 \text{ mol HBr} \times 198.70 \dfrac{J}{\text{mol HBr K}} \right) + \left(3 \text{ mol HF} \times 173.78 \dfrac{J}{\text{mol HF K}} \right) \right\} -$

$\left\{ \left(1 \text{ mol BrF}_3 \times 292.53 \dfrac{J}{\text{mol BrF}_3 \text{ K}} \right) + \left(2 \text{ mol H}_2 \times 130.68 \dfrac{J}{\text{mol H}_2 \text{ K}} \right) \right\} = 166.15 \dfrac{J}{K}$

This value matches your prediction of a positive $\Delta S°$ value.

(c) $\Delta S = \left\{ \left(1 \text{ mol FeO} \times 57.9 \dfrac{J}{\text{mol FeO K}} \right) + \left(1 \text{ mol NO} \times 210.76 \dfrac{J}{\text{mol NO K}} \right) \right\} -$

$\left\{ \left(1 \text{ mol Fe} \times 27.8 \dfrac{J}{\text{mol Fe K}} \right) + \left(1 \text{ mol NO}_2 \times 240.06 \dfrac{J}{\text{mol NO}_2 \text{ K}} \right) \right\} = 0.80 \dfrac{J}{K}$

This value is close to very zero, but is slightly positive.

(d) $\Delta S = \left\{ \left(2 \text{ mol AgCl} \times 96 \dfrac{J}{\text{mol AgCl K}} \right) + \left(1 \text{ mol Hg} \times 76 \dfrac{J}{\text{mol Hg K}} \right) \right\} -$

$\left\{ \left(1 \text{ mol HgCl}_2 \times 146 \dfrac{J}{\text{mol HgCl}_2 \text{ K}} \right) + \left(2 \text{ mol Ag} \times 43 \dfrac{J}{\text{mol Ag K}} \right) \right\} = 36 \dfrac{J}{K}$

This value matches your prediction of a positive $\Delta S°$ value. This value is only slightly positive because the entropy change from a solid to a liquid rather small.

7. (a) $\Delta G = \left\{ 1 \text{ mol H}_2\text{O(s)} \times -236.58 \dfrac{kJ}{\text{mol H}_2\text{O(s)}} \right\} - \left\{ 1 \text{ mol H}_2\text{O}(\ell) \times -237.13 \dfrac{kJ}{\text{mol H}_2\text{O}(\ell)} \right\}$

$\Delta G° = 0.55$ kJ. So, this reaction is reactant-favored at 25°C.

(b) $\Delta G = \left\{ 1 \text{ mol KOH(aq)} \times -440.5 \dfrac{kJ}{\text{mol KOH(aq)}} \right\} - \left\{ 1 \text{ mol KOH(s)} \times -379.1 \dfrac{kJ}{\text{mol KOH(s)}} \right\}$

$\Delta G° = -61.4$ kJ. So, this reaction is product-favored at 25°C.

(c) $\Delta G = \left\{ 1 \text{ mol CS}_2\text{(g)} \times 67.12 \dfrac{kJ}{\text{mol CS}_2\text{(g)}} \right\} - \left\{ 1 \text{ mol CS}_2(\ell) \times 65.27 \dfrac{kJ}{\text{mol CS}_2(\ell)} \right\}$

$\Delta G° = 1.85$ kJ. So, this reaction is reactant-favored at 25°C.

8. (a) $\Delta H = \left\{ 1 \text{ mol H}_2\text{O(s)} \times -291.84 \dfrac{kJ}{\text{mol H}_2\text{O(s)}} \right\} - \left\{ 1 \text{ mol H}_2\text{O}(\ell) \times -285.83 \dfrac{kJ}{\text{mol H}_2\text{O}(\ell)} \right\} = -6.01$ kJ

$\Delta S = \left\{ 1 \text{ mol H}_2\text{O(s)} \times 47.91 \dfrac{J}{\text{mol H}_2\text{O(s) K}} \right\} - \left\{ 1 \text{ mol H}_2\text{O}(\ell) \times 69.91 \dfrac{J}{\text{mol H}_2\text{O}(\ell) \text{ K}} \right\} = -22.00 \dfrac{J}{K}$

$\Delta G = \Delta H - T\Delta S = -6.01 \text{ kJ} - (273.15 + 250 \text{ K}) \left(-22.00 \dfrac{J}{K} \right) \left(\dfrac{1 \text{ kJ}}{10^3 \text{ J}} \right) = 5.50$ kJ

Since $\Delta G°$ is positive, this reaction is reactant-favored at 250°C.

(b) $\Delta H = \left\{ 1 \text{ mol KOH(aq)} \times -482.4 \dfrac{kJ}{\text{mol KOH(aq)}} \right\} - \left\{ 1 \text{ mol KOH(s)} \times -424.8 \dfrac{kJ}{\text{mol KOH(s)}} \right\} = -57.6$ kJ

$\Delta S = \left\{ 1 \text{ mol KOH(aq)} \times 91.6 \dfrac{J}{\text{mol KOH(aq) K}} \right\} - \left\{ 1 \text{ mol KOH(s)} \times 78.9 \dfrac{J}{\text{mol KOH(s) K}} \right\} = 12.7 \dfrac{J}{K}$

$\Delta G = \Delta H - T\Delta S = -57.6 \text{ kJ} - (273.15 + 250 \text{ K}) \left(12.7 \dfrac{J}{K} \right) \left(\dfrac{1 \text{ kJ}}{10^3 \text{ J}} \right) = -64.2$ kJ

Since $\Delta G°$ is negative, this reaction is product-favored at 250°C.

(c) $\Delta H = \left\{ 1 \text{ mol } CS_2(g) \times 117.36 \ \dfrac{kJ}{\text{mol } CS_2(g)} \right\} - \left\{ 1 \text{ mol } CS_2(\ell) \times 89.70 \ \dfrac{kJ}{\text{mol } CS_2(\ell)} \right\} = 27.66 \text{ kJ}$

$\Delta S = \left\{ 1 \text{ mol } CS_2(g) \times 237.84 \ \dfrac{J}{\text{mol } CS_2(g) \text{ K}} \right\} - \left\{ 1 \text{ mol } CS_2(\ell) \times 151.34 \ \dfrac{J}{\text{mol } CS_2(\ell) \text{ K}} \right\} = 86.50 \ \dfrac{J}{K}$

$\Delta G = \Delta H - T\Delta S = 27.66 \text{ kJ} - (273.15 + 250 \text{ K}) \left(86.50 \ \dfrac{J}{K} \right) \left(\dfrac{1 \text{ kJ}}{10^3 \text{ J}} \right) = -17.59 \text{ kJ}$

Since $\Delta G°$ is negative, this reaction is product-favored at 250°C.

9. (a) Since this reaction has a negative $\Delta H°$ value (–6.01 kJ), it will product-favored at low temperatures, and since $\Delta S°$ is negative (–22.00 J/K), this reaction will be reactant-favored at high temperatures. This reaction will have a transition temperature where the reaction changes from spontaneous to nonspontaneous. This value (which matches the melting point of ice very well) is:

$$T_{eq} = \frac{\Delta H}{\Delta S} = \frac{-6.01 \text{ kJ}}{-22.00 \text{ J/K}} \times \frac{10^3 \text{ J}}{1 \text{ kJ}} = 273 \text{ K} = 0 \text{ C}$$

(b) Since $\Delta H°$ is negative and $\Delta S°$ is positive, the reaction is product-favored at low temperatures and at high temperatures (spontaneous as written at all temperatures). Since this reaction is spontaneous at all temperatures, there is no transition temperature.

(c) Since this reaction has a positive $\Delta H°$ value (27.66 kJ), it will reactant-favored at low temperatures, and since $\Delta S°$ is positive (86.50 J/K), this reaction will be product-favored at high temperatures. This reaction will have a transition temperature where the reaction changes from spontaneous to nonspontaneous. This value matches the boiling point of carbon disulfide (46.3°C) very well.

$$T_{eq} = \frac{\Delta H}{\Delta S} = \frac{27.66 \text{ kJ}}{86.50 \text{ J/K}} \times \frac{10^3 \text{ J}}{1 \text{ kJ}} = 319.8 \text{ K} = 46.6 \text{ C}$$

10. (a) Calculate the $\Delta G°$ value at 1000°C, and then solve for $K°$ using the equation $\Delta G° = -RT \ln K°$.

$\Delta G = \Delta H - T\Delta S = \left(-718 \text{ kJ} \times \dfrac{10^3 \text{ J}}{1 \text{ kJ}} \right) - (273.15 + 1000 \text{ K}) \left(-367 \ \dfrac{J}{K} \right) = -250,754 \text{ J (unrounded)}$

$\dfrac{\Delta G}{RT} = \dfrac{-250,754 \ \dfrac{J}{1 \text{ mol } CuSO_4}}{(8.314 \text{ J/mol K})(1273.15 \text{ K})} = -23.69, \quad K = e^{-\frac{\Delta G}{RT}} = e^{23.69} = 1.94 \times 10^{10} = \dfrac{1}{P_{O_2}^2} = K_P$

(b) $\Delta G = \Delta G + RT \ln Q = -250,754 \ \dfrac{J}{\text{mol}} + \left(8.314 \ \dfrac{J}{\text{mol K}} \right)(1273.15 \text{ K}) \ln \left(\dfrac{1}{2.125^2} \right) = -267,000 \ \dfrac{J}{\text{mol}}$

$\Delta G = -267,000 \ \dfrac{J}{\text{mol } CuSO_4} \times 1 \text{ mol } CuSO_4 = -267,000 \text{ J}$

11. (a) Since $\Delta H°$ is negative and $\Delta S°$ is positive, the reaction is product-favored at low temperatures and at high temperatures (spontaneous as written at all temperatures).

(b) Even though this reaction is spontaneous (product-favored) at all temperatures, it is kinetically stable at room temperature. This means the reaction has a high activation energy, so very few molecules at room temperature have enough energy to reach the top of the 'energy hill' and react to make CO_2 and H_2O. This correlates to a small rate constant. However, if a spark is added to the reaction, it will provide enough energy to a few C_3H_8 and O_2 molecules so they can react. When they react, they will release energy (an exothermic reaction) that allows other C_3H_8 and O_2 molecules to react, releasing more energy and allowing even more molecules to react until the propane gas is burning.

ANSWERS — PRACTICE TEST

Chapter 19

1.   (a)   **Cl_2O:** Rule 4 says $2 \times O.N._{Cl} + O.N._{O} = 0$; Rule 3d says $O.N._{O} = -2$, so $2 \times O.N._{Cl} = +2$ and $O.N._{Cl} = +1$. **N_2H_4:** Rule 4 says $2 \times O.N._{N} + 4 \times O.N._{H} = 0$; Rule 3c says $O.N._{H} = +1$, so $2 \times O.N._{N} + 4(+1) = 0$, $2 \times O.N._{N} = -4$, and $O.N._{N} = -2$. **Cl^-:** Rule 4 states that $O.N._{Cl} = -1$. **N_2:** Rule 4 says that $2 \times O.N._{N} = 0$, so $O.N._{N} = 0$. Since chlorine's oxidation number decreases from +1 to –1, the Cl atoms in Cl_2O have been reduced and Cl_2O is the oxidizing agent; since nitrogen's oxidation number increases from –2 to 0, the N atoms in N_2 have been oxidized and N_2 is the reducing agent.

    (b)   (1) One of the half-equations contains Cl compounds and the other one has the N compounds. (2) In the Cl reaction, add a 2 in front of the Cl^-; in the N reaction, the N atoms are already balanced. (3) In the Cl reaction, add 1 water molecule to the right side of the equation to balance the O atoms; in the N reaction, there are no O atoms. (4) Add 2 H^+ ions to the left side of the Cl reaction and 4 H^+ ions to the right side of the N reaction to balance H atoms. (5) To balance charge, add 4 e^- to the left side of the Cl reaction and 4 e^- to the right side of the N reaction. (6) Since these equations have the same number of electrons, you do not need to multiply either equation by any number to get the electron to cancel. Remove 2 H^+(aq) from both sides of the equation. (7) Add 2 H_2O to both sides of the equation, recognizing that 2 H^+ and 2 H_2O make 2 H_3O^+. (8) Remove one water molecule from both sides.

$$Cl_2O(aq) + 2\,H^+(aq) + 4\,e^- \rightarrow 2\,Cl^-(aq) + H_2O(\ell)$$

$$N_2H_4(aq) \rightarrow N_2(g) + 4\,H^+(aq) + 4\,e^-$$

$$Cl_2O(aq) + 2\,H^+(aq) + N_2H_4(aq) \rightarrow 2\,Cl^-(aq) + H_2O(\ell) + N_2(g) + 4\,H^+(aq)$$

$$Cl_2O(aq) + N_2H_4(aq) \rightarrow 2\,Cl^-(aq) + H_2O(\ell) + N_2(g) + 2\,H^+(aq)$$

$$Cl_2O(aq) + N_2H_4(aq) + 2\,H_2O(\ell) \rightarrow 2\,Cl^-(aq) + H_2O(\ell) + N_2(g) + 2\,H_3O^+(aq)$$

$$Cl_2O(aq) + N_2H_4(aq) + H_2O(\ell) \rightarrow 2\,Cl^-(aq) + N_2(g) + 2\,H_3O^+(aq)$$

    (c)   Use the fourth equation above and steps 7' and 8': (7') Add 2 OH^- to both sides of the equation, recognizing that 2 H^+ and 2 OH^- make 2 H_2O, for a total of 3 H_2O. (8') There is nothing to remove.

$$Cl_2O(aq) + N_2H_4(aq) \rightarrow 2\,Cl^-(aq) + H_2O(\ell) + N_2(g) + 2\,H^+(aq)$$

$$Cl_2O(aq) + N_2H_4(aq) + 2\,OH^-(aq) \rightarrow 2\,Cl^-(aq) + 3\,H_2O(\ell) + N_2(g)$$

2.   (a)   The reaction with the more positive reduction potential will undergo reduction, and the reaction with the more negative (less positive) reduction potential will undergo oxidation. So, Pt^{2+} is reduced to Pt and Ag is oxidized to Ag^+.

    (b)   The reduction reaction occurs as written and the oxidation reaction must be written backwards. To get the net equation, multiply the silver reaction by 2 and add the two half-reactions.

$$Pt^{2+}(aq) + 2\,e^- \rightarrow Pt(s)$$

$$Ag(s) \rightarrow Ag^+(aq) + e^-$$

$$Pt^{2+}(aq) + 2\,Ag(s) \rightarrow Pt(s) + 2\,Ag^+(aq)$$

    (c) $E_{cell} = E_{red} - E_{ox} = +1.20\,V - 0.80\,V = +0.40\,V$.

    (d)   Since the oxidation process occurs at the Ag electrode it is the anode, and since the reduction process occurs at the Pt electrode it is the cathode.

    (e)   The electrons will leave the Ag electrode (once the Ag atoms are converted to Ag^+ ions), and flow through the external circuit into the Pt electrode (where they will convert the Pt^{2+} ions into Pt atoms).

    (f)   The Ag^+ ions flow away from the anode (and toward the cathode) as they are produced, and the Pt^{2+} ions flow toward the cathode as they react at the cathode's surface. In general, cations flow toward the cathode (and away from the anode) and anions flow toward the anode (and away from the cathode). In fact, <u>an</u>ions were named because they flow toward the <u>an</u>ode and <u>cat</u>ions were named because they flow toward the <u>cat</u>hode.

(g)   As the reaction progresses, Ag^+ ions are produced in the Ag half-cell, leading to a build-up of positive charges in this half-cell. NO_3^- ions from the salt bridge enter the solution to balance the charges (the anions enter the oxidation half-cell and flow toward the anode). And as the reaction progresses, Pt^{2+} ions are consumed in the Pt half-cell, leading to a build-up of negative charges in this half-cell (this is because the Pt^{2+} ions react and leave the solution, but their counter anions are still in the container, leaving a net negative charge). Na^+ ions from the salt bridge will enter the solution to balance these charges (the cations enter the reduction half-cell and flow toward the cathode).

(h)   The silver electrode will lose mass because some of the Ag atoms will be converted to Ag^+ ions that dissolve in the solution. The platinum electrode will gain mass because the Pt^{2+} ions that are reduced to Pt atoms will attach themselves to the electrode, causing it to weigh more as the reaction progresses.

3.   (a)   Use the equations from the standard reduction potentials and Hess's law to get the target equation. Use equation 3 as written (which makes it the reduction reaction) and use equation 2 written backwards (which makes it the oxidation reaction). Then, calculate the cell potential from the $E°$ values.

$$PbCl_2(s) + 2\,e^- \rightarrow Pb(s) + 2\,Cl^-(aq)$$

$$Pb(s) \rightarrow Pb^{2+}(aq) + 2\,e^-$$

$$E_{cell} = E_{red} - E_{ox} = -0.26\ V - (-0.12\ V) = -0.14\ V, \text{ (reactant-favored)}.$$

(b)   Use the equations from the standard reduction potentials and Hess's law to get the target equation. Use equation 1 as written and divided by 2 (which makes it the reduction reaction) and use equation 4 written backwards and divided by 2 (which makes it the oxidation reaction). Then, calculate the cell potential from the $E°$ values (note that dividing by 2 does not change the cell potential of the reaction).

$$H_3O^+(aq) + e^- \rightarrow \frac{1}{2}\,H_2(g) + H_2O(\ell)$$

$$\frac{1}{2}\,H_2(g) + OH^-(aq) \rightarrow H_2O(\ell) + e^-$$

$$E_{cell} = E_{red} - E_{ox} = +0.00\ V - (-0.83\ V) = +0.83\ V, \text{ (product-favored)}.$$

4.   (a) $\Delta G = -nFE_{cell} = -(2\ mol\ e^-)\left(\dfrac{96,500\ C}{mol\ e^-}\right)(-0.14\ V) \times \dfrac{1\ J}{1\ C \times V} = 27,000\ J$ (reactant-favored).

(b) $\Delta G = -nFE_{cell} = -(1\ mol\ e^-)\left(\dfrac{96,500\ C}{mol\ e^-}\right)(0.83\ V) \times \dfrac{1\ J}{1\ C \times V} = -8.0 \times 10^4\ J$ (product-favored).

5.   (a) $K = e^{nFE_{cell}/RT}$, $\dfrac{nFE_{cell}}{RT} = \dfrac{(2)(96,500\ C/mol\ e^-)(-0.14\ V)}{(8.314\ J/mol\ e^-\ K)(273.15 + 25.0\ K)} \times \dfrac{1\ J}{1\ C \times V} = -10.90$

$K = e^{nFE_{cell}/RT} = e^{-10.90} = 1.8 \times 10^{-5} = [Pb^{2+}][Cl^-]^2 = K_{sp}$. The book says $K_{sp}(PbCl_2) = 1.6 \times 10^{-5}$.

(b) $K = e^{nFE_{cell}/RT}$, $\dfrac{nFE_{cell}}{RT} = \dfrac{(1)(96,500\ C/mol\ e^-)(+0.83\ V)}{(8.314\ J/mol\ e^-\ K)(273.15 + 25.0\ K)} \times \dfrac{1\ J}{1\ C \times V} = 32.31$

$K = e^{nFE_{cell}/RT} = e^{32.31} = 1.1 \times 10^{14} = \dfrac{1}{[H_3O^+][OH^-]} = \dfrac{1}{K_w} = \dfrac{1}{1.0 \times 10^{-14}} = 1.0 \times 10^{14}$.

6.   This is sort of a trick question, because half-reactions (whether oxidations or reductions) cannot happen without another (opposite) half-reaction. The Al^{3+}/Al reduction reaction will be reactant-favored if it is linked to reaction with a more positive $E°$ value, but will be product-favored when linked to a reaction with a more negative (less positive) $E°$ value.

$$Fe^{2+}(aq) + 2\,e^- \rightleftharpoons Fe(s) \qquad\qquad E° = -0.44\ V$$

$$Al^{3+}(aq) + 3\,e^- \rightleftharpoons Al(s) \qquad\qquad E° = -1.68\ V$$

$$Na^+(aq) + e^- \rightleftharpoons Na(s) \qquad\qquad E° = -2.71\ V$$

$$2\,Al^{3+}(aq) + 3\,Fe(s) \rightarrow 2\,Al(s) + 3\,Fe^{2+}(aq)$$

$$E_{cell} = E_{red} - E_{ox} = -1.68\text{ V} - (-0.44\text{ V}) = -1.24\text{ V, (reactant-favored).}$$

$$Al^{3+}(aq) + 3\,Na(s) \rightarrow Al(s) + 3\,Na^{+}(aq)$$

$$E_{cell} = E_{red} - E_{ox} = -1.68\text{ V} - (-2.71\text{ V}) = +1.03\text{ V, (product-favored).}$$

7. (a) The Nernst equation can be used for standard reduction potentials of half-reactions as well as for cell potentials of redox reactions. The E values for both of these reactions are listed below.

$$E = E - \frac{RT}{nF}\ln Q = E - \frac{RT}{nF}\ln\frac{1}{[H_3O^+]^4 P_{O_2}}$$

$$E = +1.23\text{ V} - \frac{(8.314\text{ J/mol e}^- \text{ K})(298.15\text{ K})}{(4)(96,500\text{ C/mol e}^-)} \times \frac{1\text{ J}}{1\text{ C}\times\text{V}}\ln\frac{1}{[1.0\times10^{-7}]^4(0.21)} = 1.23\text{ V} - 0.42\text{ V} = +0.81\text{ V}$$

$$E = E - \frac{RT}{nF}\ln Q = E - \frac{RT}{nF}\ln P_{H_2}[OH^-]^2$$

$$E = -0.83\text{ V} - \frac{(8.314\text{ J/mol e}^- \text{ K})(298.15\text{ K})}{(2)(96,500\text{ C/mol e}^-)} \times \frac{1\text{ J}}{1\text{ C}\times\text{V}}\ln(5.0\times10^{-7})[1.0\times10^{-7}]^2 = -0.83\text{ V} + 0.60\text{ V} = -0.23\text{ V}$$

(b) For the electrolysis of water, H_2O must be the only reactant. So, use the equation 2 as written (reduction) after multiplying by 2, and use the equation 1 backwards (oxidation). The cell potential is

$$E_{cell} = E_{red} - E_{ox} = -0.23\text{ V} - 0.81\text{ V} = -1.04\text{ V}$$

(c) Electrolysis reactions can only occur if the aqueous solution conducts electricity (i.e., the solution must have ions in it). Since pure water has only 1.0×10^{-7} M H_3O^+ and OH^- ions in it, it does not conduct electricity and will not undergo electrolysis. In order for the solution to conduct electricity and undergo electrolysis, an ionic salt (that does not react) must be dissolved in the pure water.

8. In very cold conditions, the aqueous solution of sulfuric acid will freeze solid. When this liquid is frozen, the H_3O^+ and HSO_4^- ions cannot flow from one electrode to the other, and the battery cannot function.

9. Electrolysis reactions are reactant-favored. So, the products of these reactions can react spontaneously to form the reactants. If the products are not separated, they will react to reform the reactants and the energy used to cause the electrolysis reaction will have been wasted.

10. (a) Molten magnesium bromide contains Mg^{2+} and Br^- ions. So, you should only use the equations with these ions on either side of the equation. In the molten salt, only Mg^{2+} can be reduced and only Br^- can be oxidized. The net reaction and the cell potential (assuming the reduction potentials for the molten salt are the same as those for the aqueous ions at 25°C and standard conditions) appear below.

$$Br_2(\ell) + 2\,e^- \rightleftharpoons \mathbf{2\,Br^-\,(aq)} \qquad\qquad E° = +1.07\text{ V}$$

$$\mathbf{Mg^{2+}(aq)} + 3\,e^- \rightleftharpoons Mg(s) \qquad\qquad E° = -2.36\text{ V}$$

$$Mg^{2+}(\ell) + 2\,Br^-(\ell) \rightarrow Mg(s) + Br_2(\ell)$$

$$E_{cell} = E_{red} - E_{ox} = -2.36\text{ V} - 1.07\text{ V} = -3.43\text{ V}$$

(b) Aqueous magnesium bromide contains Mg^{2+} and Br^- ions and H_2O molecules. The reduction reactions containing only these ions or molecules and e^- on one side of the equation are:

$$O_2(g) + 4\,H_3O^+(aq) + 4\,e^- \rightleftharpoons \mathbf{6\,H_2O(\ell)} \qquad\qquad E° = +1.23\text{ V}$$

$$Br_2(\ell) + 2\,e^- \rightleftharpoons \mathbf{2\,Br^-\,(aq)} \qquad\qquad E° = +1.07\text{ V}$$

$$\mathbf{2\,H_2O(\ell)} + 2\,e^- \rightleftharpoons H_2(g) + 2\,OH^-(aq) \qquad\qquad E° = -0.83\text{ V}$$

$$\mathbf{Mg^{2+}(aq)} + 3\,e^- \rightleftharpoons Mg(s) \qquad\qquad E° = -2.36\text{ V}$$

In this aqueous solution, Mg^{2+} and H_2O can be reduced. Since there is a choice of oxidizing agents, the reaction with the more positive (less negative) $E°$ value undergoes reduction (H_2O). In this aqueous solution, Br^- and H_2O can be oxidized. Since there is a choice of reducing agents, the reaction with the more negative (less positive) $E°$ value undergoes oxidation (Br^-). The net reaction and cell potential for this electrolysis reaction are:

$$2\ H_2O(\ell) + 2\ Br^-(aq) \rightarrow H_2(g) + 2\ OH^-(aq) + Br_2(\ell)$$
$$E_{cell} = E_{red} - E_{ox} = -0.83\ V - 1.07\ V = -1.90\ V$$

(c) Since magnesium metal is a product only for the molten electrolysis reaction, you must use the molten method to make $Mg(s)$.

(d) Both the molten and aqueous methods produce bromine. So, either method would work. However, the aqueous method takes less energy because you don't have to melt the solid magnesium bromide, which takes a lot of energy. So, from an energy perspective, it takes less energy to use the aqueous method and it should be the preferred method for making $Br_2(\ell)$.

11. Aqueous iron(III) iodide contains Fe^{3+} and I^- ions and H_2O molecules. The reduction reactions containing only these ions or molecules and e^- on one side of the equation (with the reactants in boldface) are:

$$O_2(g) + 4\ H_3O^+(aq) + 4\ e^- \rightleftharpoons \mathbf{6\ H_2O(\ell)} \qquad E° = +1.23\ V$$

$$\mathbf{Fe^{3+}(aq)} + e^- \rightleftharpoons Fe^{2+}(aq) \qquad E° = +0.77\ V$$

$$I_2(s) + 2\ e^- \rightleftharpoons \mathbf{2\ I^-(aq)} \qquad E° = +0.54\ V$$

$$\mathbf{Fe^{3+}(aq)} + 3\ e^- \rightleftharpoons Fe(s) \qquad E° = -0.04\ V$$

$$\mathbf{2\ H_2O(\ell)} + 2\ e^- \rightleftharpoons H_2(g) + 2\ OH^-(aq) \qquad E° = -0.83\ V$$

In this solution, Fe^{3+} ions can be reduced to Fe^{2+} ions or $Fe(s)$ and H_2O can also be reduced. Since there is a choice of oxidizing agents, the reaction with the most positive (least negative) $E°$ value undergoes reduction ($Fe^{3+} \rightarrow Fe^{2+}$). In this aqueous solution, I^- and H_2O can be oxidized. Since there is a choice of reducing agents, the reaction with the more negative (less positive) $E°$ value undergoes oxidation (I^-). The net reaction and cell potential for this electrolysis reaction are:

$$2\ Fe^{3+}(aq) + 2\ I^-(aq) \rightarrow 2\ Fe^{2+}(aq) + I_2(s)$$
$$E_{cell} = E_{red} - E_{ox} = +0.77\ V - 0.54\ V = +0.23\ V$$

Since the cell potential for this reaction is positive, this reaction is spontaneous (product-favored). So, any time Fe^{3+} and I^- ions are mixed together, they will react (the Fe^{3+} ions oxidize the I^- ions to $I_2(s)$ and the I^- ions reduce the Fe^{3+} ions to Fe^{2+} ions).

12. (a) Use stoichiometry to convert the mass of Au to moles of e^-, then use the Faraday constant to determine the total charge transferred and calculate the time for this reaction using the formula $q = it$.

$$875.0\ g\ Au \times \frac{mol\ Au}{197.0\ g\ Au} \times \frac{3\ mol\ e^-}{1\ mol\ Au} \times \frac{96,500\ C}{mol\ e^-} = 1.29 \times 10^6\ C$$

$$t = \frac{q}{i} = \frac{1.29 \times 10^6\ C}{5.00\ A} \times \frac{1\ A \times s}{1\ C} = 2.57 \times 10^5\ s \times \frac{1\ min}{60\ s} \times \frac{1\ h}{60\ min} = 71\ h$$

(b) Use the formula $q = it$ to convert the time into the total electrical charge (in C) transferred. Then, use stoichiometry and the Faraday constant to determine the mass of gold produced.

$$q = it = \left(5.00\ A \times \frac{1\ C}{1\ A \times s}\right)\left(24.0\ h \times \frac{60\ min}{1\ h} \times \frac{60\ s}{1\ min}\right) = 4.32 \times 10^5\ C$$

$$4.32 \times 10^5\ C \times \frac{mol\ e^-}{96,500\ C} \times \frac{1\ mol\ Au}{3\ mol\ e^-} \times \frac{197.0\ g\ Au}{mol\ Au} = 294\ g\ Au$$

ANSWERS — PRACTICE TEST

Chapter 20

1. (a) Although the bottom numbers of the nuclide symbols are often omitted, they can be determined from the periodic table. The sum of A values and the sum of the Z values must be the same on both sides of the equation. This gives the following formulas: $A + 4 = 12$ and $Z + 2 = 6$. So, $A = 8$ and $Z = 4$ and the reactant is ^{8}Be: $^8_4\text{Be} + ^4_2\text{He} \rightarrow ^{12}_6\text{C}$.

   (b) The sum of A values and the Z values must be equal on both sides of the equation. So, $12 + 16 = A + 4$ and $6 + 8 = Z + 2$. So, $A = 24$ and $Z = 12$ and the product is ^{24}Mg: $^{12}_6\text{C} + ^{16}_8\text{O} \rightarrow ^{24}_{12}\text{Mg} + ^4_2\text{He}$.

   (c) The two equations are: $A + 4 = 13 + 1$ and $Z + 2 = 6 + 1$. So, $A = 10$ and $Z = 5$ and the reactant is ^{10}B: $^{10}_5\text{B} + ^4_2\text{He} \rightarrow ^{13}_6\text{C} + ^1_1\text{H}$.

   (d) The two equations are: $252 = 106 + A + 4 \times 1$ and $98 = 42 + Z + 4 \times 0$. So, the A value is 142 and the Z value is 56 (^{142}Ba): $^{252}_{98}\text{Cf} \rightarrow ^{106}_{42}\text{Mo} + ^{142}_{56}\text{Ba} + 4\,^1_0\text{n}^0$.

2. (a) When ^{144}Nd emits an alpha particle, its A value decreases by 4 and its Z value decreases by 2. So, the other product is ^{140}Ce: $^{144}_{60}\text{Nd} \rightarrow ^{140}_{58}\text{Ce} + ^4_2\text{He}$.

   (b) When ^{31}Si emits an beta particle, its A value stays the same and its Z value increases by 1. So, the other product is ^{31}P: $^{31}_{14}\text{Si} \rightarrow ^{31}_{15}\text{P}^+ + ^0_{-1}\text{e}^-$.

   (c) When ^{14}O emits an positron particle, its A value stays the same and its Z value decreases by 1. So, the other product is ^{14}N (and the positron will be annihilated by an electron): $^{14}_8\text{O} \rightarrow ^{14}_7\text{N}^- + ^0_1\text{e}^+$ or $^{14}_8\text{O} \rightarrow ^{14}_7\text{N} + 2\,^0_0\gamma$.

   (d) When the ^{195}Au nucleus captures an electron, its A value stays the same and its Z value decreases by 1. So, the other product is ^{105}Pt (the ^{195}Au atom is written as $^{195}\text{Au}^+ + \text{e}^-$): $^{195}_{79}\text{Au}^+ + ^0_{-1}\text{e}^- \rightarrow ^{195}_{78}\text{Pt}$.

3. (a) Since $Z = 85$ (which is greater than 83), this atom will undergo α-emission: $^{214}_{85}\text{At} \rightarrow ^{210}_{83}\text{Bi} + ^4_2\text{He}$.

   (b) Since $Z = 29$ (which is less than 83), this atom will not undergo α-emission. To determine how it will decay, you can look at the band of stability (^{59}Cu is below the band) or you can look at its molar mass (59 is less than the mass of 63.6 in the periodic table). So, this atom can undergo positron-emission or electron capture: $^{59}_{29}\text{Cu} \rightarrow ^{59}_{28}\text{Ni}^- + ^0_1\text{e}^+$, which becomes $^{59}_{29}\text{Cu} \rightarrow ^{59}_{28}\text{Ni} + 2\,^0_0\gamma$ (e$^+$-emission) or (electron capture). ^{59}Ni is a more stable isotope (closer to M in the table).

   (c) Since $Z = 9$ (which is less than 83), this atom will not undergo α-emission. Since ^{20}F is above the band of stability and its molar mass is more than the mass of 19 in the periodic table, this atom will undergo beta-emission: $^{20}_9\text{F} \rightarrow ^{20}_{10}\text{Ne}^+ + ^0_{-1}\text{e}^-$. ^{20}Ne is a stable isotope (and matches M in the table).

   (d) Since $Z = 91$ (which is greater than 83), this atom will undergo α-emission: $^{230}_{91}\text{Pa} \rightarrow ^{226}_{89}\text{Ac} + ^4_2\text{He}$.

   (e) Since $Z = 46$ (which is less than 83), this atom will not undergo α-emission. Since ^{109}Pd is above the band of stability and its molar mass is more than the mass of 106.4 in the periodic table, this atom will undergo beta-emission: $^{109}_{46}\text{Pd} \rightarrow ^{109}_{47}\text{Ag}^+ + ^0_{-1}\text{e}^-$. ^{109}Ag is stable (closer to M in the periodic table).

   (f) Since $Z = 22$ (which is less than 83), this atom will not undergo α-emission. Since ^{45}Ti is below the band of stability and its molar mass is less than the mass of 48 in the periodic table, this atom can undergo positron-emission or electron capture: $^{45}_{22}\text{Ti} \rightarrow ^{45}_{21}\text{Sc}^- + ^0_1\text{e}^+$, or $^{45}_{22}\text{Ti} \rightarrow ^{45}_{21}\text{Sc} + 2\,^0_0\gamma$ (e$^+$-emission) or (electron capture). ^{45}Sc is stable (same as M in the table).

4. (a) $\Delta m = m_f - m_i = 2\left(0.0000\ \dfrac{\text{g}}{\text{mol}}\right) - \left(0.0005\ \dfrac{\text{g}}{\text{mol}} + 0.0005\ \dfrac{\text{g}}{\text{mol}}\right) = -0.0010\ \dfrac{\text{g}}{\text{mol}}$

   $\Delta E = \Delta m c^2 = \left(-0.0010\ \dfrac{\text{g}}{\text{mol}} \times \dfrac{1\ \text{kg}}{10^3\ \text{g}}\right)\left(3.00 \times 10^8\ \dfrac{\text{m}}{\text{s}}\right)^2 \times \dfrac{1\ \text{J s}^2}{1\ \text{kg m}^2} \times \dfrac{1\ \text{kJ}}{10^3\ \text{J}} = -9.0 \times 10^7\ \dfrac{\text{kJ}}{\text{mol}}$

Remember that γ-rays are massless (0.0000 g/mol) since they are electromagnetic radiation.

(b) $\Delta m = m_f - m_i = \left(27.9769\ \dfrac{g}{mol}\right) - \left(12.0000\ \dfrac{g}{mol} + 15.9949\ \dfrac{g}{mol}\right) = -0.0180\ \dfrac{g}{mol}$

$\Delta E = \Delta mc^2 = \left(-0.0180\ \dfrac{g}{mol} \times \dfrac{1\ kg}{10^3\ g}\right)\left(3.00 \times 10^8\ \dfrac{m}{s}\right)^2 \times \dfrac{1\ J\ s^2}{1\ kg\ m^2} \times \dfrac{1\ kJ}{10^3\ J} = -1.62 \times 10^9\ \dfrac{kJ}{mol}$

(c) $\Delta m = m_f - m_i = \left(12.0000\ \dfrac{g}{mol} + 4.0026\ \dfrac{g}{mol}\right) - \left(15.9949\ \dfrac{g}{mol}\right) = 0.0077\ \dfrac{g}{mol}$

$\Delta E = \Delta mc^2 = \left(0.0077\ \dfrac{g}{mol} \times \dfrac{1\ kg}{10^3\ g}\right)\left(3.00 \times 10^8\ \dfrac{m}{s}\right)^2 \times \dfrac{1\ J\ s^2}{1\ kg\ m^2} \times \dfrac{1\ kJ}{10^3\ J} = 6.9 \times 10^8\ \dfrac{kJ}{mol}$

Since the Δm value is positive, mass is made in this reaction. In order to make mass, the reactant must absorb energy (ΔE). Because this is a large amount of energy to absorb, this reaction is unlikely.

(d) $\Delta m = m_f - m_i = \left(12.0005\ \dfrac{g}{mol} + 0.0005\ \dfrac{g}{mol}\right) - \left(12.0186\ \dfrac{g}{mol}\right) = -0.0176\ \dfrac{g}{mol}$

$\Delta E = \Delta mc^2 = \left(-0.0176\ \dfrac{g}{mol} \times \dfrac{1\ kg}{10^3\ g}\right)\left(3.00 \times 10^8\ \dfrac{m}{s}\right)^2 \times \dfrac{1\ J\ s^2}{1\ kg\ m^2} \times \dfrac{1\ kJ}{10^3\ J} = -1.58 \times 10^9\ \dfrac{kJ}{mol}$

The mass of $^{12}C^-$ is simply the mass of the ^{12}C atom plus the mass of an e^- (12.0000 + 0.0005).

5. The protons inside all atoms do repel each other. However, they are held together with the neutrons and each other by strong nuclear attractions (the binding energy of the atom). In terms of mass-energy, all atoms weigh less than the sum of their protons, neutrons, and electrons. This mass loss (negative Δm) corresponds to a release of energy (negative ΔE), and shows that the atoms are more stable (lower in energy) than their individual components.

6. (a) The ^{32}P atom decays by β-emission, so one of the products is an electron: $^{32}_{15}P \rightarrow\ ^{32}_{16}S^+ +\ ^{0}_{-1}e^-$.

   (b) Use the integrated rate law to solve for the rate constant, then use that value to find the half-life.

   $$\ln\dfrac{N}{N_0} = -kt,\ k = \dfrac{\ln\dfrac{N}{N_0}}{-t} = \dfrac{\ln\dfrac{33,000\ Ci}{91,000\ Ci}}{-21.0\ d} = 0.048\ d^{-1}$$

   $$kt_{1/2} = \ln 2,\ t_{1/2} = \dfrac{\ln 2}{k} = \dfrac{0.693}{0.048\ d^{-1}} = 14\ d$$

   (c) To determine the ^{32}P activity after 84.0 d, use the integrated rate law and the k value calculated above.

   $$\ln\dfrac{N}{N_0} = -kt,\ \ln\dfrac{N}{91,000\ Ci} = -(0.048\ d^{-1})(84.0\ d) = -4.0574\ (\text{unrounded})$$

   $$\ln\dfrac{N}{91,000\ Ci} = -4.0574,\ \dfrac{N}{91,000\ Ci} = e^{-4.0574} = 0.0173,\ N = 0.0173 \times 91,000\ Ci = 1,600\ Ci$$

   (d) $$\ln\dfrac{N}{N_0} = -kt,\ t = \dfrac{\ln\dfrac{N}{N_0}}{-k} = \dfrac{\ln\dfrac{15,000\ Ci}{91,000\ Ci}}{-0.048\ d^{-1}} = 37\ d.$$

7. (a) The ^{14}C atom decays by β-emission, so one of the products is an electron: $^{14}_{6}C \rightarrow\ ^{14}_{7}N^+ +\ ^{0}_{-1}e^-$.

   (b) Use the k-$t_{1/2}$ relationship and the half-life to find the rate constant.

   $$kt_{1/2} = \ln 2,\ k = \dfrac{\ln 2}{t_{1/2}} = \dfrac{0.693}{5730\ yr} = 1.21 \times 10^{-4}\ yr^{-1}$$

   (c) If the Shroud were created in 33 A.D., it would have been 1,955 yr old in 1988.

   $$\ln\dfrac{N}{N_0} = -kt,\ \ln\dfrac{N}{15.3\ g^{-1}\ min^{-1}} = -(1.21 \times 10^{-4}\ yr^{-1})(1955\ yr) = -0.2365\ (\text{unrounded})$$

$$\frac{N}{15.3 \text{ g}^{-1} \text{ min}^{-1}} = e^{-0.2365} = 0.7894, \quad N = 0.7894 \times 15.3 \text{ g}^{-1} \text{ min}^{-1} = 12.1 \text{ g}^{-1} \text{ min}^{-1}$$

(d) $\ln\dfrac{N}{N_0} = -kt, \quad t = \dfrac{\ln\dfrac{N}{N_0}}{-k} = \dfrac{\ln\dfrac{14.15 \text{ g}^{-1} \text{ min}^{-1}}{15.3 \text{ g}^{-1} \text{ min}^{-1}}}{-1.21 \times 10^{-4} \text{ yr}^{-1}} = 646 \text{ yr}.$ Since the Shroud of Turin was about 646 yr old in 1998, it was created around 1350 A.D.

(e) Radiocarbon dating assumes that the living object is continually replenishing the ^{14}C that decays in it by respiration of $^{14}CO_2$ from the earth's atmosphere (and the incorporation of ^{14}C into the living object). When the living object dies, it no longer incorporates additional ^{14}C into itself, and the existing ^{14}C atoms slowly decay into ^{14}N.

8. (a) The ^{222}Rn atom decays by α-emission, so one of the products is ^{4}He: $^{222}_{86}Rn \rightarrow {}^{218}_{84}Po + {}^{4}_{2}He$.

(b) First, calculate the volume of the house and use the value of 1.5 pCi/L to determine the activity (A, in Ci, then Bq = s^{-1}). This value tells you that 25,000 ^{222}Rn atoms decay every second in this house.

$$V = 2,000 \text{ ft}^2 \times 8 \text{ ft} \times \frac{28.3 \text{ L}}{1 \text{ ft}^3} = 452,800 \text{ L (unrounded)}$$

$$A = \left(1.5 \times 10^{-12} \frac{\text{Ci}}{\text{L}}\right)(452,800 \text{ L}) = 6.8 \times 10^{-7} \text{ Ci} \times \frac{3.70 \times 10^{10} \text{ s}^{-1}}{1 \text{ Ci}} = 25,000 \text{ s}^{-1}$$

9. (a) In order to determine the number of chest X-rays that provide the same amount of background radiation as smoking one pack of cigarettes a day (900 mrem), use the conversion factor for chest X-rays and mrem (1 chest X-ray = 10 mrem).

$$1 \text{ pack per day} \times \frac{900 \text{ mrem}}{1 \text{ pack per day}} \times \frac{1 \text{ chest X-ray}}{10 \text{ mrem}} = 90 \text{ chest X-rays}$$

(b) To determine the number of hours of TV watching that provide the same amount of background radiation as smoking one pack of cigarettes a day (900 mrem), use the conversion factor for watching TV and mrem (1 h TV watching = 0.15 mrem).

$$1 \text{ pack per day} \times \frac{900 \text{ mrem}}{1 \text{ pack per day}} \times \frac{1 \text{ h TV watching}}{0.15 \text{ mrem}} = 6,000 \text{ h TV watching}$$

(c) To determine the fraction of the year that must be spent in front of the TV, convert 6,000 hours of TV watching into years.

$$6,000 \text{ h} \times \frac{1 \text{ d}}{24 \text{ h}} \times \frac{1 \text{ yr}}{365.25 \text{ d}} = 0.68 \text{ yr}$$

The value of 0.68 yr means that you would have to spend 68% of the entire year in front of the TV to get the same background radiation as smoking one pack of cigarettes a day for the entire year.

ANSWERS — PRACTICE TEST

Chapter 21

1. (a) Although the bottom numbers of the nuclide symbols are often omitted, they can be determined from the periodic table. The sum of A values and the sum of the Z values must be the same on both sides of the equation. This gives the following formulas: $13 + 1 = A + 4$ and $6 + 1 = Z + 2$. So, $A = 10$ and $Z = 5$ and the reactant is ^{10}B: $^{13}_{6}C + ^{1}_{1}H \rightarrow ^{10}_{5}B + ^{4}_{2}He$.

   (b) The sum of A values and the Z values must be equal on both sides of the equation. So, $13 + 4 = A + 1$ and $6 + 2 = Z + 0$. So, $A = 16$ and $Z = 8$ and the product is ^{16}O: $^{13}_{6}C + ^{4}_{2}He \rightarrow ^{16}_{8}O + ^{1}_{0}n^{0}$.

   (c) The two equations are: $2 \times 12 = A + 4$ and $2 \times 6 = Z + 2$. So, $A = 20$ and $Z = 10$ and the product is ^{20}Ne: $2\,^{12}_{6}C \rightarrow ^{20}_{10}Ne + ^{4}_{2}He$.

   (d) The two equations are: $2 \times 20 = A + 16$ and $2 \times 10 = Z + 8$. So, $A = 24$ and $Z = 12$ and the product is ^{24}Mg: $2\,^{20}_{10}Ne \rightarrow ^{24}_{12}Mg + ^{16}_{8}O$.

2. (a) Chapter 20 stated that $^{56}_{26}Fe$ is the most stable atom in terms of binding energy per nucleon. So, atoms with lower Z values will become more stable if they fuse together into bigger atoms. Therefore, when $^{12}_{6}C$ and $^{16}_{8}O$ fuse into a larger atom, this reaction should be exothermic (you calculated $\Delta E = -1.62 \times 10^{9}$ kJ/mol for this reaction in practice test 4b in Chapter 20, indicating this reaction is exothermic).

   (b) Even though this is an exothermic reaction, a very large temperature is needed to make this reaction occur because it requires huge amounts of energy to overcome this reaction's tremendously high activation energy. This reaction has a huge activation energy because it takes a lot of energy to force the +6 C nucleus and the +8 O nucleus together so that they can fuse together into ^{28}Si).

3. (a) Since cristobalite has the formula SiO_2, it is a three-dimensional silicate.

   (b) Kaolinite contains Al^{3+} ions and OH^- ions. The two Al^{3+} ions have a total charge of +6 and the four OH^- ions have a total charge of –4. So, the (Si_2O_5) ion must have a charge of –2, and this silicate has $Si_2O_5^{2-}$ ions. Compounds containing $Si_2O_5^{2-}$ ions are two-dimensional silicates.

   (c) Spodumene contains Li^+ ions and Al^{3+} ions. The Li^+ ion and the Al^{3+} ions have a total charge of +4, so the (Si_2O_6) ion must have a charge of –4 $(Si_2O_6^{4-})$. The empirical formula of this silicate ion is SiO_3^{2-}, which means that this compound is a one-dimensional silicate.

   (d) Talc contains Mg^{2+} ions and OH^- ions. The three Mg^{2+} ions have a total charge of +6 and the two OH^- ions have a total charge of –2. So, the (Si_4O_{10}) ion must have a charge of –4 $(Si_4O_{10}^{4-})$. The empirical formula of this silicate ion is $Si_2O_5^{2-}$, which means that this compound is a two-dimensional silicate.

   (e) Tremolite contains Ca^{2+} ions, Mg^{2+} ions, and OH^- ions. The two Ca^{2+} ions and the five Mg^{2+} ions have a total charge of +14 and the two OH^- ions have a total charge of –2. So, the (Si_8O_{22}) ion must have a charge of –12 $(Si_8O_{22}^{12-})$. The empirical formula of this silicate ion is $Si_4O_{11}^{6-}$, which means that this compound is a one-dimensional silicate.

   (f) Zircon contains Zr ions. Since Zr is a transition metal in column 4B, you might expect the Zr ion to be Zr^{4+}. This means that the silicate ion is SiO_4^{4-} and this is a zero-dimensional silicate.

4. In order to purify nitrogen from the air, it must be liquefied (condensed). So, pure air is pressurized and cooled (causing a P increase and a T decrease). Once the air is liquefied, it is distilled to separate liquid N_2 from liquid O_2 and liquid Ar. To purify sulfur from the inside the earth, it must also be liquefied (melted). The solid sulfur in the ground is heated using superheated steam ($T = 165°C$), which melts the sulfur into a liquid and condenses the steam into liquid water. Then, pressurized air is injected into the ground to force the liquid sulfur/water mixture up to the surface. This mixture is allowed to cool, and the sulfur resolidifies.

5. (a) This mixture of $NaCl(\ell)$ and $CaCl_2(\ell)$ contains $Na^+(\ell)$, $Ca^{2+}(\ell)$, and $Cl^-(\ell)$ ions.

   (b) Since there is only one ion present that can be oxidized (Cl^-), the Cl^- ions will be converted to $Cl_2(g)$. However, there are two ions that can be reduced (Na^+ or Ca^{2+} ions). Since there is a choice of two oxidizing agents, the reaction with the more positive (less negative) $E°$ value undergoes reduction. So, Na^+ ions are reduced to Na metal preferentially over the reduction of Ca^{2+} ions.

A 63

$$Cl_2(g) + 2\,e^- \rightleftharpoons 2\,Cl^-(aq) \qquad E° = +1.36\ V$$

$$Na^+(aq) + e^- \rightleftharpoons Na(s) \qquad E° = -2.71\ V$$

$$Ca^{2+}(aq) + 2\,e^- \rightleftharpoons Ca(s) \qquad E° = -2.87\ V$$

(c)   When there is a choice of oxidizing agents, the reduction reaction with the more positive (or less negative) $E°$ value will be preferentially reduced. So, sodium metal will be reduced before calcium metal because it takes less energy (potential) to reduce sodium metal.

6.   The reactants in this reaction are solid $AlCl_3$ (made of Al^{3+} and Cl^- ions) and solid Na. One of the products is solid Al metal, so the other product must be sodium chloride (containing Na^+ and Cl^- ions). Since the charge of Al changes from +3 to 0, it is reduced and the Al atom in $AlCl_3$ must be the oxidizing agent. Since the charge of Na changes from 0 to +1, it is oxidized and Na must be the reducing agent. Since the charge of the Cl^- atom does not change in this reaction, it is not oxidized or reduced.

$$3\,Na(s) + AlCl_3(s) \rightarrow 3\,NaCl(s) + Al(s)$$

7.   The Hall-Heroult process is a very expensive method for producing aluminum metal. These expenses include: (1) The purchase of cryolite (Na_3AlF_6) which is not involved in the process, but is used as an electrolytic solvent for the reaction; (2) The energy needed to melt cryolite into a liquid at a temperature of 1000°C; (3) The continual purchase and purification of bauxite (Al_2O_3), the ore that is reduced to produce aluminum metal; (4) The continual purchase and replacement of the carbon anodes that react with the O^{2-} ions and are oxidized to $CO_2(g)$; and (5) The purchase of electricity needed to force the nonspontaneous electrolysis reaction to occur.

8.   (a)   Calcium phosphate is the source of P atoms that are converted to elemental phosphorus, $P_4(g)$.
    (b)   The coke (graphite) is the chemical that allows the P atoms to be reduced from a +5 oxidation number in calcium phosphate to an oxidation number of 0 in elemental phosphorus (i.e., C is the reducing agent that is oxidized to carbon monoxide as the reaction proceeds).
    (c)   The textbook does not explain why silicon dioxide is added to this reaction. If the silicon dioxide were not present, the Ca atoms in calcium phosphate would end up as calcium oxide (CaO). However, CaO has a very high melting point (2614°C) and would remain a solid in the reaction container (which is at a temperature of 1400-1500°C). Silicon dioxide is added to produce calcium silicate (*slag*), which has a much lower melting point and will exist as a liquid in this reaction. Since the slag is a liquid, it is easier to remove from the reaction container (by draining) than a solid would be.

$$2\,Ca_3(PO_4)_2(\ell) + 10\,C(s) \rightarrow 6\,CaO(s) + P_4(g) + 10\,CO(g)$$
$$2\,Ca_3(PO_4)_2(\ell) + 10\,C(s) + 6\,SiO_2(\ell) \rightarrow 6\,CaSiO_3(\ell) + P_4(g) + 10\,CO(g)$$

    (d)   Water is not part of the reactants that are used to make elemental phosphorus. Instead, it is used to separate the two gaseous products of this reaction. Carbon monoxide has a very low melting point (−199°C) so it will remain a gas when it is bubbled through water, and since it is not very soluble in water (because it is rather nonpolar) it will bubble out of the water. However, elemental phosphorus has a much higher melting point (44°C) so when it is bubbled into liquid water at room temperature, the water cools the gaseous P_4, converting it into an insoluble (nonpolar) solid. This solid sinks to the bottom of the tank of water since $P_4(s)$ has a higher density than water (1.82 g/mL).

9.   (a)   Lithium metal forms Li^+ ions and magnesium metal forms Mg^{2+} ions. Since nitride ions are N^{3-}, the formulas of these compounds are $Li_3N(s)$ and $Mg_3N_2(s)$. These reactions are combination reactions.

$$6\,Li(s) + N_2(g) \rightarrow 2\,Li_3N(s)$$
$$3\,Mg(s) + N_2(g) \rightarrow Mg_3N_2(s)$$

    (b)   When the N^{3-} ion reacts with water, it steals H^+ ions to form $NH_3(g)$, leaving behind OH^- ions to react with the metal ions. These reactions are acid-base reactions (they are also hydrolysis reactions).

$$Li_3N(s) + 3\,H_2O(\ell) \rightarrow 3\,LiOH(aq) + NH_3(g)$$
$$Mg_3N_2(s) + 6\,H_2O(\ell) \rightarrow 3\,Mg(OH)_2(s) + 2\,NH_3(g)$$

ANSWERS — PRACTICE TEST

Chapter 22

1. (a) The name tells you that the lead atom is present as a Pb^{2+} ion, so chromate must be CrO_4^{2-}. Use this ion to find the oxidation number for Cr: $O.N._{Cr} + 4 \times O.N._O = -2$, $O.N._{Cr} = -2 - 4(-2) = +6$.

    (b) Since lead is a main group metal and chromium is a transition element, the yellow color probably comes from Cr. This is consistent with the fact that $PbSO_4(s)$ is white and $K_2CrO_4(s)$ is bright yellow.

2. (a) To determine the electron configuration for Cu^{2+}, you need to start with the electron configuration for the Cu atom—$[Ar]\,4s^1 3d^{10}$. To get the electron configuration for Cu^{2+}, remove two e^- (one from the $4s$ orbital and one from the $3d$ orbital).

    Cu $[Ar]$ $4s$ [↑] $3d$ [↑↓|↑↓|↑↓|↑↓|↑↓]   **Cu^{2+}** $[Ar]$ $4s$ [ ] $3d$ [↑↓|↑↓|↑↓|↑↓|↑]

    (b) The electron configuration for the Co atom is $[Ar]\,4s^2 3d^7$. To get the electron configuration for Co^{2+}, remove two e^- (from the $4s$ orbital).

    Co $[Ar]$ $4s$ [↑↓] $3d$ [↑↓|↑↓|↑|↑|↑]   **Co^{2+}** $[Ar]$ $4s$ [ ] $3d$ [↑↓|↑↓|↑|↑|↑]

    (c) The electron configuration for the Ru atom is $[Kr]\,5s^1 4d^7$. To get the electron configuration for Ru^{3+}, remove three e^- (one from the $5s$ orbital and two from the $4d$ orbital). You would have obtained the same answer if you had used the predicted electron configuration of the Ru atom (i.e., $[Kr]\,5s^2 4d^6$).

    Ru $[Kr]$ $5s$ [↑] $4d$ [↑↓|↑↓|↑|↑|↑]   **Ru^{3+}** $[Kr]$ $5s$ [ ] $4d$ [↑|↑|↑|↑|↑]

    (d) The electron configuration for the Ti atom is $[Ar]\,4s^2 3d^2$. To get the electron configuration for Ti^{4+}, remove two e^- from the $4s$ orbital and two e^- from the $3d$ orbitals.

    Ti $[Ar]$ $4s$ [↑↓] $3d$ [↑|↑| | |]   **Ti^{4+}** $[Ar]$ $4s$ [ ] $3d$ [| | | |]

3. (a) This compound contains Ba^{2+} and O^{2-} ions. Using the charges of these ions, you can find the charge (oxidation number) of Ru: $(+2) + O.N._{Ru} + 3 \times (-2) = 0$, $O.N._{Ru} = 0 - 2 + 6 = +4$.

    (b) Cesium forms Cs^+ ions and chlorine forms Cl^- ions. So, the oxidation number for the Cu atom is: $2 \times O.N._{Cs} + O.N._{Cu} + 4 \times O.N._{Cl} = 0$, $O.N._{Cu} = 0 - 2 \times (+1) - 4(-1) = 0 - 2 + 4 = +2$.

    (c) Lithium forms Li^+ ions (+1 oxidation number) and oxygen forms O^{2-} ions (–2 oxidation number). The oxidation number for the Nb atoms is: $O.N._{Li} + 3 \times O.N._{Nb} + 8 \times O.N._O = 0$, $3 \times O.N._{Nb} = 0 - 1 - 8(-2) = -1 + 16 = +15$, $O.N._{Nb} = +5$.

    (d) This compound contains NO_3^- and OH^- ions. Using the charges of these ions, you can find the charge (oxidation number) of Pt: $O.N._{Pt} + 2 \times (-1) + 2 \times (-1) = 0$, $O.N._{Pt} = 0 + 2 + 2 = +4$.

4. (a) Since S atoms form S^{2-} ions, the Cu and Fe atoms must have a total charge of +4 to balance the two S^{2-} ions. Cu commonly forms Cu^+ and Cu^{2+} ions and Fe commonly forms Fe^{2+} or Fe^{3+} ions. So, this compound either contains a Cu^+ ion and an Fe^{3+} ion or a Cu^{2+} and an Fe^{2+} ion.

    (b) Since O atoms form O^{2-} ions, the Mn and Cr atoms must have a total charge of +8. Mn commonly forms Mn^{2+} and Mn^{3+} ions and Cr commonly forms Cr^{2+} or Cr^{3+} ions. So, this compound either contains an Mn^{2+} ion and two Cr^{3+} ions or it contains an Mn^{3+} ion, an Cr^{2+} ion, and an Cr^{3+} ion.

5. (a) The magnetite will react with gaseous carbon monoxide (produced from solid carbon and oxygen gas).

$$4\,C(s) + 2\,O_2(g) \rightarrow 4\,CO(g)$$
$$\underline{Fe_3O_4(s) + 4\,CO(g) \rightarrow 3\,Fe(\ell) + 4\,CO_2(g)}$$
$$Fe_3O_4(s) + 4\,C(s) + 2\,O_2(g) \rightarrow 3\,Fe(\ell) + 4\,CO_2(g)$$

    (b) Since O atoms form O^{2-} ions, the Fe atoms must have a total charge of +8 to balance the four O^{2-} ions. Since Fe commonly forms Fe^{2+} and Fe^{3+} ions, this compound contains one Fe^{2+} ion and two Fe^{3+} ions.

6.   (a)   Metals with standard reduction potentials less than or equal to +0.34 V will be oxidized at the anode. So, the aqueous solution would contain Fe^{2+} ions after the reaction is completed.

      (b)   Metals with standard reduction potentials less than or equal to +0.34 V will be oxidized at the anode. So, the aqueous solution would contain Ni^{2+} ions after the reaction is completed.

      (b)   Metals with standard reduction potentials greater than +0.34 V will not be oxidized at the anode and these metals will appear in the anode sludge. So, the anode sludge will contain Pt atoms.

      (d)   Metals with standard reduction potentials greater than +0.34 V will not be oxidized at the anode and these metals will appear in the anode sludge. So, the anode sludge will contain Rh atoms.

7.   (a)   The CN^- (cyanide) ligands are named 'cyano', and the charge on the Ag ion is +1. So, the $[Ag(CN)_2]^-$ ion is named the dicyanoargentate(I) ion. Because this is an anion, use the Latin name + '-ate' for silver. You may also leave the (I) out of the ion name because silver forms only one ion: Ag^+.

      (b)   The charge for Au in this complex is also +1. So, $[Au(CN)_2]^-$ is called the dicyanoaurate(I) ion. Use the Latin name + '-ate' for gold since this is an anion. So, $Na[Au(CN)_2]$ is sodium dicyanoaurate(I).

      (c)   The charge on the Zn ion is +2. So, the $[Zn(CN)_4]^{2-}$ ion is named the tetracyanozincate(II) ion. Because this is an anion, use the '-ate' ending for zinc. You may also leave the (II) out of the ion name because zinc forms only one ion: Zn^{2+}.

8.   Use the rules introduced in Chapter 19 to balance these half-reactions. (1) One of the half-equations has the Cr-containing compounds and the other one has the S-containing compounds. (2) In the Cr reaction, place a 2 in front of the Cr^{3+} ions to balance Cr atoms; in the S reaction, the S atoms are already balanced. (3) In the Cr reaction, add 7 water molecules to the right side of the equation to balance the O atoms; in the S reaction, add one water molecule to the left side of the equation to balance the O atoms. (4) Add 14 H^+ ions to the left side of the Cr reaction and 4 H^+ ions to the right side of the S reaction to balance H atoms. (5) To balance charge, add 6 e^- to the left side of the Cr reaction and 2 e^- to the right side of the S reaction. (6) Multiply the S equation by 3 so that there will be the same number of e^- on both sides of the equation. Cancel the 6 e^-, 12 H^+(aq), and 3 $H_2O(\ell)$ from both sides of the equation. (7) Add 2 water molecules to both sides of the equation, recognizing that 2 H^+ and 2 H_2O make 2 H_3O^+. (8) There are no chemical on both sides to cancel.

$$Cr_2O_7{}^{2-}(aq) + 14\,H^+(aq) + 6\,e^- \rightarrow 2\,Cr^{3+}(aq) + 7\,H_2O(\ell)$$

$$H_2SO_3(aq) + H_2O(\ell) \rightarrow SO_4{}^{2-}(aq) + 4\,H^+(aq) + 2\,e^-$$

$$Cr_2O_7{}^{2-}(aq) + 14\,H^+(aq) + 3\,H_2SO_3(aq) + 3\,H_2O(\ell) \rightarrow 2\,Cr^{3+}(aq) + 7\,H_2O(\ell) + 3\,SO_4{}^{2-}(aq) + 12\,H^+(aq)$$

$$Cr_2O_7{}^{2-}(aq) + 2\,H^+(aq) + 3\,H_2SO_3(aq) \rightarrow 2\,Cr^{3+}(aq) + 4\,H_2O(\ell) + 3\,SO_4{}^{2-}(aq)$$

$$Cr_2O_7{}^{2-}(aq) + 2\,H_3O^+(aq) + 3\,H_2SO_3(aq) \rightarrow 2\,Cr^{3+}(aq) + 6\,H_2O(\ell) + 3\,SO_4{}^{2-}(aq)$$

9.   You can balance this equation using a variation of the rules used to balance redox equations. To balance the Cr atoms, place a coefficient of 2 in front of the $CrO_4{}^{2-}$ ion. To balance the O atoms, add 1 water to the left side of the equation. Then, add 2 H^+(aq) ions to the right side to balance H atoms. Now, the equation is balanced for acidic conditions. To convert this reaction to basic conditions, add 2 OH^- to both sides of the equation, recognizing that 2 H^+ and 2 OH^- make 2 H_2O, then remove 1 H_2O from both sides of the equation.

$$Cr_2O_7{}^{2-}(aq) + H_2O(\ell) \rightarrow 2\,CrO_4{}^{2-}(aq) + 2\,H^+(aq)$$

$$Cr_2O_7{}^{2-}(aq) + 2\,OH^-(aq) \rightarrow 2\,CrO_4{}^{2-}(aq) + H_2O(\ell)$$

10.   (a)   The $CO_3{}^{2-}$ (carbonate) ligands are named 'carbonato'; the charge on the Co ion is +3 (+3 + 3(–2) = –3). So, the $[Co(CO_3)_3]^{3-}$ ion is named the tricarbonatocobaltate(III) ion.

      (b)   The O^{2-} (oxide) ligands are named 'oxo' and the Cl^- ions (chloride) ligands are named 'chloro'; the charge on the W atom in $[WOCl_5]^{2-}$ is $O.N._W = -2 - 1 \times (-2) - 5 \times (-1) = +5$ and $H_2[WOCl_5]$ is named hydrogen pentachlorooxowulfrate(V). As an acid, it could be named pentachlorooxowulfric(V) acid.

      (c)   The CN^- (cyanide) ligands are name 'cyano'; the ion is Hg^{2+} (+1 + 2 + 3 × (–1) = 0). So, $K[Hg(CN)_3]$ is named potassium tricyanohydrargyrate(II).

      (d)   The en ligand is ethylenediamine and the H_2O ligands are called 'aqua'; the charge on Pt is +2 (+2 + 0 + 2 × (0) + 2 × (–1) = 0). So, $[Pt(en)(H_2O)_2]Cl_2$ is named diaquaethylenediamineplatinum(II) chloride.

(e)   The H_2O (water) ligands are called 'aqua' ligands; the complex ion is $[V(H_2O)_6]^{2+}$ and the charge on V atom is +2 (to balance the 2 NO_3^-). So, $[V(H_2O)_6](NO_3)_2$ is named hexaaquavanadium(II) nitrate.

(f)   The NH_3 (ammonia) ligands are 'ammine' ligands and the Cl^- ions are 'chloro' ligands. This compound has two coordination ions; since Zn forms only +2 ions, the two coordination ions are $[Zn(NH_3)_4]^{2+}$ and $[ZnCl_4]^{2-}$, and this compound is named tetramminezinc tetrachlorozincate.

11.   (a)   The 'ethylenediaminetetraacetato' ligand is the $EDTA^{4-}$ ion when it is bound to a metal. The charge of the Co-EDTA complex ion is –2 (+2 + (–4) = –2). So, the complex ion is $[Co(EDTA)]^{2-}$ and the ionic compound is $Co^{2+}[Co(EDTA)]^{2-}$ or $Co[Co(EDTA)]$.

(b)   The 'carbonyl' ligand is carbon monoxide molecule (CO), so the charge of the Mo atom is 0. The 'hexa-' prefix means there are 6 CO ligands, and the formula of this compound is $[Mo(CO)_6]$.

(c)   The 'oxo' ligand is an O^{2-} ion (oxo = oxide) and there are 4 of them ('tetra-'). Since the complex contains Mn^{7+}, the ion is $[MnO_4]^-$ (+7 + 4 × (–2) = –1) and the formula is $K[MnO_4]$ (the MnO_4^- ion is also called the 'permanganate' ion and $KMnO_4$ is called 'potassium permanganate').

(d)   The 'aqua' ligand is water and 'fluoro' is the fluoride (F^-) ion; there are three H_2O molecules and three F^- ions. The metal is Fe^{3+} and the compound is neutral $[Fe(H_2O)_3F_3]$ (+3 + 3 × (0) + 3 × (–1) = 0).

12.   (a)   Since there are three H_2O and three ONO ligands, this compound will be a *fac* or *mer* isomer. Since two pairs of the ONO (and H_2O) ligands are 90° apart and one pair is 180° apart, this is a *mer* isomer.

(b)   Since this compound has only one Cl^- ion and five CO molecules, it is not a *cis*, *trans*, *fac*, or *mer* isomer (none of these).

(c)   There are three H_2O ligands and three other coordination sites (an ammine ligand and an oxalate ion). Since the three H_2O ligands are all 90° apart, this is a *fac* isomer.

(d)   Since there are two CN^- ligands (and the en ligand with two coordination sites), this compound will be a *cis* or *trans* isomer. Since the two CN^- ions are 90° apart, this is the *cis* isomer.

(e)   Since this compound has one I^- and three NH_3 ligands, it is not a *cis* or *trans* isomer (none of these).

(f)   There are four NH_3 ligands and two other ligands (a Br^- ion and an OH^- ion). So, this will be a *cis* or *trans* isomer. Since the two other ligands are 180° apart, this is a *trans* isomer.

13.   (a)   This complex contains an Mn^{2+} ion (+2 + 4 × (–1) = –2). The electron configuration of Mn is $[Ar]$ $4s^2 3d^5$; Mn^{2+} is $[Ar]\,3d^5$. This ion is tetrahedral (two *e* orbitals at lower energy and three t_2 at higher energy); since this is a high-spin complex its configuration is $e^2 t_2^3$ and has five unpaired electrons.

(b)   This low-spin complex contains a Fe^{2+} ion (+2 + 4 × (0) = +2) with an electron configuration of $[Ar]$ $3d^6$. Since it is tetrahedral and low-spin, its configuration is $e^4 t_2^2$ and has two unpaired electrons.

(c)   This complex contains a Cr^0 atom, which has an electron configuration of $[Ar]\,4s^1 3d^5$ (6 valence e^- which are all added to the *d* orbitals). Since this ion is octahedral (with three t_{2g} at lower energy and two e_g orbitals at higher energy) and low-spin, its configuration is t_{2g}^6 and has no unpaired electrons.

(d)   This complex contains a Co^{2+} ion. The electron configuration of Co^{2+} is $[Ar]\,3d^7$. Since this ion is octahedral and high-spin, its configuration is $t_{2g}^5 e_g^2$ and has three unpaired electrons.

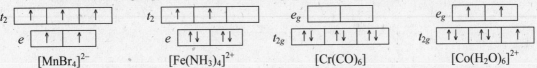

14.   (a)   These ions all have Co^{3+} in them. The name for the complex $[Co(NH_3)_6]Cl_3$ is hexaamminecobalt(III) chloride, the name for the complex $[Co(NH_3)_5Cl]Cl_2$ is pentaamminechlorocobalt(III) chloride, and the complex $[Co(NH_3)_4Cl_2]Cl$ is tetraamminedichlorocobalt(III) chloride.

(b)   Looking at the color wheel in the textbook (p. 1069), the yellow complex absorbs blue-violet light, the purple complex absorbs yellow-green light, and the green complex absorbs red-violet light.

(c)   The spectrochemical series says that NH_3 is a stronger-field ligand than the Cl^- ion. So, replacing a NH_3 ligand with a Cl^- ligand will cause the crystal-field splitting energy (Δ_o) to become slightly smaller. The compound with six NH_3 ligands absorbs blue-violet light (the highest energy light), but replacing one of the NH_3 ligands with a Cl^- ion causes this compound to absorb lower-energy light (yellow-green) and replacing two of the NH_3 ligands with Cl^- ions causes this compound to absorb even lower-energy light (red-violet, the lowest energy light).